ATTAINING HIGH PERFORMANCE COMMUNICATIONS

A VERTICAL APPROACH

ATTAINING HIGH PERFORMANCE COMMUNICATIONS

A VERTICAL APPROACH

EDITED BY

ADA GAVRILOVSKA

CRC Press
Taylor & Francis Group
Boca Raton London New York

CRC Press is an imprint of the
Taylor & Francis Group, an **informa** business

A CHAPMAN & HALL BOOK

CRC Press
Taylor & Francis Group
6000 Broken Sound Parkway NW, Suite 300
Boca Raton, FL 33487-2742

First issued in hardback 2019

© 2010 by Taylor and Francis Group, LLC
Chapman & Hall/CRC is an imprint of Taylor & Francis Group, an Informa business

No claim to original U.S. Government works

ISBN-13: 978-1-4200-9308-7 (hbk)

Library of Congress Cataloging-in-Publication Data

Attaining high performance communications : a vertical approach / editor, Ada
   Gavrilovska.
      p. cm.
   Includes bibliographical references and index.
   ISBN 978-1-4200-9308-7 (hardcover : alk. paper)
   1. High performance computing. I. Gavrilovska, Ada.

QA76.88.A77 2010
004.1'1--dc22
                                           2009027266

Visit the Taylor & Francis Web site at
http://www.taylorandfrancis.com

and the CRC Press Web site at
http://www.crcpress.com

To my family.

Contents

5 Network Interfaces for High Performance Computing   113

Keith Underwood, Ron Brightwell, and Scott Hemmert

6 Network Programming Interfaces for High Performance Computing   149

Ron Brightwell and Keith Underwood

9 Accelerating Communication Services on Multi-Core Platforms   **209**

Ada Gavrilovska

10 Virtualized I/O   **229**

Ada Gavrilovska, Adit Ranadive, Dulloor Rao, and Karsten Schwan

11 The Message Passing Interface (MPI)   251

Jeff Squyres

12 High Performance Event Communication   281

Greg Eisenhauer, Matthew Wolf, Hasan Abbasi, and Karsten Schwan

15 Network Simulation   **353**

George Riley

List of Figures

List of Tables

Preface

As technology pushes the Petascale barrier, and the next generation Exascale Computing requirements are being defined, it is clear that this type of computational capacity can be achieved only by accompanying advancement in communication technology. Next generation computing platforms will consist of hundreds of thousands of interconnected computational resources. Therefore, careful consideration must be placed on the choice of the communication fabrics connecting everything, from cores and accelerators on individual platform nodes, to the internode fabric for coupling nodes in large-scale data centers and computational complexes, to wide area communication channels connecting geographically disparate compute resources to remote data sources and application end-users.

The traditional scientific HPC community has long pushed the envelope on the capabilities delivered by existing communication fabrics. The computing complex at Oak Ridge National Laboratory has an impressive 332 Gigabytes/sec I/O bandwidth and a 786 Terabyte/sec global interconnection bandwidth [305], used by extreme science applications from "subatomic to galactic scales" domains [305], and supporting innovation in renewable energy sources, climate modeling, and medicine. Riding on a curve of a thousandfold increase in computational capabilities over the last five years alone, it is expected that the growth in both compute resources, as well as the underlying communication infrastructure and its capabilities, will continue to climb at mind-blowing rates.

Other classes of HPC applications have equally impressive communication needs beyond the main computing complex, as they require data from remote sources — observatories, databases, remote instruments such as particle accelerators, or large scale data-intensive collaborations across many globally distributed sites. The computational grids necessary for these applications must by enabled by communication technology capable of moving terabytes of data in a timely manner, while also supporting the interactive nature of the remote collaborations.

More unusual, however, are the emerging extreme scale communication needs outside of the traditional HPC domain. First, the confluence of the multi-core nature of emerging hardware resources, coupled with the renaissance of virtualization technology, is creating exceptional consolidation possibilities, and giving rise to *compute clouds* of virtualized multi-core resources. At the same time, the management complexity and operating costs associated with today's IT hardware and software stacks are pushing a broad range of enterprise class applications into such clouds. As clouds spill outside of data center boundaries and move towards becoming Exascale platforms across globally distributed facilities [191,320], their communication needs — for computations distributed across inter- and intra-cloud resources, for management, provisioning and QoS, for workload migration, and for end-user interactions — become exceedingly richer.

Next, even enterprise applications which, due to their critical nature, are not likely candidates for prime cloud deployment, are witnessing a "skyrocketing" increase in communication needs. For instance, financial market analyses are forecasting as much as 128 billion electronic trades a day just by 2010 [422], and similar trends toward increased data rates and lower latency requirements are expected in other market segments, from ticketing services, to parcel delivery and distribution, to inventory management and forecasting.

Finally, the diversity in commodity end-user application, from 3D distributed games, to content sharing and rich multimedia based applications, to telecommunication and telepresence types of services, supported on everything from high-end workstations to cellphones and embedded devices, is further exacerbating the need for high performance communication services.

Objectives

No silver bullet will solve all of the communications-related challenges faced by the emerging types of applications, platforms, and usage scenarios described above. In order to address these requirements we need an ecosystem of solutions along a stack of technology layers: (i) efficient interconnection hardware; (ii) scalable, robust, end-to-end protocols; and (iii) system services and tools specifically targeting emerging multi-core environments.

Toward this end, this book is a unique effort to discuss technological advances and challenges related to high performance communications, by addressing each layer in the vertical stack — the low-level architectural features of the hardware interconnect and interconnection devices; the selection of communication protocols; the implementation of protocol stacks and other operating system features, including on modern, homogeneous or heterogeneous multi-core platforms; the middleware software and libraries used in applications with high-performance communication requirements; and the higher-level application services such as file systems, monitoring services, and simulation tools. The rationale behind this approach is that no single solution, applied at one particular layer, can help applications address all performance-related issues with their communication services. Instead, a coordinated effort is needed to eliminate bottlenecks and address optimization opportunities at each layer — from the architectural features of the hardware, through the protocols and their implementation in OS kernels, to the manner in which application services and middleware are using the underlying platforms.

This book is an edited collection of chapters on these topics. The choice to organize this title as a collection of chapters contributed by different individuals is a natural one — the vertical scope of the work calls for contributions from researchers with many different areas of expertise. Each of the chapters is organized in a manner which includes historic perspective, discussion of state-of-the-art technology solutions and current trends, summary of major research efforts in the community, and, where appropriate, greater level of detail on a particular effort that the chapter contributor is mostly affiliated with.

The topics covered in each chapter are important and complex, and deserve a separate book to cover adequately and in-depth all technical challenges which surround them. Many such books exist [71, 128, 181, 280, 408, etc.]. Unique about this title, however, is that it is a more comprehensive text for a broader audience, spanning a community with interests across the entire stack. Furthermore, each chapter abounds in references to past and current technical papers and projects, which can guide the reader to sources of additional information.

Finally, it is worth pointing out that even this type of text, which touches on many different types of technologies and many layers across the stack, is by no means complete. Many more topics remain only briefly mentioned throughout the book, without having a full chapter dedicated to them. These include topics related to physical-layer technologies, routing protocols and router architectures, advanced communications protocols for multimedia communications, modeling and analysis techniques, compiler and programming language-related issues, and others.

Target audience

The relevance of this book is multifold. First, it is targeted at academics for instruction in advanced courses on high-performance I/O and communication, offered as part of graduate or undergraduate curricula in Computer Science or Computer Engineering. Similar courses have been taught for several years at a number of universities, including Georgia Tech (High-Performance Communications, taught by this book's editor), The Ohio State University (Network-Based Computing, by Prof. Dhabaleswar K. Panda), Auburn University (Special Topics in High Performance Computing, by Prof. Weikuan Yu), or significant portions of the material are covered in more traditional courses on High Performance Computing in many Computer Science or Engineering programs worldwide.

In addition, this book can be relevant reference and instructional material for students and researchers in other science and engineering areas, in academia or at the National Labs, that are working on problems with significant communication requirements, and on high performance platforms such as supercomputers, high-end clusters or large-scale wide-area grids. Many of these scientists may not have formal computer science/computer engineering education, and this title aims to be a single text which can help steer them to more easily identify performance improvement opportunities and find solutions best suited for the applications they are developing and the platforms they are using.

Finally, the text provides researchers who are specifically addressing problems and developing technologies at a particular layer with a single reference that surveys the state-of-the-art at other layers. By offering a concise overview of related efforts from "above" or "below," this book can be helpful in identifying the best ways to position one's work, and in ensuring that other elements of the stack are appropriately leveraged.

Organization

The book is organized in 15 chapters, which roughly follow a vertical stack from bottom to top.

- **Chapter 1** discusses design alternatives for interconnection networks in massively parallel systems, including examples from current Cray, Blue Gene, as well as cluster-based supercomputers from the Top500 list.

- **Chapter 2** provides in-depth discussion of the InfiniBand interconnection technology, its hardware elements, protocols and features, and includes a number of case studies from the HPC and enterprise domains, which illustrate its suitability for a range of applications.

- **Chapter 3** contrasts the traditional high-performance interconnection solutions to the current capabilities offered by Ethernet-based interconnects, and demonstrates several convergence trends among Ethernet and EtherNOT technologies.

- **Chapter 4** describes the key board- and chip-level interconnects such as PCI Express, HyperTransport, and QPI; the capabilities they offer for tighter system integration; and a case study for a service enabled by the availability of low-latency integrated interconnects — global partitioned address spaces.

- **Chapter 5** discusses a number of considerations regarding the hardware and software architecture of network interface devices (NICs), and contrasts the design points present in NICs in a number of existing interconnection fabrics.

- **Chapter 6** complements this discussion by focusing on the characteristics of the APIs natively supported by the different NIC platforms.

- **Chapter 7** focuses on IP-based transport protocols and provides a historic perspective on the different variants and performance optimization opportunities for TCP transports, as well as a brief discussion of other IP-based transport protocols.

- **Chapter 8** describes in greater detail Remote Direct Memory Access (RDMA) as an approach to network communication, along with iWARP, an RDMA-based solution based on TCP transports.

- **Chapter 9** more explicitly discusses opportunities to accelerate the execution of communication services on multi-core platforms, including general purpose homogeneous multi-cores, specialized network processing accelerators, as well as emerging heterogeneous many-core platforms, comprising both general purpose and accelerator resources.

- **Chapter 10** targets the mechanisms used to ensure high-performance communication services in virtualized platforms by discussing VMM-level techniques as well as device-level features which help attain near-native performance.

- **Chapter 11** provides some historical perspective on MPI, the de facto communication standard in high-performance computing, and outlines some of the challenges in creating software and hardware implementations of the MPI standard. These challenges are directly related to MPI applications and their understanding can impact the programmers' ability to write better MPI-based codes.

- **Chapter 12** looks at event-based middleware services as a means to address the high-performance needs of many classes of HPC and enterprise applications. It also gives an overview of the EVPath high-performance middleware stack and demonstrates its utility in several different contexts.

- **Chapter 13** first describes an important class of applications with high performance communication requirements — electronic trading platforms used in the financial industry. Next, it provides implementation detail and performance analysis for the authors' approach to leverage the capabilities of general purpose multi-core platforms and to attain impressive performance levels for one of the key components of the trading engine.

- **Chapter 14** describes the data-movement approaches used by a selection of commercial, open-source, and research-based storage systems used by massively parallel platforms, including Lustre, Panasas, PVFS2, and LWFS.

- **Chapter 15** discusses the overall approach to creating the simulation tools needed to design, evaluate and experiment with a range of parameters in the interconnection technology space, so as to help identify design points with adequate performance behaviors.

Acknowledgments

This book would have never been possible without the contributions from the individual chapter authors. My immense gratitude goes out to all of them for their unyielding enthusiasm, expertise, and time.

Next, I would like to thank my editor Alan Apt, for approaching and encouraging me to write this book, and the extended production team at Taylor & Francis, for all their help with the preparation of the manuscript.

I am especially grateful to my mentor and colleague, Karsten Schwan, for supporting my teaching of the High Performance Communications course at Georgia Tech, which was the basis for this book. He and my close collaborators, Greg Eisenhauer and Matthew Wolf, were particularly accomodating during the most intense periods of manuscript preparation in providing me with the necessary time to complete the work.

Finally, I would like to thank my family, for all their love, care, and unconditional support. I dedicate this book to them.

Ada Gavrilovska
Atlanta, 2009

About the Editor

Ada Gavrilovska is a Research Scientist in the College of Computing at Georgia Tech, and at the Center for Experimental Research in Computer Systems (CERCS). Her research interests include conducting experimental systems research, specifically addressing high-performance applications on distributed heterogeneous platforms, and focusing on topics that range from operating and distributed systems, to virtualization, to programmable network devices and communication accelerators, to active and adaptive middleware and I/O. Most recently, she has been involved with several projects focused on development of efficient virtualization solutions for platforms ranging from heterogeneous multi-core systems to large-scale cloud enviroments.

In addition to research, Dr. Gavrilovska has a strong commitment to teaching. At Georgia Tech she teaches courses on advanced operating systems and high-performance communications topics, and is deeply involved in a larger effort aimed at upgrading the Computer Science and Engineering curriculum with multicore-related content.

Dr. Gavrilovska has a B.S. in Electrical and Computer Engineering from the Faculty of Electrical Engineering, University Sts. Cyril and Methodius, in Skopje, Macedonia, and M.S. and Ph.D. degrees in Computer Science from Georgia Tech, both completed under the guidance of Dr. Karsten Schwan. Her work has been supported by the National Science Foundation, the U.S. Department of Energy, and through numerous collaborations with industry, including Intel, IBM, HP, Cisco Systems, Netronome Systems, and others.

List of Contributors

Hasan Abbasi
College of Computing
Georgia Institute of Technology
Atlanta, Georgia

Virat Agarwal
IBM TJ Watson Research Center
Yorktown Heights, New York

David Bader
College of Computing
Georgia Institute of Technology
Atlanta, Georgia

Pavan Balaji
Argonne National Laboratory
Argonne, Illinois

Ron Brightwell
Sandia National Laboratories
Albuquerque, New Mexico

Dennis Dalessandro
Ohio Supercomputer Center
Columbus, Ohio

Lin Duan
IBM TJ Watson Research Center
Yorktown Heights, New York

Greg Eisenhauer
College of Computing
Georgia Institute of Technology
Atlanta, Georgia

Wu-chun Feng
Departments of Computer Science
    and Electrical & Computer
    Engineering
Virginia Tech
Blacksburg, Virginia

Ada Gavrilovska
College of Computing
Georgia Institute of Technology
Atlanta, Georgia

Scott Hemmert
Sandia National Laboratories
Albuquerque, New Mexico

Matthew Jon Koop
Department of Computer Science
    and Engineering
The Ohio State University
Columbus, Ohio

Todd Kordenbrock
Hewlett-Packard
Albuquerque, New Mexico

Lurng-Kuo Liu
IBM TJ Watson Research Center
Yorktown Heights, New York

Ron A. Oldfield
Sandia National Laboratories
Albuquerque, New Mexico

Scott Pakin
Los Alamos National Laboratory
Los Alamos, New Mexico

Dhabaleswar K. Panda
Department of Computer Science
  and Engineering
The Ohio State University
Columbus, Ohio

Davide Pasetto
IBM Computational Science Center
Dublin, Ireland

Michaele Perrone
IBM TJ Watson Research Center
Yorktown Heights, New York

Fabrizio Petrlni
IBM TJ Watson Research Center
Yorktown Heights, New York

Adit Ranadive
College of Computing
Georgia Institute of Technology
Atlanta, Georgia

Dulloor Rao
College of Computing
Georgia Institute of Technology
Atlanta, Georgia

George Riley
School of Electrical and Computer
  Engineering
Georgia Institute of Technology
Atlanta, Georgia

Karsten Schwan
College of Computing
Georgia Institute of Technology
Atlanta, Georgia

Jeff Squyres
Cisco Systems
Louisville, Kentucky

Sayantan Sur
IBM TJ Watson Research Center
Hawthorne, New York

Keith Underwood
Intel Corporation
Albuquerque, New Mexico

Patrick Widener
University of New Mexico
Albuquerque, New Mexico

Matthew Wolf
College of Computing
Georgia Institute of Technology
Atlanta, Georgia

Pete Wyckoff
Ohio Supercomputer Center
Columbus, Ohio

Sudhakar Yalamanchili
School of Electrical and Computer
  Engineering
Georgia Institute of Technology
Atlanta, Georgia

Jeffrey Young
School of Electrical and Computer
  Engineering
Georgia Institute of Technology
Atlanta, Georgia

Chapter 1

High Performance Interconnects for Massively Parallel Systems

Scott Pakin
Los Alamos National Laboratory

1.1 Introduction

If you were intent on building the world's fastest supercomputer, how would you design the interconnection network? As with any architectural endeavor, there are a suite of trade-offs and design decisions that need to be considered:

Maximize performance Ideally, *every* application should run fast, but would it be acceptable for a smaller set of "important" applications or application classes to run fast? If so, is the overall performance of those applications dominated by communication performance? Do those applications use a known communication pattern that the network can optimize?

Minimize cost Cheaper is better, but how much communication performance are you willing to sacrifice to cut costs? Can you exploit existing hardware components, or do you have to do a full-custom design?

Maximize scalability Some networks are fast and cost-efficient at small scale but quickly become expensive as the number of nodes increases. How can you ensure that you will not need to rethink the network design from scratch when you want to increase the system's node code? Can the network grow incrementally? (Networks that require a power-of-two number of nodes quickly become prohibitively expensive, for example.)

Minimize complexity How hard is it to reason about application performance? To observe good performance, do parallel algorithms need to be constructed specifically for your network? Do an application's processes need to be mapped to nodes in some non-straightforward manner to keep the network from becoming a bottleneck? As for some more mundane complexity issues, how tricky is it for technicians to lay the network

1

cables correctly, and how much rewiring is needed when the network size increases?

Maximize robustness The more components (switches, cables, etc.) compose the network, the more likely it is that one of them will fail. Would you utilize a naturally fault-tolerant topology at the expense of added complexity or replicate the entire network at the expense of added cost?

Minimize power consumption Current estimates are that a sustained megawatt of power costs between US$200,000–1,200,000 per year [144]. How much power are you willing to let the network consume? How much performance or robustness are you willing to sacrifice to reduce the network's power consumption?

There are no easy answers to those questions, and the challenges increase in difficulty with larger and larger networks: Network performance is more critical; costs may grow disproportionately to the rest of the system; incremental growth needs to be carefully considered; complexity is more likely to frustrate application developers; fault tolerance is both more important and more expensive; and power consumption becomes a serious concern. In the remainder of this chapter we examine a few aspects of the network-design balancing act. Section 1.2 quantifies the measured network performance of a few massively parallel supercomputers. Section 1.3 discusses network topology, a key design decision and one that impacts all of performance, cost, scalability, complexity, robustness, and power consumption. In Section 1.4 we turn our attention to some of the "bells and whistles" that a network designer might consider including when trying to balance various design constraints. We briefly discuss some future directions in network design in Section 1.5 and summarize the chapter in Section 1.6.

Terminology note In much of the system-area network literature, the term *switch* implies an indirect network while the term *router* implies a direct network. In the wide-area network literature, in contrast, the term *router* typically implies a switch plus additional logic that operates on packet contents. For simplicity, in this chapter we have chosen to consistently use the term *switch* when describing a network's switching hardware.

1.2   Performance

One way to discuss the performance of an interconnection network is in terms of various mathematical characteristics of the topology including — as a function at least of the number of nodes — the network diameter (worst-case distance in switch hops between two nodes), average-case communication distance, bisection width (the minimum number of links that need to be removed to partition the network into two equal halves), switch count, and network capacity (maximum number of links that can carry data simultaneously). However, there are many nuances of real-world networks that are not captured by simple mathematical expressions. The manner in which applications, messaging layers, network interfaces, and network switches interact can be complex and unintuitive. For example, Arber and Pakin demonstrated that the address in memory at which a message buffer is placed can have a significant impact on communication performance and that different systems favor different buffer alignments [17]. Because of the discrepancy between the expected performance of an idealized network subsystem and the measured performance of a physical one, the focus of this section is on actual measurements of parallel supercomputers containing thousands to tens of thousands of nodes.

1.2.1 Metrics

The most common metric used in measuring network performance is half of the round-trip communication time ($\frac{1}{2}$RTT), often measured in microseconds. The measurement procedure is as follows: Process A reads the time and sends a message to process B; upon receipt of process A's message, process B sends an equal-sized message back to process A; upon receipt of process B's message, process A reads the time again, divides the elapsed time by two, and reports the result as $\frac{1}{2}$RTT. The purpose of using round-trip communication is that it does not require the endpoints to have precisely synchronized clocks. In many papers, $\frac{1}{2}$RTT is referred to as *latency* and the message size divided by $\frac{1}{2}$RTT as *bandwidth* although these definitions are far from universal. In the LogP parallel-system model [100], for instance, latency refers solely to "wire time" for a zero-byte message and is distinguished from *overhead*, the time that a CPU is running communication code instead of application code. In this section, we use the less precise but more common definitions of latency and bandwidth in terms of $\frac{1}{2}$RTT for an arbitrary-size message. Furthermore, all performance measurements in this section are based on tests written using MPI [408], currently the de facto standard messaging layer.

1.2.2 Application Sensitivity to Communication Performance

An underlying premise of this entire book is that high-performance applications require high-performance communication. Can we quantify that statement in the context of massively parallel systems? In a 2006 paper, Kerbyson used detailed, application-specific, analytical performance models to examine the impact of varying the network latency, bandwidth, and number of CPUs per node [224]. Kerbyson's study investigated two large applications (SAGE [450] and Partisn [24]) and one application kernel (Sweep3D [233])

TABLE 1.1:   Network characteristics of Purple, Red Storm, and Blue Gene/L [187]

Metric	Purple	Red Storm	Blue Gene/L
CPU cores	12,288	10,368	65,536
Cores per node	8	1	2
Nodes	1,536	10,368	32,768
NICs per node	2	1	1
Topology (cf. §1.3)	16-ary 3-tree	$27 \times 16 \times 24$ mesh in x,y; torus in z	$32 \times 32 \times 32$ torus in x,y,z
Peak link bandwidth (MB/s)	2,048	3,891	175
Achieved bandwidth (MB/s)	2,913	1,664	154
Achieved min. latency (μs)	4.4	6.9	2.8

at three different network sizes. Of the three applications, Partisn is the most sensitive to communication performance. At 1024 processors, Partisn's performance can be improved by 7.4% by reducing latency from 4 μs to 1.5 μs, 11.8% by increasing bandwidth from 0.9 GB/s to 1.6 GB/s, or 16.4% by decreasing the number of CPUs per node from 4 to 2, effectively halving the contention for the network interface controller (NIC). Overall, the performance difference between the worst set of network parameters studied (4 μs latency, 0.9 GB/s bandwidth, and 4 CPUs/node) and the best (1.5 μs latency, 1.6 GB/s bandwidth, and 2 CPUs/node) is 68% for Partisn, 24% for Sweep3D, and 15% for SAGE, indicating that network performance is in fact important to parallel-application performance.

Of course, these results are sensitive to the particular applications, input parameters, and the selected variations in network characteristics. Nevertheless, other researchers have also found communication performance to be an important contributor to overall application performance. (Martin et al.'s analysis showing the significance of overhead [277] is an oft-cited study, for example.) The point is that high-performance communication is needed for high-performance applications.

1.2.3   Measurements on Massively Parallel Systems

In 2005, Hoisie et al. [187] examined the performance of three of the most powerful supercomputers of the day: IBM's Purple system [248] at Lawrence Livermore National Laboratory, Cray/Sandia's Red Storm system [58] at Sandia National Laboratories, and IBM's Blue Gene/L system [2] at Lawrence Livermore National Laboratory. Table 1.1 summarizes the node and network characteristics of each of these three systems at the time at which Hoisie et al. ran their measurements. (Since the data was acquired all three systems have had substantial hardware and/or software upgrades.)

Purple is a traditional cluster, with NICs located on the I/O bus and

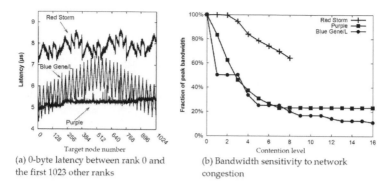

(a) 0-byte latency between rank 0 and
the first 1023 other ranks

(b) Bandwidth sensitivity to network
congestion

FIGURE 1.1:   Comparative communication performance of Purple,
Red Storm, and Blue Gene/L.

connecting to a network fabric. Although eight CPUs share the two NICs in a
node, the messaging software can stripe data across the two NICs, resulting in a
potential doubling of bandwidth. (This explains how the measured bandwidth
in Table 1.1 can exceed the peak link bandwidth.) Red Storm and Blue Gene/L
are massively parallel processors (MPPs), with the nodes and network more
tightly integrated.

Figure 1.1 reproduces some of the more interesting measurements from
Hoisie et al.'s paper. Figure 1.1(a) shows the $\frac{1}{2}$RTT of a 0-byte message sent
from node 0 to each of nodes 1–1023 in turn. As the figure shows, Purple's
latency is largely independent of distance (although plateaus representing
each of the three switch levels are in fact visible); Blue Gene/L's latency
clearly matches the Manhattan distance beween the two nodes, with peaks
and valleys in the graph representing the use of the torus links in each 32-node
row and 32×32-node plane; however, Red Storm's latency appears erratic.
This is because Red Storm assigns ranks in the computation based on a node's
physical location on the machine room floor — aisle (0–3), cabinet within an
aisle (0–26), "cage" (collection of processor boards) within a cabinet (0–2),
board within a cage (0–7), and socket within a board (0–3) — not, as expected,
by the node's x (0–26), y (0–15), and z (0–23) coordinates on the mesh.
While the former mapping may be convenient for pinpointing faulty hardware,
the latter is more natural for application programmers. Unexpected node
mappings are one of the subtleties that differentiate *measurements* of network
performance from *expectations* based on simulations, models, or theoretical
characteristics.

Figure 1.1(b) depicts the impact of link contention on achievable bandwidth.
At contention level 0, two nodes are measuring bandwidth as described in
Section 1.2.1. At contention level 1, two other nodes lying between the first pair
exchange messages while the first pair is measuring bandwidth; at contention

level 2, another pair of coincident nodes consumes network bandwidth, and so on. On Red Storm and Blue Gene/L, the tests are performed across a single dimension of the mesh/torus to force all of the messages to overlap. Each curve in Figure 1.1(b) tells a different story. The Blue Gene/L curve contains plateaus representing messages traveling in alternating directions across the torus. Purple's bandwidth degrades rapidly up to the node size (8 CPUs) then levels off, indicating that contention for the NIC is a greater problem than contention within the rest of the network. Relative to the other two systems, Red Storm observes comparatively little performance degradation. This is because Red Storm's link speed is high relative to the speed at which a single NIC can inject messages into the network. Consequently, it takes a relatively large number of concurrent messages to saturate a network link. In fact, Red Storm's overspecified network links largely reduce the impact of the unusual node mappings seen in Figure 1.1(a).

Which is more important to a massively parallel architecture: faster processors or a faster network? The tradeoff is that faster processors raise the peak performance of the system, but a faster network enables more applications to see larger fractions of peak performance. The correct choice depends on the specifics of the system and is analogous to the choice of eating a small piece of a large pie or a large piece of a small pie. On an infinitely fast network (relative to computation speed), all parallelism that can be exposed — even, say, in an arithmetic operation as trivial as $(a + b) \cdot (c + d)$ — leads to improved performance. On a network with finite speed but that is still fairly fast relative to computation speed, trivial operations are best performed sequentially although small blocks of work can still benefit from running in parallel. On a network that is much slower than what the processors can feed (e.g., a wide-area computational Grid [151]), only the coarsest-grained applications are likely to run faster in parallel than sequentially on a single machine. As Hoisie et al. quantify [187], Blue Gene/L's peak performance is 18 times that of the older ASCI Q system [269] but its relatively slow network limits it to running SAGE [450] only 1.6 times as fast as ASCI Q. In contrast, Red Storm's peak performance is only 2 times ASCI Q's but its relatively fast network enables it to run SAGE 2.45 times as fast as ASCI Q. Overspecifying the network was in fact an important aspect of Red Storm's design [58].

1.3 Network Topology

One of the fundamental design decisions when architecting an interconnection network for a massively parallel system is the choice of network topology. As mentioned briefly at the start of Section 1.2, there are a wide variety of metrics one can use to compare topologies. Selecting a topology also involves a large

TABLE 1.2: Properties of a ring vs. a fully connected network of n nodes

Metric	Ring	Fully connected
Diameter	$n/2$	1
Avg. dist.	$n/4$	1
Degree	**2**	$n-1$
Bisection	2	$\left(\frac{n}{2}\right)^2$
Switches	n	n
Minimal paths	1	1
Contention	blocking	**nonblocking**

number of tradeoffs that can impact performance, cost, scalability, complexity, robustness, and power consumption:

Minimize diameter Reducing the hop count (number of switch crossings) should help reduce the overall latency. Given a choice, however, should you minimize the *worst-case* hop count (the *diameter*) or the *average* hop count?

Minimize degree How many incoming/outgoing links are there per node? Low-radix networks devote more bandwidth to each link, improving performance for communication patterns that map nicely to the topology (e.g., nearest-neighbor 3-space communication on a 3-D mesh). High-radix networks reduce average latency — even for communication patterns that are a poor match to the topology — and the cost of the network in terms of the number of switches needed to interconnect a given number of nodes. A second degree-related question to consider is whether the number of links per node is constant or whether it increases with network size.

Maximize bisection width The minimum number of links that must be removed to split the network into two disjoint pieces with equal node count is called the *bisection width*. (A related term, *bisection bandwidth*, refers to the aggregate data rate between the two halves of the network.) A large bisection width may improve fault tolerance, routability, and global throughput. However, it may also greatly increase the cost of the network if it requires substantially more switches.

Minimize switch count How many switches are needed to interconnect n nodes? $O(n)$ implies that the network can cost-effectively scale up to large numbers of nodes; $O(n^2)$ topologies are impractical for all but the smallest networks.

Maximize routability How many minimal paths are there from a source node to a destination node? Is the topology naturally deadlock-free or does it require extra routing sophistication in the network to maintain deadlock freedom? The challenge is to ensure that concurrently transmitted messages cannot mutually block each other's progress while simultaneously allowing the use of as many paths as possible between a source and a destination [126].

Minimize contention What communication patterns (mappings from a set of source nodes to a — possibly overlapping — set of destination nodes) can proceed with no two paths sharing a link? (Sharing implies reduced bandwidth.) As Jajszczyk explains in his 2003 survey paper, a network can be categorized as *nonblocking in the strict sense, nonblocking in the wide sense, repackable, rearrangeable,* or *blocking,* based on the level of effort (roughly speaking) needed to avoid contention for arbitrary mappings of sources to destinations [215]. But how important is it to support *arbitrary* mappings — vs. just a few common communication patterns — in a contention-free manner? Consider also that the use of packet switching and adaptive routing can reduce the impact of contention on performance.

Consider two topologies that represent the extremes of connectivity: a ring of n nodes (in which nodes are arranged in a circle, and each node connects only to its "left" and "right" neighbors) and a fully connected network (in which each of the n nodes connects directly to each of the other $n-1$ nodes). Table 1.2 summarizes these two topologies in terms of the preceding characteristics. As that table shows, neither topology outperforms the other in all cases (as is typical for any two given topologies), and neither topology provides many minimal paths, although the fully connected network provides a large number of non-minimal paths. In terms of scalability, the ring's primary shortcomings are its diameter and bisection width, and the fully connected network's primary shortcoming is its degree. Consequently, in practice, neither topology is found in massively parallel systems.

1.3.1 The "Dead Topology Society"

What topologies *are* found in massively parallel systems? Ever since parallel computing first started to gain popularity (arguably in the early 1970s), the parallel-computing literature has been rife with publications presenting a virtually endless menagerie of network topologies, including — but certainly not limited to — 2-D and 3-D meshes and tori [111], banyan networks [167], Beneš networks [39], binary trees [188], butterfly networks [251], Cayley graphs [8], chordal rings [18], Clos networks [94], crossbars [342], cube-connected cycles [347], de Bruijn networks [377], delta networks [330], distributed-loop networks [40], dragonflies [227], extended hypercubes [238], fat trees [253],

	June 1993	June 1994	June 1995	June 1996	June 1997	June 1998	June 1999	June 2000	June 2001	June 2002	June 2003	June 2004	June 2005	June 2006	June 2007	June 2008
1	F X	M X	X X	H H	M M	M M	M M	M M	F F	F X	X X	X X	M M	M M	M M	M F
2	F F	X M	M M	X X	H M	M M	H H	H H	M F	F F	F F	F F	H M	M M	M M	M M
3	F F	F F	M M	M H	X M	M M	M M	H H	H M	F F	F F	F F	X H	F F	M M	C M
4	F F	F F	M M	M M	H H	H M	H H	H M	F H	H M	F F	F F	M F	X H	H F	M O F
5	F	X X	X X	M M	M M	M M	M M	M H	H F	H H	F F	F F	F F	F F	F F	M F M
6	X X	F F	M M	M H	H H	H H	M H	F H	F F	F F	F F	F F	M M	F F	F M M
7	F	X X	X F	X M	M X	M M	M M	H M	H F	H M	F X	F X	F F	X F	F M M	C
8	M M	F F	F F	F X	M X	M H	H M	M F	X H	F F	F F	M M	M F	M H	F M O
9	M	F F	M	F X	M M	M M	M M	M M	H H	F F	F H	F F	F F	M M	M F	F M M
10	F		M F	X M	M M	M M	F M	M F	F F	F F	F F	F F	M M	X M	H M	C
Most procs.	C C	M M	M M	M M	M M	M M	M M	M M	M M	M M	M M	M M	M M	M M	M M	M M

Legend and statistics

M	Mesh or torus	(35.8%)		No network	(3.2%)
F	Fat tree	(34.8%)	C	Hypercube	(1.0%)
H	Hierarchical	(13.2%)	O	Other	(0.6%)
X	Crossbar	(11.3%)			

FIGURE 1.2: Network topologies used in the ten fastest supercomputers and the single most parallel supercomputer, June and November, 1993–2008.

flattened butterflies [228], flip networks [32], fractahedrons [189], generalized hypercube-connected cycles [176], honeycomb networks [415], hyperbuses [45], hypercubes [45], indirect binary n-cube arrays [333], Kautz graphs [134], KR-Benes networks [220], Moebius graphs [256], omega networks [249], pancake graphs [8], perfect-shuffle networks [416], pyramid computers [295], recursive circulants [329], recursive cubes of rings [421], shuffle-exchange networks [417], star graphs [7], X-torus networks [174], and X-Tree structures [119]. Note that many of those topologies are general classes of topologies that include other topologies on the list or are specific topologies that are isomorphic to other listed topologies.

Some of the entries in the preceding list have been used in numerous parallel-computer implementations; others never made it past a single publication. Figure 1.2 presents the topologies used over the past fifteen years both in the ten fastest supercomputers and in the single most parallel computer (i.e., the supercomputer with the largest number of processors, regardless of its performance).

The data from Figure 1.2 were taken from the semiannual Top500 list

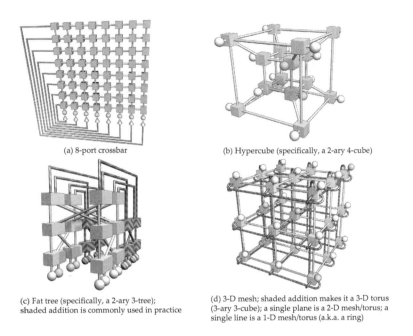

(a) 8-port crossbar

(b) Hypercube (specifically, a 2-ary 4-cube)

(c) Fat tree (specifically, a 2-ary 3-tree);
shaded addition is commonly used in practice

(d) 3-D mesh; shaded addition makes it a 3-D torus
(3-ary 3-cube); a single plane is a 2-D mesh/torus; a
single line is a 1-D mesh/torus (a.k.a. a ring)

FIGURE 1.3:   Illustrations of various network topologies.

(`http://www.top500.org/`), and therefore represent achieved performance on
the HPLinpack benchmark [123].[1] Figure 1.3 illustrates the four topologies
that are the most prevalent in Figure 1.2: a crossbar, a hypercube, a fat tree,
and a mesh/torus.

One feature of Figure 1.2 that is immediately clear is that the most parallel
system on each Top500 list since 1994 has consistently been either a 3-D mesh
or a 3-D torus. (The data are a bit misleading, however, as only four different
systems are represented: the CM-200 [435], Paragon [41], ASCI Red [279],
and Blue Gene/L [2].) Another easy-to-spot feature is that two topologies —
meshes/tori and fat trees — dominate the top 10. More interesting, though,
are the trends in those two topologies over time:

- In June 1993, half of the top 10 supercomputers were fat trees. Only
  one mesh/torus made the top 10 list.

---

[1]HPLinpack measures the time to solve a system of linear equations using LU decomposition
with partial pivoting. Although the large, dense matrices that HPLinpack manipulates are
rarely encountered in actual scientific applications, the decade and a half of historical HPLinpack performance data covering thousands of system installations is handy for analyzing
trends in supercomputer performance.

TABLE 1.3:   Hypercube conference: length of proceedings

Year	Conference title	Pages
1986	First Conference on Hypercube Multiprocessors	286
1987	Second Conference on Hypercube Multiprocessors	761
1988	Third Conference on Hypercube Concurrent Computers and Applications	2682
1989	Fourth Conference on Hypercubes, Concurrent Computers, and Applications	1361

- From November 1996 through November 1999 there were *no* fat trees in the top 10. In fact, in June 1998 every one of the top 10 was either a mesh or a torus. (The entry marked as "hierarchical" was in fact a 3-D mesh of crossbars.)

- From November 2002 through November 2003 there were no meshes or tori in the top 10. Furthermore, the November 2002 list contained only a single non-fat-tree.

- In November 2006, 40% of the top 10 were meshes/tori and 50% were fat trees.

Meshes/tori and fat trees clearly have characteristics that are well suited to high-performance systems. However, the fact that the top 10 is alternately dominated by one or the other indicates that changes in technology favor different topologies at different times. The lesson is that the selection of a topology for a massively parallel system must be based on what current technology makes fast, inexpensive, power-conscious, etc.

Although hypercubes did not often make the top 10, they do make for an interesting case study in a topology's boom and bust. Starting in 1986, there were enough hypercube systems, users, and researchers to warrant an entire conference devoted solely to hypercubes. Table 1.3 estimates interest in hypercubes — rather informally — by presenting the number of pages in this conference's proceedings. (Note though that the conference changed names almost every single year.) From 1986 to 1988, the page count increased almost tenfold! However, the page count halved the next year even as the scope broadened to include non-hypercubes, and, in 1990, the conference, then known as the Fifth Distributed Memory Computing Conference, no longer focused on hypercubes.

How did hypercubes go from being the darling of massively parallel system design to topology non grata? There are three parts to the answer. First, topologies that match an expected usage model are often favored over those that don't. For example, Sandia National Laboratories have long favored 3-D meshes because many of their applications process 3-D volumes. In contrast, few scientific methods are a natural match to hypercubes. The second reason that hypercubes lost popularity is that the topology requires a doubling of processors (more for non-binary hypercubes) for each increment in processor count. This limitation starts to become impractical at larger system sizes. For example, in 2007, Lawrence Livermore National Laboratory upgraded their Blue Gene/L [2] system (a 3-D torus) from 131,072 to 212,992 processors; it was too costly to go all the way to 262,144 processors, as would be required by a hypercube topology.

The third, and possibly most telling, reason that hypercubes virtually disappeared is because of techology limitations. The initial popularity of the hypercube topology was due to how well it fares with many of the metrics listed at the start of Section 1.3. However, Dally's PhD dissertation [110] and subsequent journal publication [111] highlighted a key fallacy of those metrics: They fail to consider wire (or pin) limitations. Given a fixed number of wires into and out of a switch, an n-dimensional binary hypercube (a.k.a. a 2-ary n-cube) divides these wires — and therefore the link bandwidth — by n; the larger the processor count, the less bandwidth is available in each direction. In contrast, a 2-D torus (a.k.a. a k-ary 2-cube), for example, provides 1/4 of the total switch bandwidth in each direction regardless of the processor count. Another point that Dally makes is that high-dimension networks in general require long wires — and therefore high latencies — when embedded in a lower-dimension space, such as a plane in a typical VLSI implementation.

In summary, topologies with the best mathematical properties are not necessarily the best topologies to implement. When selecting a topology for a massively parallel system, one must consider the real-world features and limitations of the current technology. A topology that works well one year may be suboptimal when facing the subsequent year's technology.

1.3.2   Hierarchical Networks

A *hierarchical network* uses an "X of Y" topology — a base topology in which every node is replaced with a network of a (usually) different topology. As Figure 1.2 in the previous section indicates, many of the ten fastest supercomputers over time have employed hierarchical networks. For example, Hitachi's SR2201 [153] and the University of Tsukuba's CP-PACS [49] (#1, respectively, on the June and November 1996 Top500 lists) were both $8 \times 17 \times 16$ meshes of crossbars (a.k.a. 3-D hypercrossbars). Los Alamos National Laboratory's ASCI Blue Mountain [268] (#4 on the November 1998 Top500 list) used a deeper topology hierarchy: a HIPPI [433] 3-D torus of "hierarchical fat bristled hypercubes" — the SGI Origin 2000's network of eight crossbars, each of which connected to a different node within a set of eight 3-D hypercubes and where each node contained two processors on a bus [246]. The topologies do not need to be different: NASA's Columbia system [68] (#2 on the November 2004 Top500 list) is an InfiniBand [207, 217] 12-ary fat tree of NUMAlink 4-ary fat trees. (NUMAlink is the SGI Altix 3700's internal network [129].)

Hierarchical topologies can be useful for exploiting packaging hierarchies [227] — one topology within a socket, another within a circuit board, another across a backplane, another within a cabinet, and another across cabinets — with each topology optimized for the physical features and limitations of the corresponding technology. A related advantage of hierarchical topologies is that they can more easily control tradeoffs between cost and performance. For example, a higher-performance but higher-cost topology may be used closer to the processors while a lower-performance but lower-cost topology may be used further away. Assuming that applications exhibit good locality and that they relatively infrequently use the upper layers of the network, this can be a worthwhile tradeoff. Also, consider that massively parallel systems are typically space-shared, with different applications running in different partitions. This usage model effectively increases locality even further.

In fact, *most* new parallel computers today use a hierarchy of topologies. At a minimum, one topology (e.g., a crossbar) interconnects processor cores within a socket, another (e.g., a 2-D mesh) connects sockets within a node, and a third (e.g., a fat tree) connects nodes. Los Alamos National Laboratory's Roadrunner supercomputer [28] (#1 on the June 2008 Top500 list and the first supercomputer in the #1 spot to use a commodity interconnect for internode communication) integrates its 122,400 processor cores using a *five*-level network hierarchy as shown in Figure 1.4. Roadrunner's hierarchy comprises a 12-ary fat tree of InfiniBand[2] [217], a 1-D mesh (i.e., a line) of HyperTransport [3], and point-to-point PCI Express [281] and FlexIO [88] connections. This hierarchy continues into the CPU sockets. Roadrunner is a hybrid architecture with two processor types. The PowerXCell 8i [180] sockets connect nine cores (eight "SPE" and one "PPE") using an Element Interconnect Bus [230], which is in fact a set of four unidirectional rings, two in each direction. The Opteron sockets connect two cores using a "system request interface" [97], which is a simple point-to-point network. While complex, Roadrunner's topology hierarchy represents a collection of price:performance tradeoffs. For example, a commodity interconnect may deliver less communication performance than a custom interconnect, but if it is significantly less costly than a custom interconnect then the money saved can go towards purchasing more processor cores, which may lead to a gain in overall system performance.

1.3.3   Hybrid Networks

While some massively parallel systems use a hierarchy of networks, others connect each node to more than one network. We call these *hybrid networks*. The idea is to recognize that different topologies have different strengths

---

[2]In fact, Roadrunner effectively contains *two* 12-ary InfiniBand fat trees: one used for communication within each 180-node compute unit (CU) and a separate fat tree for communication among compute units [28].

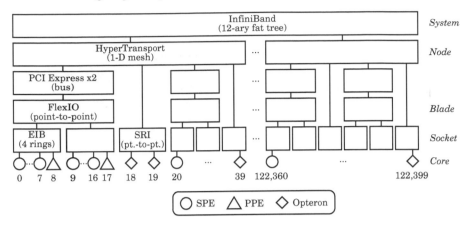

FIGURE 1.4:   The network hierarchy of the Roadrunner supercomputer.

and different weaknesses. Hence, rather than trying to construct a single topology that is good in all cases it may be more cost-effective to include multiple networks in a single system. For example, the Thinking Machines CM-5 utilized a fat-tree network for data communication but an ordinary (skinny) tree network for collective-communication operations [254]. The latter network implements a wide variety of collectives within the tree itself: quiescence detection, barriers, global interrupts, broadcasts, parallel prefix and suffix operations (a.k.a. forward and backward scans), and asynchronous OR operations. The CM-5 targeted both message-passing and data-parallel programming, so including a network optimized for low-latency collectives was a wise design decision. We further discuss optimizations for specific programming models in Section 1.4.1.

The IBM Blue Gene/P [12] is a more recent example of a massively parallel system interconnected using multiple networks. Blue Gene/P contains three networks, each making a different tradeoff between performance and generality.[3] The first network is a 3-D torus that provides point-to-point communication with high bisection bandwidth. This is the most general of Blue Gene/P's networks as any communication operation can be implemented atop point-to-point messages. The second network uses a tree topology. While this network can technically perform point-to-point communication (albeit with low bisection bandwidth), it has hardware support for low-latency broadcasts and reductions, similar to the CM-5's tree network. The third network, a barrier/interrupt network implemented as a wired OR, is the least general of the three. It can perform only a couple of operations (global barriers and interrupts), but it can perform them with extremely low latency. An

---

[3]Blue Gene/P in fact has five networks if one includes the 10 Gigabit Ethernet maintenance network and the JTAG diagnostic network.

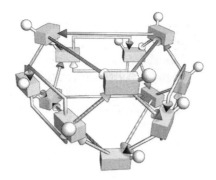

FIGURE 1.5:   Kautz graph (specifically, $\mathcal{K}_2^3$ — a degree 2, dimension 3 Kautz graph).

interesting point is that all three networks grow linearly with the number of nodes — scalability was a key design decision — yet by implementing different topologies and specialized features Blue Gene/P is able simultaneously to deliver high-throughput point-to-point communication and low-latency collective communication.

1.3.4   Novel Topologies

Figure 1.2 in Section 1.3.1 illustrated how two basic topologies have dominated the top of the Top500 list since its inception: meshes/tori and fat trees. Does this imply that there is little opportunity for future innovation in topology design? On the contrary, new technologies are continually being introduced, and different technologies are likely to favor different network topologies.

A recent massively parallel system that utilizes a unique network topology is the SiCortex SC5832 [413]. The SC5832 organizes 972 six-core nodes (5832 total cores) into a degree 3, dimension 6 Kautz graph (sometimes notated as $\mathcal{K}_3^6$) [134]. A Kautz graph is an interesting topology for a tightly integrated massively parallel system because it contains the largest number of nodes for a given degree and dimension and because it provides multiple disjoint paths between any two nodes. However, the Kautz graph is also an unusual topology because it uses unidirectional network links and is asymmetric. That is, node A may be one hop away from node B, but node B may be more than one hop away from node A. (Consider how this impacts the use of $\frac{1}{2}$RTT for measuring the per-hop time, as described in Section 1.2.1.) Finding optimal mappings of even fairly basic communication patterns (e.g., nearest-neighbor communication in two-dimensional space) onto an arbitrary Kautz graph without relying on brute-force search is currently an open research problem.

Fortunately, like the Red Storm system discussed in Section 1.2.3, the SC5832's network is fast relative to compute performance [30] so the need for optimal task placement is greatly reduced.

Figure 1.5 illustrates a 12-node Kautz graph with degree 2 and dimension 3. Note that each switch contains exactly two (the degree minus one) outgoing and exactly two incoming links; it is possible to route a message from any node to any other node in no more than three (the dimension) switch hops; and there are two (the degree) disjoint — but not necessarily minimal — paths from any node to any other node. On the SC5832 no two nodes are more than six hops apart, and there are three disjoint paths between any pair of nodes.

1.4   Network Features

Besides topology, another important design choice when constructing a high-performance interconnect for a massively parallel system is the set of features that the NICs and switches support. A design with more features may imply higher power demands, greater costs, and even a slower NIC or switch than a less featureful design. However, if these features reduce the average *amount* of work needed to communicate, then there may be an overall gain in performance. For example, a network that uses source-based routing (i.e., one in which the source NIC specifies every switch that each packet traverses) may reduce switch complexity and therefore switching time versus a network that uses adaptive routing (i.e., one in which routing decisions are made dynamically at each switch, typically based on link availability). However, adaptive routing reduces contention within the network, which can produce an overall gain in communication throughput. Then again, adaptive routing leads to out-of-order delivery, and if the NICs do not automatically reorder packets then a software messaging layer must do so, assuming a messaging layer such as MPI [408] that provides ordering guarantees. As Karamcheti and Chien observed, relegating packet reordering to software can add almost 100% to the base messaging time [222]. This example motivates the significance of careful selection of network features: the absence of a feature such as adaptive routing can lead to poor performance, but so can its inclusion if not considered in the larger picture of NICs plus switches plus communication layers plus applications.

The interconnects found in different massively parallel systems represent a spectrum of feature sets. Interconnects may include sophisticated communication support in the NICs, switches, both, or neither. The following are examples of each of the four alternatives:

Simple NICs, simple switches In Blue Gene/L, the CPU cores are responsible for injecting data into the network 16 bytes at a time, and the torus network does little more than routing [10]. Separate networks provide support for collectives, as discussed in Section 1.3.3 in the context of Blue Gene/P. (Blue Gene/P uses essentially the same network structure as Blue Gene/L but includes a DMA engine for communication offloading [12].)

Smart NICs, simple switches Myricom's Myrinet network [48] dominated the Top500 list from June 2002 through November 2004 — 38.6% of the systems in the November 2003 list used Myrinet as their interconnection network. Myrinet switches — 16- or 32-port crossbars typically organized as an 8-ary fat tree — simply route data from an input port to an output port designated in the packet itself (and specified by the NIC). The Myrinet NIC contains a fully programmable 32-bit processor, 2–4 MB of SRAM, and support for DMA transfers between the NIC's SRAM and either the network or main memory [300]. Programmability implies that parts of the messaging layer can be offloaded to the NIC to free up the main CPU for computation. In fact, if necessary, the program running on the NIC can be customized for various usage models (e.g., running MPI [408] versus TCP [344] applications). Furthermore, because Myrinet uses static, source-based routing, the source NIC can dynamically select a path to a given destination.

Simple NICs, smart switches QLogic's (formerly PathScale's) InfiniPath network attaches very simple NICs to standard InfiniBand switches [60]. The philosophy behind InfiniPath is to provide nearly complete CPU onload. That is, rather than incorporating an expensive CPU or a slow embedded processor into the NIC, InfiniPath dictates that the main CPU perform virtually all aspects of communication itself including flow control, error handling, send-side data transfer (via programmed I/O), and all communication progress mechanisms. While the InfiniPath NICs are designed to be simple but fast, the InfiniBand switches include support for multicasts, virtual lanes (i.e., message priority levels), programmable routing tables, and other features [217].

Smart NICs, smart switches The Quadrics QsNet[II] network [36], which is used in Bull's Tera-10 supercomputer at the Commissariat à l'Énergie Atomique [72], is an example of a network that contains both smart NICs *and* smart switches. The NICs include a fully programmable, multithreaded, 64-bit main processor, a separate, specialized command processor for local memory movement and synchronization, 64 MB of SDRAM memory, and various communication-specific logic. The NICs support DMA operations between NIC memory and main memory, remote data transfers between local and remote NIC memory (with optimizations for both large transfers and transfers as small as a single word), communication to and from demand-paged virtual memory, remote operations (boolean tests, queue management, and synchronization), and an event mechanism that enables the completion of one communication operation to trigger another communication operation. QsNet[II]'s switches, which are organized as a 4-ary fat tree, support adaptive routing (which can be limited or disabled on a per-packet basis), reliable hardware multicast with acknowledgment combining — an important but often overlooked feature for avoiding ACK implosion [112] — network conditionals ("Is condition P true on all nodes?," which can be used for barrier synchronization), and tagged acknowledgment tokens [337].

A network designer always needs to be mindful, however, of the tradeoffs — in terms of common-case performance, power, cost, etc. — involved in including any given feature. Does the added feature benefit all communication patterns or only a select few? Does a NIC with a dedicated but slow processor provide better or worse performance than a simple NIC controlled by a state-of-the-art CPU running a full software stack? Is the flexibility provided by extra programmability beneficial, or can all of the interesting cases be implemented in dedicated hardware?

1.4.1   Programming Models

An appropriate way to reason about the interconnect features that should be included or excluded is to consider the system's usage model, in particular how the system is expected to be programmed. The following are some examples of parallel-programming models and the sorts of network features that may benefit implementations of those models:

Data-parallel A data-parallel program uses sequential control flow but works with arrays of data as single entities. A typical data-parallel operation involves moving data collectively between all processors and all memory banks (and/or among all processors, depending on the architecture) in any of a variety of permutations. A topology that can support commonly used permutations (e.g., shift) in a contention-free manner may offer improved performance over other topologies. Not surprisingly, in the early days of parallel computing, when SIMD-style parallelism [148] dominated, topology support for contention-free implementations of specific permutations was the focus of a great deal of research [32, 249, 333, 416, etc.].

Thread-based shared memory In this model of programming, a set of independent threads concurrently accesses a common memory. In practice, for scalability reasons, the memory is distributed, data is allowed to reside in local caches, and a cache-coherency protocol presents applications with an illusion of a single memory. Supporting cache-coherency in hardware generally involves optimizing the network for cache-line-sized transfers, including coherency bits as message metadata, and integrating the NIC close to the memory subsystem (as opposed to placing it further away on an I/O bus). These are the basic network features that have been included in scalable cache-coherent systems from the early Stanford Dash [257] through the more recent Altix line of parallel computers from SGI [460].

Two-sided message passing Two-sided message passing expresses communication in terms of matched sends and receives between pairs of concurrently executing processes. More generally, all processes involved in any communication must explicitly indicate their participation. MPI [408] is currently the de facto standard for two-sided message passing — in fact, *most* parallel programs today are written in MPI — so it is worthwhile to engineer a network for efficient support of MPI's specific features. For example, MPI allows for selective reception of messages based on their "communicator" (a communication context that encapsulates a subset of processes), tag (an arbitrary integer associated with a message), and source process, with support for wildcards on tags and source processes; it provides ordering guarantees on message delivery; and it uses a reliable-delivery model in that a dropped message is considered a catastrophic event, not a common occurrence. (An MPI implementation may not casually drop a message to simplify buffer management, for example.)

Underwood et al. demonstrated that supporting MPI-specific message matching in hardware can lead to substantial performance benefits [436]. MPI provides a wealth of collective-communication operations including barriers, broadcasts, reductions, scatters, and gathers. If these are expected to dominate performance for key applications, the interconnect can include hardware support for them, as do the CM-5 [254] and Blue Gene/P [12] (Section 1.3.3). Finally, messages used in two-sided message-passing programs tend to be larger than the messages used in data-parallel or thread-based shared-memory programs, so it may also be useful to optimize the network for data rate, not just for startup latency.

One-sided communication A comparatively recent development in message passing is the use of one-sided communication, also known as *remote memory access* (RMA) or *remote direct memory access* (RDMA) communication. In contrast to the two-sided model, one-sided communication requires the explicit participation of only one process. The basic operations in one-sided communication are PUT (transfer data from the initiating node's memory to a target node's memory) and GET (transfer data from a target node's memory to the initiating node's memory). However, implementations of one-sided communication (e.g., ARMCI [310]) may also include atomic operations, synchronization operations, locks, and memory-management operations. Network support for one-sided communication may include a shared address space across nodes, demand-paged virtual memory, remote atomic operations, and of course hardware

PUT/GET primitives. Because one-sided data transfers tend to be small, low-latency/low-overhead communication is also advantageous.

As a case study, consider the Cray T3E [388], a supercomputer that represented at least 50% of the top 10 systems in the Top500 list from June 1997 through June 1999 — and all but two of the top 10 systems in June 1998. The T3E targeted data-parallel programming (High-Performance Fortran [270]), two-sided message passing (MPI [408]), and one-sided communication (Cray's SHMEM library [31]). Because of this broad scope, the T3E provided a wealth of hardware features in its interconnect. A set of "E-registers" located externally to the CPU are used both for address mapping (i.e., mapping the global address space into E-register space with highly flexible support for data striping) and for data transfer (i.e., single- or multiple-word GETs and PUTs, with the ability to pack strided data into cache lines). Once an address mapping is established, reading and writing E-registers translates into GET and PUT operations, and the issue rate is sufficiently fast that even single-word GETs and PUTs achieve high bandwidth. The T3E's network supports atomic fetch&increment, fetch&add, compare&swap, and masked-swap operations on arbitrary memory locations. Although PUT/GET and atomic operations can be used to implement two-sided message passing in software, the T3E additionally provides hardware support for message queues to help reduce the latency and overhead of messaging. Finally, the T3E supports eureka notification and barrier synchronization, including fuzzy barriers (i.e., separate entry and completion phases with intervening computation possible) and optional interrupt triggering upon completion of a eureka/barrier. Unlike the CM-5 [254] and Blue Gene/P [12], the T3E does not use a separate synchronization network but rather embeds a logical tree (configurable) within the 3-D torus data network. With all of these network features, the T3E is a prime example of how a network can be designed to support a variety of different programming models and what features may be useful to include for each model.

1.5   Future Directions

What does the future hold in store for interconnects for massively parallel systems? While it is impossible to know for certain, we can examine a few technologies that are currently being developed.

There has been some recent research into the use of high-radix switches [229] — crossbars with large numbers of ports — for constructing interconnection networks. The insight is that recent technological advances have substantially increased the achievable bandwidth per wire so a reduced number of wires per port can still deliver acceptable per-port bandwidth while an increased number of ports significantly reduces latency and cost (number of pins, connectors, and switches). While the crossbar chips used in interconnection networks have long utilized modest port counts (Myrinet [48]: 16 or 32; Mellanox's InfiniBand [217] switches: 24; Quadrics's QsNet [338]: 24), researchers are actively considering how to build and utilize switches containing significantly more ports than that [227, 229].

A recently demonstrated technology that may have deep implications for interconnection networks is Sun Microsystems's *proximity communication* [125]. Instead of connecting neighboring chips with wires, proximity communication overlaps the chips' surfaces and uses capacitive coupling, inductive coupling, optical transmission, or another technology to transmit signals between surfaces. The result is an estimated 60-fold improvement in signaling density over conventional, area-ball-bonded I/O as well as high signaling rates, low bit-error rates, and low power utilization per bit. Already, researchers have proposed techniques for utilizing proximity communication to construct very high-radix, single-stage crossbars, including a 1728-port crossbar comprising only 16 chips [131].

While proximity communication targets delivering very high performance across short distances, another technology, optical communication — transmitting beams of light through fiber or free space instead of transmitting electrical signals through copper or another conductor — targets high communication performance across longer distances. As an example of a recently proposed form of optical communication, optical circuit switching [29] uses arrays of mirrors based on micro electro-mechanical systems (MEMS) that rotate to reflect optical data from an input port to an output port [328]. The advantages of optical circuit-switched interconnects are that they do not require high-latency conversions between electrical and optical signals at each switch; they can deliver low-latency, high-bandwidth communication effectively independently of physical distance; they are likely to be less expensive and less power-hungry than interconnects based on electrical switches; and they can support high-radix (~1024-port) switches to keep the network diameter low. The primary disadvantage of optical circuit-switched interconnects is that connection setup is extremely slow: Rotating a MEMS mirror can take several milliseconds. Barring radical technological improvements or extremely large mirror arrays that can support simultaneous connections from each node to every other node, the key is to consider programming models, as discussed in Section 1.4.1. Nodes may require connections to only a few other nodes, thereby making optical circuit switching a practical technology in such instances [29]. While optical circuit switching in particular is not used in any existing massively parallel system, there has already been much research on optical interconnects in general, including entire workshops and conferences devoted to the subject (e.g., MPPOI [203]).

Finally, with transistor counts still increasing — if not faithfully following Moore's Law [298] — technology is enabling more functionality than ever before to appear on a single chip. It therefore stands to reason that NICs and perhaps also switches integrated onto a chip will soon become commonplace. However, one can do better than simply to combine an existing CPU with an existing NIC. As a historical example, the Intel/CMU iWarp processor [51] integrated communication into the CPU's *register* space to support exceptionally fine-grained communication. Special "gate" registers could be associated with message queues. Writing to a gate register caused data to be injected into the network, and reading from a gate register caused data to be read from the network. Consequently, two single-word receives, an arithmetic operation on those words, and a single-word send of the result could be performed with a single machine instruction. The iWarp integrated this low-overhead word-by-word communication with a DMA engine that enabled overlap of communication and computation; a single message transfer could in fact utilize either or both mechanisms. Perhaps future interconnects for massively parallel systems will revisit the iWarp's approach of integrating communication into the deepest parts of the processor core to reduce communication overhead and thereby increase parallel efficiency.

1.6   Chapter Summary

This chapter examined some of the issues involved in designing an interconnection network for a massively parallel supercomputer. An important, recurring theme throughout the chapter is that there are numerous tradeoffs to be made at every stage of network design. Section 1.1 listed various tradeoffs for the network as a whole; Section 1.2 showed how various tradeoffs in network design affect certain aspects of communication performance; Section 1.3 focused on tradeoffs in the topology alone; and Section 1.4 discussed how adding specialized features to the network can help performance if chosen well, hurt performance if not, and merely increase costs if not applicable to the target programming model. Even the forthcoming technologies presented in Section 1.5 improve some aspects of network communication while negatively impacting other aspects.

Another message that runs through this chapter is that one needs to beware of differences between theory and practice when architecting a large-scale interconnect: massive parallelism is where interconnection-network theory and practice diverge. Although meshes, tori, and fat trees have well-understood theoretical properties, Section 1.2 showed that actual, massively parallel systems that employ those topologies may exhibit unexpected performance characteristics because of subtle implementation details such as node mappings or bandwidth ratios between NICs and switches. While the networking literature presents an abundance of interesting network topologies, many of which are provably optimal for some particular metric, Section 1.3 showed that only a few basic topologies have ever been represented by the world's fastest or largest-scale supercomputers. Even hypercubes, long admired by many for their nice mathematical properties, have little representation in current massively parallel systems. Sections 1.4 and 1.5 emphasize that usage models and technology trends play a key role in interconnect design. A network that is optimized for data-parallel programming may prove to be suboptimal when running message-passing programs and vice versa. A network that exploits switch chips containing only a small number of low-bandwidth connectors may be the wrong network to use once switch chips with large numbers of high-bandwidth connectors become feasible.

In short, there is no one, right way to design a high-performance interconnect for a massively parallel system. Costs, usage models, and technological constraints dictate the tradeoffs that must be made to produce an interconnection network that best matches its intended purpose.

Chapter 2

Commodity High Performance Interconnects

Dhabaleswar K. (DK) Panda
The Ohio State University
Pavan Balaji
Argonne National Laboratory
Sayantan Sur
IBM TJ Watson Research Center
Matthew Jon Koop
The Ohio State University

2.1 Introduction

Clusters with commodity processors and commodity interconnects are gaining momentum to provide cost-effective processing power for a range of applications in multiple disciplines. Examples of such applications include: high-end computing, parallel file systems, multi-tier web datacenters, web servers, visualization servers, multimedia servers, database and data-mining systems, etc. As the performance and features of commodity processors and commodity networking continue to grow, clusters are becoming popular to design systems with scalability, modularity and upgradability. According to the TOP500 list [434], clusters entered the high-end computing market in a very small manner (only 6%) in 2001. However, as of November 2008, 82% of TOP500 systems are clusters.

These commodity clusters require cost-effective commodity interconnects with varying requirements: Systems Area Networks with good inter-processor communication performance (low latency, high bandwidth and low CPU utilization), Storage Area Networks with high performance I/O, good Wide Area Network connectivity, Quality of Service (QoS) and support for Reliability, Availability and Serviceability (RAS). Keeping these requirements in mind, many different commodity interconnects have emerged during the last ten years. We provide an overview of these interconnects and their associated features in the next section. From all these commodity interconnects, InfiniBand architecture is gaining a lot of momentum in recent years. In the latest TOP500 list (November '08), 28.2% clusters use InfiniBand. In the following sections of this chapter, we provide a detailed overview of InfiniBand architecture. We start the discussion with switches and adapters, and gradually move up the InfiniBand software stack, all the way up to designing MPI, Parallel file systems and Enterprise data-centers. Finally, we discuss current adoptions and community usage of InfiniBand and its future trends.

2.2   Overview of Past Commodity Interconnects, Features and Trends

As the commodity clusters started getting popular around mid-nineties, the common interconnect used for such clusters was Fast Ethernet (100 Mbps). These clusters were identified as Beowulf clusters [35]. Even though it was cost-effective to design such clusters with Ethernet/Fast Ethernet, the communication performance was not very good because of the high overhead associated with the standard TCP/IP communication protocol stack. The high overhead was because of the TCP/IP protocol stack being completely executed on the host processor. The Network Interface Cards ((NICs) or commonly known as network adapters) on those systems were also not intelligent. Thus, these adapters didn't allow any overlap of computation with communication or I/O. This led to high latency, low bandwidth and high CPU requirement for communication and I/O operations on those clusters.

To alleviate the communication performance bottlenecks, researchers from academia and industry started exploring the design of intelligent NICs to offload some part of the communication protocols to the NICs. They also started designing user-level communication protocols with Operating Systems (OS)-Bypass [44,47,446]. The basic idea was to register user communication buffers (send and receive) with the network adapters. This step involves the help from OS. After the buffers are registered, communication can take place directly by the user processes. Intelligence in the network adapters also allowed data to be fetched from or stored to the registered memory by Direct Memory Access (DMA) operations. The adapters also started providing additional logic (and sometimes programmability) to implement checksum computation/verification, reliability through data retransmission in case of packet drops, etc. Such offload together with the concept of user-level communication protocols led to low latency, high bandwidth, low CPU utilization and overlap of computation with communication and I/O.

Leading projects focusing on these novel designs were: U-Net on ATM Network [446] and Fast Message (FM) on Myrinet network [325]. These designs allowed clusters to use commodity interconnects (such as Myrinet [48] and ATM [20]) for clusters. The ATM standard originally came from the telephony world and didn't sustain long among the clusters for high-end computing.

However, Myrinet became a very popular interconnect for commodity clusters. During the late nineties, Myricom (the designer of the proprietary Myrinet interconnect) kept on introducing faster and intelligent adapters, faster switches and faster links to match with the faster computing platforms. This led to large-scale deployment of Myrinet clusters. The June 2002 ranking of TOP500 systems reported 32.4% of the systems using Myrinet.

As Myrinet was gaining popularity, several other organizations also started designing high performance interconnects for clusters. Quadrics [353] led to the design of extremely high performance QSNet adapters and switches. Many large-scale clusters used Quadrics to obtain best performance and scalability. A new industry consortium consisting of Intel, Compaq and Microsoft attempted to standardize the concepts of user-level communication and intelligent network adapters into an *open standard* called *Virtual Interface Architecture* (VIA). The most prominent cluster interconnect supporting VIA in hardware was GigaNet [162].

As the above interconnects were used in commodity clusters to achieve good inter-processor communication, the popular interconnect for storage remained as Fibre Channel [375]. The Fibre Channel was standardized in 1994 and worked as a successor to the HIPPI protocol [92]. The Fibre Channel remained successful as it allowed SCSI commands to be transported over Fibre Channel networks and allowed the design of scalable storage systems with SCSI disks.

During late nineties, major commodity clusters had three kinds of interconnects: Myrinet or Quadrics for inter-processor communication, Fibre Channel for I/O, and Ethernet (Fast Ethernet or Gigabit Ethernet) for communication over Wide Area Network (WAN). The use of three different interconnects was pushing the cost of clusters (both for high-end computing and commercial data-centers) significantly higher. The management of all three interconnects was also complex, demanding expertise from a set of network administrators. The popular inter-processor interconnects (Myrinet and Quadrics) were also *proprietary*, not open standards. Thus, it was difficult to add new features to these interconnects.

The above limitations led to a goal for designing a new and converged interconnect with open standard. Seven industry leaders (Compaq, Dell, IBM, Intel, Microsoft and Sun) formed a new InfiniBand Trade association [207]. The charter for this association was "to design a scalable and high performance communication and I/O architecture by taking an integrated view of computing, networking and storage technologies." Many other companies participated in this consortium later. The members of this consortium deliberated and defined the InfiniBand architecture specification. The first specification (Volume 1, Version 1.0) was released to the public on October 24, 2000. Since then this standard is becoming enhanced periodically. The latest version is 1.2.1, released during January 2008.

The word "InfiniBand" is coined from two words "Infinite Bandwidth." The architecture is defined in such a manner that as the speed of computing and networking technologies improve over time, the InfiniBand architecture should be able to deliver higher and higher bandwidth. During the inception in 2001, InfiniBand was delivering a link speed of 2.5 Gbps (payload data rate of 2 Gbps, due to the underlying 8/10 encoding and decoding). Now in 2009, it is able to deliver a link speed of 40 Gbps (payload data rate of 32 Gbps).

In the next few sections, we discuss in depth the InfiniBand architecture and its features.

2.3   InfiniBand Architecture

The InfiniBand Architecture (IBA) describes a switched interconnect technology for inter-processor communication and I/O. The architecture is independent of the host operating system and the processor platform. However, several implementations of the InfiniBand hardware and software stacks are more closely tied to the OS and various system components.

2.3.1   IB Communication Model

IB's communication model comprises an interaction of multiple components in the interconnect system. In this chapter, we briefly discuss some of these components.

2.3.1.1   IB Topology and Network Components

At a high-level, IBA serves as an interconnection of nodes, where each node can be a processor node, an I/O unit, a switch or a router to another network, as illustrated in Figure 2.1. Processor nodes or I/O units are typically referred to as "end nodes," while switches and routers are referred to as "intermediate nodes" (or sometimes as "routing elements"). An IBA network is subdivided into subnets interconnected by routers. Overall, the IB fabric is comprised of four different components: (a) channel adapters, (b) links and repeaters, (c) switches and (d) routers.

Channel Adapters. A processor or I/O node connects to the fabric using channel adapters connecting them to the IB fabric. These channel adapters consume and generate IB packets. Most current channel adapters are equipped with programmable direct memory access (DMA) engines with protection features implemented in hardware. Each adapter can have one or more physical ports connecting it to either a switch/router or another adapter. Each physical port itself internally maintains two or more virtual channels (or virtual lanes) with independent buffering for each of them. The channel adapters

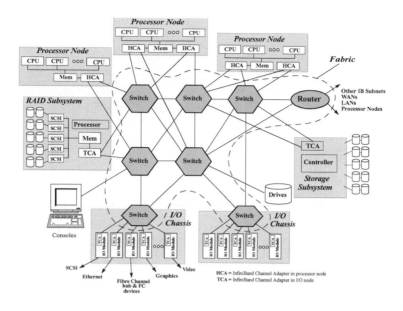

FIGURE 2.1: Typical InfiniBand cluster (Courtesy InfiniBand Standard).

also provide a memory translation and protection (MTP) mechanism that translates virtual addresses to physical addresses and validates access rights. Each channel adapter has a globally unique identifier (GUID) assigned by the channel adapter vendor. Additionally, each port on the channel adapter has a unique port GUID assigned by the vendor as well.

Links and Repeaters. Links interconnect channel adapters, switches, routers and repeaters to form an IB network fabric. Different forms of IB links are currently available, including copper links, optical links and printed circuit wiring on a backplane. Repeaters are transparent devices that extend the range of a link. Links and repeaters themselves are not accessible in the IB architecture. However, the status of a link (e.g., whether it is up or down) can be determined through the devices which the link connects.

Switches. IBA switches are the fundamental routing components for intra-subnet routing. Switches do not generate or consume data packets (they generate/consume management packets). Every destination within the subnet is configured with one or more unique local identifiers (LIDs). Packets contain a destination address that specifies the LID of the destination (DLID). The switch is typically configured out-of-band with a forwarding table that allows it to route the packet to the appropriate output port. This out-of-band configuration of the switch is handled by a separate component within IBA called the subnet manager, as we will discuss later in this chapter. It is to be

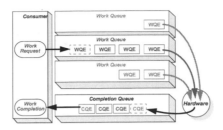

FIGURE 2.2:   Consumer queuing model (Courtesy InfiniBand Standard).

noted that such LID-based forwarding allows the subnet manager to configure multiple routes between the same two destinations. This, in turn, allows the network to maximize availability by re-routing packets around failed links through reconfiguration of the forwarding tables.

IBA switches support unicast forwarding (delivery of a single packet to a single destination) and may optionally support multicast forwarding (delivery of a single packet to multiple destinations) as well.

Routers. IBA routers are the fundamental routing component for inter-subnet routing. Like switches, routers do not generate or consume data packets; they simply pass them along. Routers forward packets based on the packet's global route header and replace the packet's local route header as the packet passes from one subnet to another. The primary difference between a router and a switch is that routers are not completely transparent to the end-nodes since the source must specify the LID of the router and also provide the global identifier (GID) of the destination. IB routers use the IPv6 protocol to derive their forwarding tables.

2.3.1.2   IB Messaging

IBA operations are based on the ability of a consumer to queue up a set of instructions that the hardware executes. In the IB terminology, this facility is referred to as a work queue. In general, work queues are created in pairs, referred to as Queue Pairs (QPs), one for send operations and one for receive operations. The consumer process submits a work queue element (WQE) to be placed in the appropriate work queue. The channel adapter executes WQEs in the order in which they were placed. After executing a WQE, the channel adapter places a completion queue entry (CQE) in a completion queue (CQ); each CQE contains or points to all the information corresponding to the completed request.

Send Operation. There are three classes of send queue operations: (a) Send, (b) Remote Direct Memory Access (RDMA) and (c) Memory Binding.

For a send operation, the WQE specifies a block of data in the consumer's memory. The IB hardware uses this information to send the associated data to the destination. The send operation requires the consumer process to pre-post a receive WQE, which the consumer network adapter uses to place the incoming data in the appropriate location.

For an RDMA operation, together with the information about the local buffer and the destination end point, the WQE also specifies the address of the remote consumer's memory. Thus, RDMA operations do not need the consumer to specify the target memory location using a receive WQE. There are four types of RDMA operations: RDMA write, RDMA write with immediate data, RDMA read, and Atomic. In an RDMA write operation, the sender specifies the target location to which the data has to be written. In this case, the consumer process does not need to post any receive WQE at all. In an RDMA write with immediate data operation, the sender again specifies the target location to write data, but the receiver still needs to post a receive WQE. The receive WQE is marked complete when all the data is written. An RDMA read operation is similar to an RDMA write operation, except that instead of writing data to the target location, data is read from the target location. Finally, there are two types of atomic operations specified by IBA: (i) atomic compare and swap and (ii) atomic fetch and increment. Both operations operate only on 64-bit integral datatypes, and occur within a critical section. It is to be noted that these operations are atomic only as seen by the network adapter, i.e., if the network adapter receives two atomic operation requests for the same location simultaneously, the second operation is not started till the first one completes. However, such atomicity is not guaranteed when another device is operating on the same data. For example, if the CPU or another InfiniBand adapter modifies data while one InfiniBand adapter is performing an atomic operation on this data, the target location might get corrupted.

Memory binding does not directly perform any data communication, but allows a process to specify which portions of memory it shares with other nodes. For example, it can specify which portions of memory, a remote process can write to, or read from, or both. This operation produces a memory key (called a remote key or R_KEY) that the consumer process can provide to other processes allowing them to access its memory.

Receive Operations. Unlike the send operations, there is only one type of receive operation. For a send operation, the corresponding receive WQE specifies where to place data that is received, and if requested places the receive WQE in the completion queue. For an RDMA write with immediate data operation, the consumer process does not have to specify where to place the data (if specified, the hardware will ignore it). When the message arrives, the receive WQE is updated with the memory location specified by the sender, and the WQE placed in the completion queue, if requested.

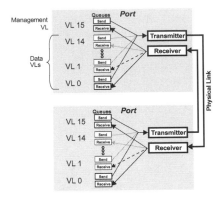

FIGURE 2.3:   Virtual lanes (Courtesy InfiniBand Standard).

2.3.2   Overview of InfiniBand Features

IBA describes a multi-layer network protocol stack. Each layer provides several features that are usable by applications. In this section, we present an overview of some of the features offered by IB. We divide up the features into different layers of the network stack: (i) link layer features, (ii) network layer features and (iii) transport layer features.

2.3.2.1   Link Layer Features

CRC-based data integrity. IBA provides two forms of CRC-based data integrity to achieve both early error detection as well as end-to-end reliability: invariant CRC and variant CRC.

Invariant CRC (ICRC) covers fields that do not change on each network hop. This includes routing headers (which do not change unless the packet passes through any routers), transport headers and the data payload. This is a strong 32-bit CRC that is compatible with Ethernet CRC. ICRC is only calculated once at the sender and is verified at the receiver since it does not change within the network fabric. Thus, it provides an end-to-end reliability. Note that, here, end-to-end only refers to communication from the source adapter to the destination adapter and does not include the remaining system components such as the I/O bus and the memory bus.

Variant CRC (VCRC), on the other hand, covers the entire packet including the variant as well as the invariant fields. This is a weaker 16-bit CRC to allow for faster calculation, since this CRC needs to be calculated on each network hop. This also allows the network fabric to drop erroneous packets early, without having to transmit them to their final destination.

Buffering and Flow Control. IBA provides an absolute credit-based flow-control where the receiver guarantees that it has enough space allotted to receive N blocks of data. The sender sends only N blocks of data before waiting for an acknowledgment from the receiver. The receiver occasionally updates the sender with an acknowledgment as and when the receive buffers get freed up. Note that this link-level flow control has no relation to the number of messages sent, but only to the total amount of data that has been sent.

Virtual Lanes, Service Levels and QoS. Virtual Lanes (VLs) are a mechanism that allow the emulation of multiple virtual links within a single physical link as illustrated in Figure 2.3. Each port provides at least 2 and up to 16 virtual lanes (VL0 to VL15). VL15 is reserved exclusively for subnet management traffic. All ports support at least one data VL (VL0) and may provide VL1 to VL15. Each VL is independent with respect to its buffering capability as well as its flow control. That is, traffic on one VL does not block traffic on another VL.

Together with VLs, IBA also defines different QoS levels called service levels (SLs). Each packet has a SL which is specified in the packet header. As the packet traverses the fabric, its SL determines which VL will be used on each link. However, IBA does not define this mapping and leaves it open to be determined by the subnet manager as configured by the system administrator.

Together with VL-based QoS provisioning, SLs also allow for fabric partitioning. Fabric administration can assign SLs for particular platforms. This allows the network to allow traffic only in certain parts of the networks. Note that this mechanism does not provide complete electrical segregation of network partitions, but instead relies on SL partition tables configured on switches to allow or disallow traffic in different parts of the network.

Hardware Multicast. IBA optionally provides a feature allowing the network switches and routers to perform hardware multicast of data packets. The architecture specifies a detailed group management scheme including algorithms to create, join, leave, delete and prune multicast groups.

Multicast groups are identified by unique GIDs. Processing requests to join or leave multicast groups uses management routines, that are discussed later in this section. When a process joins a multicast group, the management infrastructure updates all the switches and routers participating in the multicast with information about this new process; thus, when any of these switches or routers gets a packet for that multicast group, a copy is forwarded to the new process as well. The management infrastructure ensures that there are no loops by forming a single spanning tree of links.

The sender process uses the multicast LID and GID as the destination LID for the packet. When the switch receives this packet, it replicates the packet in hardware and sends it out to each of the ports designated by the management infrastructure, except the arrival port. On a destination node, if multiple processes have registered for the same multicast group, the channel adapter replicates and distributes the packet amongst the different QPs.

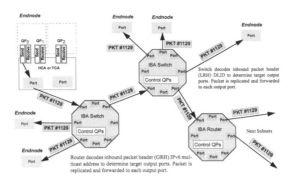

FIGURE 2.4: Example of unreliable multicast operation (Courtesy Infini-Band Standard).

Congestion Control. Together with flow-control, IBA also defines a multistep congestion control mechanism for the network fabric. Specifically, congestion control when used in conjunction with the flow-control mechanism is intended to alleviate head-of-line blocking for non-congested flows in a congested environment. To achieve effective congestion control, IB switch ports need to first identify whether they are the root or the victim of a congestion. A switch port is a root of a congestion if it is sending data to a destination faster than it can receive, thus using up all the flow-control credits available on the switch link. On the other hand, a port is a victim of a congestion if it is unable to send data on a link because another node is using up all of the available flow-control credits on the link.

In order to identify whether a port is the root or the victim of a congestion, IBA specifies a simple approach. When a switch port notices congestion, if it has no flow-control credits left, then it assumes that it is a victim of congestion. On the other hand, when a switch port notices congestion, if it has flow-control credits left, then it assumes that it is the root of a congestion. This approach is not perfect: it is possible that even a root of a congestion may not have flow-control credits remaining at some point of the communication (for example, if the receiver process is too slow in receiving data); in this case, even the root of the congestion would assume that it is a victim of the congestion. Thus, though not required by the IBA, in practice, IB switches are configured to react with the congestion control protocol irrespective of whether they are the root or the victim of the congestion or not.

The IBA congestion control protocol consists of three steps. In the first step, a port that detects congestion marks a subset of the packets it is transmitting with a congestion bit known as the Forward Explicit Congestion Notification (FECN). When the target channel adapter receives a packet with a FECN bit set, it informs the sender about the congestion by setting a return packet (either a data or an acknowledgment packet) with a different congestion bit known as the Backward Explicit Congestion Notification (BECN). Finally, when the sender receives a packet with the BECN bit set, it temporarily slows down its transmission rate hoping to alleviate the congestion in the fabric.

IB Multipathing. While IBA itself does not specify any generic dynamic multipathing capability, the architecture allows end nodes to use multiple paths simultaneously. Each port within an IB subnet is configured with one or more unique LIDs. A switch can be configured to use a different route for each of the LIDs. Thus, a source can pick different routes by using different DLIDs to reach the required destination.

Static Rate Control. IBA defines a number of different link bit rates, such as 1x SDR (2.5 Gbps), 4x SDR (10 Gbps), 12x SDR (30 Gbps), 4x DDR (20 Gbps) and so on. It is to be noted that these rates are signaling rates; with an 8b/10b data encoding, the actual maximum data rates would be 2 Gbps (1x SDR), 8 Gbps (4x SDR), 24 Gbps (12x SDR) and 16 Gbps (4x DDR), respectively. To simultaneously support multiple link speeds within a fabric, IBA defines a static rate control mechanism that prevents faster links from overrunning the capacity of ports with slower links. Using a static rate control mechanism, each destination has a timeout value based on the ratio of the destination and source link speeds. If the links are homogeneous, there is no timeout; if the source link is faster than the destination link, the source sends out packets only every time the timeout expires, thus pacing its data injection rate.

2.3.2.2 Network Layer Features

IB Routing. IB network-layer routing is similar to the link-layer switching, except for two primary differences. First, network-layer routing relies on a global routing header (GRH) that allows packets to be routed between subnets. Second, VL15 (the virtual lane reserved for management traffic) is not respected by IB routers since management traffic stays within the subnet and is never routed across subnets.

Flow Labels. Flow labels identify a single flow of packets. For example, these can identify packets belonging to a single connection that need to be delivered in order, thus allowing routers to optimize traffic routing when using multiple paths for communication. IB routers are allowed to change flow labels as needed, for example to distinguish two different flows which have been given the same label. However, during such relabeling, routers ensure that all packets corresponding to the same flow will have the same label.

2.3.2.3 Transport Layer Features

IB Transport Services. IBA defines four types of transport services (reliable connection, reliable datagram, unreliable connection and unreliable datagram), and two types of raw communication services (raw IPv6 datagram

Attribute	Reliable Connection	Reliable Datagram	Unreliable Datagram	Unreliable Connection	Raw Datagram (both IPv6 & ethertype)
Scalability (M processes on N Processor nodes communicating with all processes on all nodes)	M^2*N QPs required on each processor node, per CA	M QPs required on each processor node, per CA.	M QPs required on each processor node, per CA.	M^2*N QPs required on each processor node, per CA.	1 QP required on each end node, per CA.

Reliability

Corrupt data detected	Yes				
Data delivery guarantee	Data delivered exactly once		No guarantees		
Data order guaranteed	Yes, per connection	Yes. packets from any one source QP are ordered to multiple destination QPs.	No	Unordered and duplicate packets are detected.	No
Data loss detected	Yes		No	Yes	No
Error recovery	**Reliable**. Errors are detected at both the requestor and the responder. The requestor can transparently recover from errors (retransmission, alternate path, etc.) without any involvement of the client application. QP processing is halted only if the destination is inoperable or all fabric paths between the channel adapters have failed.	**Unreliable**. Packets with some types of errors may not be delivered. Neither source nor destination QPs are informed of dropped packets.	**Unreliable**. Packets with errors, including sequence errors, are detected and may be logged by the responder. The requestor is not informed.	**Unreliable**. Packets with errors are not delivered. The requestor and responder are not informed of dropped packets.
RDMA and ATOMIC Operations	Yes	Yes	No	Yes: RDMA WRITEs No: RDMA READs & ATOMICs	No
Bind Memory Window	Yes	Yes	No	Yes	No
IBA Unreliable Multicast Support	No	No	Yes	No	No
Raw Multicast	No	No	No	No	Yes
Message Size	Message size 0 to 2^{31} bytes. Smaller max size may be negotiated by Connection Management. A message may consist of multiple packets.	**Single PMTU packet** datagrams - 0 to 4096 bytes.	Message size 0 to 2^{31} bytes. Smaller max size may be negotiated by Connection Management. A message may consist of multiple packets.	**Single PMTU packet** datagrams - 0 to 4096 bytes.
Connection Oriented?	**Connected**. The client connects the local QP to one and only one remote QP. No other traffic flows over these QPs.	**Connectionless**. Appears connectionless to the client - uses one or more End-to-End contexts per CA to provide reliability service.	**Connectionless**. No prior connection is needed for communication.	**Connected**. The client connects the local QP to one and only one remote QP. No other traffic flows over these QPs.	**Connectionless**. No prior connection is needed for communication.

FIGURE 2.5: IB transport services (Courtesy InfiniBand Standard).

and raw Ethertype datagram) to allow for encapsulation of non-IBA protocols. The transport service describes the degree of reliability and how the data is communicated.

As illustrated in Figure 2.5, each transport service provides a trade-off in the amount of resources used and the capabilities provided. For example, while the reliable connection provides the most features, it has to establish a QP for each peer process it communicates such, thus requiring a quadratically increasing number of QPs with the number of processes. Reliable and Unreliable datagram services, on the other hand, utilize fewer resources since a single QP can be used to communicate with all peer processes.

Further, for the reliable and unreliable connection models, InfiniBand allows message-based communication. That is, processes do not have to segment their messages into the size of the maximum transmission unit or frame size that is sent on the network. The network hardware handles this. The hardware also maintains message boundaries; thus, the send WQE is marked complete after the entire message is sent out, and the receive WQE posted by the consumer process is marked complete after the entire message is received. This is especially beneficial for large message communication, where the host processor does not have to be interrupted to continue transmitting or receiving the message at regular time intervals.

Automatic Path Migration. The Automatic Path Migration (APM) feature of InfiniBand allows a connection end-point to specify a fallback path, together with a primary path for a data connection. All data is initially sent over the primary path. However, if the primary path starts throwing excessive errors or is heavily loaded, the hardware can automatically migrate the connection to the new path, after which all data is only sent over the new path. Such migration follows a multi-step protocol which is handled entirely in hardware without any intervention from either the source or the destination process. Once the migration is complete, another alternate path can be loaded for the channel adapter to utilize, but this requires intervention from the management infrastructure. In summary, the APM feature allows the first migration to be completely in hardware, but the later migrations require software intervention.

Transaction Ordering. IBA specifies several ordering rules for data communication, as described below:

- A requester always transmits request messages in the order in which the WQEs are posted. Send and RDMA write WQEs are completed in the order in which they are transmitted. RDMA read and atomic WQEs, though started in order, might be completed out-of-order with the send and RDMA write WQEs.

- Receiving of messages (on a send or RDMA write with immediate data) is strictly ordered as they were transmitted. If messages arrive out-of-order, they are either buffered or dropped to maintain the receive order.

- Receive WQEs always complete in the order in which they are posted.

- If a send WQE describes multiple data segments, the data payload will use the same order in which the segments are specified.

- The order in which a channel adapter writes data bytes to memory is unspecified by IBA. For example, if an application sets up data buffers that overlap for separate data segments, the resultant buffer content is non-deterministic.

- A WQE can be set with a fence attribute which ensures that all prior requests are completed before this request is completed.

Message-level Flow Control. Together with the link-level flow-control, for reliable connection, IBA also defines an end-to-end message level flow-control mechanism. This message-level flow-control does not deal with the number of bytes of data being communicated, but rather with the number of messages being communicated. Specifically, a sender is only allowed to send as many messages that use up receive WQEs (i.e., send or RDMA write with immediate data messages) as there are receive WQEs posted by the consumer. That is, if the consumer has posted 10 receive WQEs, after the sender sends out 10 send or RDMA write with immediate data messages, the next message is not communicated till the receiver posts another receive WQE.

Since message-level flow control does not deal with the number of bytes of data that are being communicated, it is possible that the receiver post a receive WQE for fewer bytes than the incoming message. However, this flow control does not disallow such behavior. It is up to the consumer to ensure that such erroneous behavior does not take place.

Shared Receive Queue (SRQ). Introduced in the InfiniBand 1.2 specification, Shared Receive Queues (SRQs) were added to help address scalability issues with InfiniBand memory usage.

When using the RC transport of InfiniBand, one QP is required per communicating peer. To prepost receives on each QP, however, can have very high memory requirements for communication buffers. To give an example, consider a fully-connected MPI job of 1K processes. Each process in the job will require $1K - 1$ QPs, each with n buffers of size s posted to it. Given a conservative setting of $n = 5$ and $s = 8KB$, over 40MB of memory per process would be required simply for communication buffers that may not be used. Given that current InfiniBand clusters now reach 60K processes, maximum memory usage would potentially be over 2GB per process in that configuration.

Recognizing that such buffers could be pooled, SRQ support was added, so instead of connecting a QP to a dedicated RQ, buffers could be shared across QPs. In this method, a smaller pool can be allocated and then refilled as needed instead of pre-posting on each connection.

Extended Reliable Connection (XRC). eXtended Reliable Connection (XRC) provides the services of the RC transport, but defines a very different connection model and method for determining data placement on the receiver in channel semantics. This mode was mostly designed to address multi-core clusters and lower memory consumption of QPs.

For one process to communicate with another over RC each side of communication must have a dedicated QP for the other. There is no distinction as to the node in terms of allowing communication. In XRC, a process no longer needs to have a QP to every process on a remote node. Instead, once one QP to a node has been set up, messages can be routed to the other processes by giving the address/number of a Shared Receive Queue (SRQ). In this model the number of QPs required is based on the number of nodes in the job rather than the number of processes.

2.3.3 InfiniBand Protection and Security Features

InfiniBand provides a detailed set of security and protection features to allow for secure communication capabilities for applications. Some of these features are described in this section.

Virtual Memory Management. For security and ease-of-use, IBA requires that consumers directly use virtual address spaces for communication. However, like all other hardware devices, InfiniBand itself performs its data movement between physical address pages. Thus, it requires that the translation from virtual address space to physical address space happen in a reliable and secured manner without much overhead in data communication. In order to handle such translation, IBA requires that the buffers that are used for communication be *registered* before the actual communication starts.

Memory registration provides mechanisms that allow IBA consumer processes to describe a virtual address space that would be used for communication. Apart from the advantage that this allows consumer processes to perform communication without having to deal with the virtual-to-physical address translation, and the benefit of avoiding unauthorized access to a process' memory, memory registration also allows the IB device and its associated software to perform several other tasks. First, the IB driver can pin the communication buffers in the kernel. This ensures that the kernel does not page out these buffers during communication, thus corrupting either the data being communicated or other memory buffers. Second, this can potentially also allow the IB device to cache such translation on the device itself; thus, during communication the appropriate physical page lookup can happen efficiently.

Unlike some other high-speed networks, such as Quadrics [353], IBA does not specify a page fault mechanism in hardware. That is, if a buffer being used for communication has not been registered and the hardware device cannot find a virtual-to-physical address translation for the buffer, this is considered an error event.

Memory Protection (Keys and Protection Domains). IBA provides various keys to allow for isolation and protection. While these keys are used for security and protection, in the current IBA standard, the keys themselves are not secure because they are not encrypted on the physical link. Thus, a network snooper can get access to these keys. Here, we discuss two types of keys, namely, *Local Memory Key (L_KEY)* and *Remote Memory Key (R_KEY)*. These two keys allow IB consumers to control access to their memory. When a consumer registers its memory, the IB device and its associated software returns an *L_KEY* and an *R_KEY* to the consumer. The *L_KEY* is used in WQEs to describe local memory permissions to the QP. The consumer can also pass the *R_KEY* to its peer processes, which in turn can use this key to access the corresponding remote memory using RDMA and atomic operations.

In some cases, memory protection keys alone are not sufficient for appropriate management of permissions. For example, a consumer might not want to let all peer processes have the same access to its registered memory. To handle such scenarios, IBA also specifies the concept of a *protection domain*. Protection domains allow a consumer to control which QPs can access which memory regions. Protection domains are created before registering memory and are associated with the memory registration call. Thus, the memory keys such as *L_KEY* and *R_KEY* are only valid in the context of a protection domain and are not globally usable.

Partition Isolation. IBA provides partition isolation (or partitioning) capability that allows a set of nodes to behave as a separate cluster disconnected from the remaining nodes in the fabric. Such partition isolation is not provided through any electrical decoupling, but rather through key-based lookups on IB channel adapters and optionally on IB switches and routers as well.

Each port of an end-node is a member of at least one partition (but could be a member of multiple partitions as well). Partitions are differentiated using a separate key known as the *P_KEY*, which is used in all communication packets. If a channel adapter receives a packet for a partition of which it is not a member, this packet is discarded. Partitioning enforced by channel adapters is the most commonly used method, but is inefficient since a packet has to traverse the network till the end node before the channel adapter identifies that it has to be discarded. Configuring switches and routers to perform such partition isolation is more effective, but requires more intervention from the IB management infrastructure to set it up.

2.3.4 InfiniBand Management and Services

Together with regular communication capabilities, IBA also defines an elaborate management semantics and several services. The basic management messaging is handled through special packets called management datagrams (MADs). The IBA management model defines several management classes: (i) subnet management, (ii) subnet administration, (iii) communication management, (iv) performance management, (v) device management, (vi) baseboard management, (vii) SNMP tunneling, (viii) vendor specific and (ix) application specific. Brief overviews of each of these classes are provided in this section.

Subnet Management. The subnet management class deals with discovering, initializing, and maintaining an IB subnet. It also deals with interfacing with diagnostic frameworks for handling subnet and protocol errors. For subnet management, each IB subnet has at least one subnet manager (SM). An SM can be a collection of multiple processing elements, though they appear as a single logical management entity. These processes can internally make management decisions, for example to manage large networks in a distributed fashion. Each channel adapter, switch and router, also maintains a subnet management agent (SMA) that interacts with the SM and handles management of the specific device on which it resides as directed by the SM. An SMA can be viewed as a worker process, though in practice it could be implemented as hardware logic.

While a subnet can have more than one SM, only one will be active (called the master SM) and the rest must be in standby to take over in case the master SM fails. The master SM is responsible for several tasks such as the following:

1. Discovering the physical topology of the subnet.

2. Assigning LIDs to each port in the system (including channel adapters, switches and routers).

3. Setting up forwarding tables for switches, thus configuring routes based on the DLID.

4. Sweeping the network subnet and dynamically discovering topology changes.

Subnet Administration. Subnet administration (SA) deals with providing consumers with access to information related to the subnet through the subnet management interface (SMI). Most of this information is collected by the SM, and hence the SA works very closely with the SM. In fact, in most implementations, a single logical SM processing unit performs the tasks of both the SM and the SA. The SA typically provides information that cannot be locally computed to consumer processes (such as data paths, SL-to-VL mappings, partitioning information). Further, it also deals with inter-SM management such as handling standby SMs.

The SA performs most of its interaction using special types of MAD packets known as SA MADs. Consumers can use these MADs to query the SA, or request it to perform subnet administration tasks such as reconfiguring the topology through the subnet manager. The SA management class defines an elaborate set of communication methods including raw management datagrams as well as a reliable multi-packet transaction protocol for some tasks. Most communication between consumers and the SA happens on a special QP reserved for subnet management traffic (QP0).

One of the most widely used capabilities provided by the SA is multicast group management. This includes creating and deleting groups, joining and leaving a group, querying a group, etc. The actual multicast data traffic, however, does not go through the SA. Once the group is formed, all data forwarding is directly handled by the appropriate switches and routers.

Communication Management. Communication management deals with the protocols and mechanisms used to establish, maintain and release channels for RC, UC and RD transport services. At creation, QPs are not ready for communication. The CM infrastructure is responsible for preparing the QPs for communication by exchanging appropriate information.

Each IB channel adapter hosts a communication manager (CM), which performs the relevant communication management tasks for that device. The CM is a single logical entity and may internally be composed of multiple processing units. Datagrams related to this management class are exchanged using the General Services Interface (GSI) and are directed to a special QP reserved for general services traffic (QP1). Once a CM request arrives at a channel adapter, the CM can request a GSI redirection which is a response to the requester asking it to use a different QP instead of QP1, thus offloading tasks from the master CM process to other processing units.

Performance Management. The performance management class allows performance management entities to retrieve performance and error statistics from IB components. There are two classes of statistics: (a) mandatory statistics that have to be supported by all IB devices and (b) optional statistics that are specified to allow for future standardization. There are several mandatory statistics supported by current IB devices, including the amount of bytes/packets sent and received, the transmit queue depth at various intervals, the number of ticks during which the port had data to send but had no flow-control credits, etc.

Device Management. Device management is an optional class that mainly focuses on devices that do not directly connect to the IB fabric, such as I/O devices and I/O controllers. An I/O unit (IOU) containing one or more IOCs is attached to the fabric using a channel adapter. The channel adapter is responsible for receiving packets from the fabric and delivering them to the relevant devices and vice versa. The device management class does not deal with directly managing the end I/O devices, but rather focuses on the communication with the channel adapter. Any other communication between the adapter and the I/O devices is unspecified by the IB standard and depends on the device vendor.

Baseboard Management. The baseboard management (BM) class focuses on low-level system management operations. This class essentially encapsulates BM messages into management datagrams and tunnels them through the IB fabric to the appropriate device. Each device has a baseboard management agent (BMA) which removes the encapsulation and passes the message to the device-specific module management entity (MME). Similarly, it re-encapsulates the response generated by the MME and passes it back to the requesting node.

SNMP Tunneling. Simple Network Management Protocol (SNMP) comprises of a set of standards for uniform network management. The SNMP tunneling management class is similar to the baseboard management class, except that it encapsulates and tunnels SNMP messages over the IB fabric. Devices advertise their support for SNMP through the SA. Once advertised, the device is queried via the GSI to access the SNMP service. This is an optional class and is mainly retained to allow for backward compatibility for network management software infrastructure that has been developed for traditional Ethernet fabrics.

Vendor and Application Specific: Vendor and application specific management classes, as the name suggests, are specific management datagrams that can be used by specific vendor-provided devices or a specific application. These datagrams are not interpreted by the rest of the IB fabric. Both these classes are optional for IB devices to provide.

2.4 Existing InfiniBand Adapters and Switches

2.4.1 Channel Adapters

Channel adapters are the method of connecting an endpoint to the fabric. There are many vendors and adapters available, but they can be classified into two main categories:

- *Onloading*: In this mode more operations are controlled and performed by the host CPU and often there is interaction with the OS kernel. The QLogic [351] adapters are an example of this category.

- *Offloading*: The style of adapter hardware contains significant "intelligence" and can perform most operations without any interaction with the host CPU or OS kernel. This style of adapter can achieve higher overlap of communication and computation. Mellanox [283] and IBM [202] designs for InfiniBand adapters are examples of this offloaded approach.

To achieve high performance, adapters are continually updated with the latest in speeds (SDR, DDR and QDR) and I/O interface technology. Earlier InfiniBand adapters supported only PCI-X I/O interface technology, but have now moved to PCI-Express (x8 and x16) and Hyper-Transport. More recently, QDR cards have moved to the PCI-Express 2.0 standard to further increase I/O bandwidth.

In 2008, Mellanox Technologies released the fourth generation of their adapter called "ConnectX." This was the first InfiniBand adapter to support

QDR speed. This adapter, in addition to increasing speeds and lowering latency, included support for 10 Gigabit Ethernet. Depending on the cable connected to the adapter port, it either performs as an Ethernet adapter or an InfiniBand adapter.

Although InfiniBand adapters have traditionally been add-on components in expansion slots, they have recently started moving onto the motherboard. Many boards including some designed by Tyan and SuperMicro include adapters directly on the motherboard. These design are made possible by "mem-free" adapter designs. Earlier InfiniBand adapters had up to 128MB of memory directly on the adapter, but with the advent of PCI-Express, these communication contexts began to be stored in main memory without significant performance degradation. Such "mem-free" adapters are becoming quite common in current generation IB clusters.

2.4.2   Switches

There have been many switch devices over the history of InfiniBand. Infini-Band switches have been designed and sold by a variety of vendors including Mellanox, Flextronics, QLogic, Voltaire and Sun Microsystems. For many years virtually all InfiniBand switches were made with Mellanox-based silicon switching chips.

SDR switches were the first and came in sizes from 8 to 144 ports with 4x widths and some 12x ports available for inter-switch connectivity. These switches were designed with 8-port switch chips. DDR switches were available next in sizes from 8 to 288 ports and were based on 24-port switch chips. Smaller scale switches (such as 8 ports) with 12x widths are also available in the market.

Most recently manufacturers have introduced QDR products. Mellanox has announced a 36-port "InfiniScale IV" switch chip that will allow higher density switches to be built. QLogic has also announced the intention to ship their own 36-port switch silicon termed "TrueScale."

The largest InfiniBand switch currently available is the Sun Microsystems "Magnum" switch with a total of 3,456 ports. This switch was used in the Texas Advanced Computing Center (TACC) "Ranger" system. A smaller "nano Magnum" switch has also been created with 72 ports.

InfiniBand switches are often integrated directly into blade-based systems developed by many manufacturers. They have also been integrated directly into proprietary systems (such as the Cray CX1 system). Switches have also been designed with InfiniBand-to-Fiber Channel and InfiniBand-to-Ethernet gateways. Such gateway switches provide flexibility for upgrading some parts of a larger cluster/data-center to InfiniBand while providing interoperability with other existing parts using Fibre Channel or Ethernet.

2.4.3 Wide Area Networks (WAN) and Routers

Long-distance networking over InfiniBand has recently gained increased prominence along the directions of remote storage, remote visualization, etc. For these applications, it is not possible to connect clusters thousands of miles apart using InfiniBand. These devices can either appear as switches and create a single subnet or as routers and link two subnets together. Initial products along this direction are available from Obsidian Research [313] and Bay Microsystems [34].

2.5 Existing InfiniBand Software Stacks

The InfiniBand specification defines a set of "verbs" to describe the functionality an InfiniBand implementation should provide to be compliant. Although it describes the functionality required, it does not specify any particular API to be used. Given this flexibility, multiple interfaces for interacting with the InfiniBand hardware have been created. Some of them interact at a very low level, following the specification and adding additional tuning parameters, such as VAPI and OpenFabrics "Gen2," while others such as PSM provide a much higher-level interface for the developer.

2.5.1 Low-Level Interfaces

The original method of interacting with InfiniBand hardware was the Verbs API (VAPI). Each vendor had their own particular closed-source version of VAPI. As a result, supporting each of these vendor-specific flavors created a problem for developing cross-vendor solutions.

Given this lack of a single interface, the OpenIB Alliance was created in 2004 with the goal of creating a common verbs-level interface for InfiniBand devices within Linux. The U.S. Department of Energy was strongly interested in this area and provided the initial funding. The initial interface proposed by OpenIB was a combination of the vendor VAPI stacks and was termed "Gen1." A next-generation "Gen2" interface was later created from scratch and is the basis for all current development. Since then the group has expanded its focus to include both the Windows platform and the iWARP [370] standard. It was renamed OpenFabrics [321] in 2006 to reflect the growing scope of the group.

The Direct Access Programming Library (DAPL) was created by the Direct Access Transport (DAT) Collaborative [113] as a transport independent software interface to RDMA-enabled devices. The initial meeting of the group was in 2001. The goal of this effort is to provide a layer to support VIA, InfiniBand and iWARP through a common interface. Interfaces for InfiniBand, iWARP, Quadrics and Myrinet have been created.

Given the expanded scope of OpenFabrics, many libraries and components that want to access low-level features of InfiniBand and iWARP opt to use the OpenFabrics verbs instead of DAPL. Since DAPL strives to be more transport independent, the features exposed are often at a higher level and do not offer as many features as OpenFabrics. For example, InfiniBand provides reliable connected, unreliable connected, and unreliable unconnected modes, but only the reliable connected mode is exposed through the DAPL interface.

2.5.2   High-Level Interfaces

The low-level access methods including OpenFabrics and DAPL expose a view that is very similar to the InfiniBand specification verbs. Other modes since then have been proposed that expose a different feature set, either for performance or compatibility. These modes include PSM and IP over InfiniBand.

The PathScale Messaging (PSM) interface was introduced by PathScale (now QLogic [351]) in 2006. This interface is vastly different than the lower-level verbs interface. Instead of exposing the low-level InfiniBand building blocks of opening connections, setting permissions, and carrying out necessary operations, it supplies an interface that is similar to MPI-1 semantics. It also provides an integrated shared memory communication. This interface more closely matches the more on-loaded approach of the QLogic HCAs and shows higher performance. It also exposes some additional features that were added into the QLogic HCAs to perform tag matching, which is not an InfiniBand specified feature.

IP over IB (IPoIB) is a upper-level protocol encapsulation of IP packets over the InfiniBand fabric. The particular protocols and encapsulation of IP traffic over InfiniBand has been standardized by the IETF. This mode allows standard IP traffic to work seamlessly over the InfiniBand fabric and is interoperable between vendors. IPoIB cannot attain the same levels of performance as the lower-level interfaces (such as OpenFabrics Gen2) or PSM since it was designed as a compatibility layer rather than a new layer to expose higher performance.

2.5.3   Verbs Capabilities

The InfiniBand verbs, which are closely modeled in the "Gen2" interface, provide the functional specification for the operations that should be allowed on an InfiniBand compliant adapter. There are a few dozen verbs that are specified and provide the functional description of opening the adapter, creating endpoints for communication, registering memory for communication and sending and receiving data.

Sample examples of InfiniBand verbs are:

- *Register Memory Region*: All memory used for communication in Infini-Band must be registered before use. This makes sure that the appropriate security is provided as well as making sure that the HCA knows the physical address of the memory. OpenFabrics API: `ibv_reg_mr` is used to carry out this operation.

- *Post Receive Request*: For channel semantics a receive must be explicitly posted. To post a receive the local key from the memory region must be provided. OpenFabrics API: `ibv_post_recv` is used to carry out this operation.

- *Post Send Request*: This operation sends a buffer to the remote destination. To post a send operation the local key from the memory region must be provided. This same verb can be used for both channel and memory semantics. When using the memory semantics the remote virtual address and remote key from the memory region must be provided. OpenFabrics API: `ibv_post_send` is used to carry out this operation.

2.6 Designing High-End Systems with InfiniBand: Case Studies

In this section, we consider a set of practical case studies to demonstrate how the novel features of InfiniBand can be used in designing the next generation high-end systems with significant gains in performance and scalability. First, we consider the scenario of designing high performance Message Passing Interface (MPI) for high-end clusters. Then we consider the design of parallel file systems like Lustre. Finally, we consider the design of enterprise data-centers.

2.6.1 Case Study: Message Passing Interface

The Message Passing Interface (MPI) is widely regarded as the de facto message passing standard. The majority of high-performance scientific applications are written in MPI. The MPI standard is supported on a wide variety of platforms. Programming in MPI requires the programmer to decompose the parallel programming task into local computation and communication. Communication can be performed using point-to-point primitives or collectives. MPI standard versions 1 and 2 are available, and MPI-3 is under discussion. The reader is encouraged to refer to [289, 290] to learn more about MPI and its semantics. In this chapter, we will discuss about MPI implementation strategies over InfiniBand. Our case study will delve into the design of MVA-PICH [309] and MVAPICH2 [265], which are implementations of MPI over InfiniBand. They are based on MPICH1 [171] and MPICH2 [219], respectively.

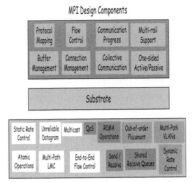

FIGURE 2.6: Layered design of MVAPICH/MVAPICH2 over IB.

MVAPICH/MVAPICH2 Software. MVAPICH/MVAPICH2 stacks are high-performance MPI library for InfiniBand and 10GigE/iWARP. As of March 2009, these stacks are used by more than 855 organizations in 44 countries around the world. There have been more than 27,000 downloads from the OSU site directly. These stacks are empowering many powerful computers, including "Ranger," which is the 6th ranked cluster according to the November 2008 TOP500 rankings. It has 62,976 cores and is located at the Texas Advanced Computing Center (TACC). MVAPICH/MVAPICH2 work with many existing IB software layers, including the OpenFabrics Gen2 interface. It also supports uDAPL device to work with any network supporting uDAPL. Both MVAPICH and MVAPICH2 are part of the Open Fabrics Enterprise Distribution (OFED) stack.

A high-level schematic of MVAPICH/MVAPICH2 is shown in Figure 2.6. The design components are separated into two layers. The layer above substrate deals with higher level design issues and the lower layer takes advantage of specific InfiniBand verbs and network-level features. We will discuss each of these components in detail and observe through experimental results how these design components affect overall performance and scalability of the MPI library.

2.6.1.1   Higher-Level MPI Design components

Protocol Mapping. The MPI library supports arbitrarily sized message transfer. In order to optimally transmit messages from one process to another, several protocol level decisions have to be made. There are two main protocols for transmitting messages over the network. One is the "Eager" protocol and the other is the "Rendezvous" Protocol. In the Eager protocol, a message send when initiated, is immediately sent over to the remote side. This assumes that there is enough buffer space available at the remote side to temporarily store the message until the MPI application asks for it. Since this protocol requires buffering, and memory copies (in order to copy messages into internal buffers), this protocol is appropriate only for small messages. Larger messages follow the Rendezvous protocol. In this protocol, messages are not immediately sent to the receiver. Instead, the sender asks the receiver about buffer availability. Only when buffers are available is the message sent. This protocol can also achieve zero-copy, i.e. no memory copies are required to transmit the message to receiving MPI application.

Besides message size, there are other considerations as well for choice of protocol. For example, the message may be small, but the remote side may have no buffer availability due to a flood of small messages. In that situation, the Protocol Mapping component has to fall back on the Rendezvous protocol. Thus, this component takes into account all these factors while deciding on which protocol to use.

Basic MPI level performance numbers on a range of IB adapters are shown in Figure 2.7. Three generation Mellanox adapters (InfiniHost III with DDR speed, ConnectX with DDR speed, and ConnectX with QDR speed) and two generation QLogic adapters (SDR and DDR) are considered. The latest MVAPICH stack is used across all these adapters to obtain these numbers. It can be observed that Mellanox ConnectX QDR adapter with PCI-Express Gen2 interface delivers the best one-way latency (1.06 microsec) for 1-byte message. QLogic DDR adapter delivers best latency for 128-1024 bytes of messages. As message size increases, ConnectX QDR adapter with PCI-Express Gen2 interface delivers the best one-way latency. These adapters also deliver different unidirectional and bidirectional bandwidth numbers. The Mellanox ConnectX DDR adapter with PCI-Express Gen2 interface is able to deliver the highest unidirectional bandwidth (2570 Million Bytes/sec) and highest bidirectional bandwidth (5012 Million Bytes/sec).

Flow Control. Flow control is an important aspect of any network library. This component ensures that healthy levels of buffer space are available on the sender and receiver ends. Buffer space is an expensive resource (in terms of memory consumption), so neither should it be wasted nor should applications starve for buffers due to flood conditions. The major task of this component is to prevent a situation where there are no buffers available to send and receive control messages. In such a case, the whole MPI application may deadlock. This component detects such cases through monitoring buffer usage (through piggy-backed information in packets), as well as by utilizing IB device features. Finally, this component feeds information to the Protocol Mapping component so that the MPI library can recover from flood conditions and avoid buffer wastage.

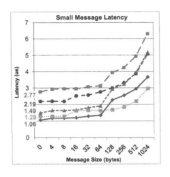

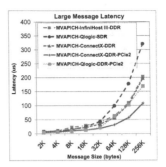

(a) One-way latency (small messages)

(b) One-way latency (large messages)

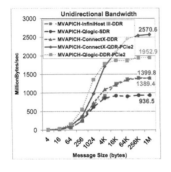

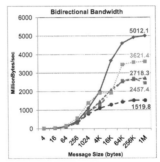

(c) Unidirectional bandwidth

(d) Bidirectional bandwidth

FIGURE 2.7: Two-sided point-to-point performance over IB on a range of adapters.

Communication Progress. This component is responsible for making progress on all outstanding communication requests. Under many situations, it may not be possible to complete communication immediately. For example, large message sends over rendezvous protocol, communication over unreliable transports (when it is required to acknowledge packets), etc. Communication progress is important since it has a fairly major impact on overall performance. Two main techniques may be employed for implementing communication progress. The first is polling based progress, in which progress is made whenever the MPI application makes a call into the MPI library. This mode is the most lightweight method of making progress, however, it can only make weak progress guarantees. The second method of making progress is through dedicated communication threads. In this approach, one or more threads are tasked with constantly polling the network and keeping track of communication requests. Using this approach, one can make strong progress guarantees, however, this mode is not light-weight. There has been some research in providing a hybrid (of polling and interrupts) in the MVAPICH stack.

Multi-rail Support. Over the past few years, computing nodes have become increasingly powerful due to the advent of multi-core computing. As a result, computation on multiple-cores is bound to generate more data and an increased demand for faster I/O. The multi-rail support component in MVAPICH attempts to improve communication bandwidths by simultaneously leveraging multiple InfiniBand HCAs and ports [267]. It performs many optimizations, including message stripping and other advanced mechanisms to optimally choose policies for multi-rail communication that result in alleviating network congestion and even usage of all network devices.

Buffer Management. The buffer management component is responsible for all MPI internal buffers used to shuttle user (and control) messages to and from processes. This component sets policies on when new buffers may be allocated and de-allocated. It leverages specific IB features (like Low watermark events on a Shared Receive Queue) to get input from the device about when buffers are running low. It manages other meta-data like buffer registration handles and storing/retrieving buffers from free/busy lists. The buffer management component gives feedback to the Flow-control component which eventually advises the Protocol mapping component to supervise overall communication flow through the MPI library.

Connection Management. As mentioned in Section 2.3.2.3, InfiniBand requires Queue Pairs (or connection structures) to communicate with remote end points. Unfortunately, these IB connections consume physical memory and associated memory in the form of message buffers (e.g. receive buffers posted into a receive queue). Since MPI provides a logical all-to-all connectivity, the MPI library is bound to provide a method by which any process can communicate with any other process. However, in the view of the memory usage concerns, the MPI library has to be intelligent about how it uses open connections. The connection management component provides this ability. It can leverage different types of connection models, all-to-all reliable connections (for small scale clusters), on-demand reliable connections (for medium to large clusters) and unreliable datagram (for very large/ultra scale clusters). This component also impacts how buffer management may be done, and that ties it to the other inter-dependent components in the MPI library.

Collective Communication. MPI provides the programmer the flexibility of using collective communications to aggregate bits of data from several processes (defined by a MPI communicator). Not only does MPI collective communication provide the user a simple way to gather data from different remote processes address space, but it also employs powerful algorithms that do this collection in clever manners so as to effectively utilize network bandwidth, alleviate congestion, and use device specific features. Collective communication algorithms are a bona-fide research area that has been active for several decades now. This component is responsible for choosing the appropriate collective algorithm based on the number of processes involved (i.e., group size), message size, and network topology. Often, this component uses the point-to-point communication primitives which are already implemented in MPI, but when optimal hardware features (like IB Hardware Multicast) are available, it may choose to use them [218].

One-sided Active and Passive Synchronization. The MPI-2 standard provides the programmer with one-sided communication primitives. Using these primitives, the programmer can issue remote get and put operations on MPI "windows" (which are opaque objects representing a particular address space in processes memory). These operations map naturally on to IB RDMA Read and Write operations. However, the synchronization models of the MPI-2 one-sided communications are somewhat complicated [197]. There are two modes of synchronization, Active and Passive [378]. In the active mode, the target process participates in the synchronization. As opposed to that, in the passive mode, the target does not participate in the synchronization. Thus, all processes remotely accessing a window have to synchronize amongst themselves without participation of the target process. In this context, the IB atomic operations provide an excellent mechanism to implement passive side synchronization. Implementation of one-sided operations in MVAPICH2 was recently enhanced with the atomic operations. This results in significant improvement at the application level as shown in Figure 2.8a. Compared to the two-sided-based design, the new direct-passive scheme shows significant performance improvement at the applications-level.

2.6.1.2   Network-Level MPI Design Components

Static Rate Control. InfiniBand provides static rate control features for the user application to "reserve" a specified bandwidth on a particular Queue Pair (i.e., per connection). This value, which cannot exceed the max link rate, is provided at the time when the queue pair is created (hence, it is "static"). However, the MPI library may make intelligent use of this feature to provide bandwidth guarantees on specific connections. Additionally, it may also employ this feature to enable a certain amount of congestion control by limiting the bandwidth a QP can attain. This component is primarily used by the multi-rail component, as described in the previous section.

Unreliable Datagram. The Unreliable Datagram (UD) is the most scalable IB transport available. While it makes no guarantees of message delivery or order, it provides huge savings on memory utilization by MPI processes. Using this transport, the receiver MPI process no longer has to provide for message buffers for each connection separately, and only needs only N QPs (for N processes), as opposed to N^2 QPs for Reliable Connected mode. In addition to huge memory savings, it also turns out that packet drop rates for InfiniBand

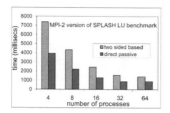

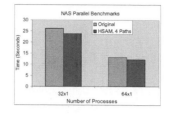

(a) MPI-2 one-sided passive synchronization

(b) Network-congestion alleviation using HSAM

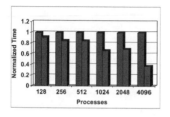

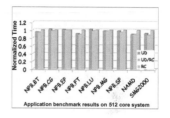

(c) SMG2000 performance with
MVAPICH-UD (progress mode)

(d) MVAPICH-Aptus normalized performance with various applications

FIGURE 2.8: Application-level evaluation of MPI over InfiniBand design
components.

networks are very low (even negligible) even at a large scale. Thus, memory
savings can be utilized while not paying a heavy cost for re-transmission and
reliability. MVAPICH-Aptus [235] is specially designed to leverage this feature
on large and ultra-large scale clusters. Figures 2.8c and 2.8d show the benefits
of this mode on the scalability and performance of MVAPICH. This component
feeds back into the protocol selection, flow-control, and buffer management
higher-level components.

Hardware Multicast. Multicast is supported at the hardware level by
IB. This feature can be utilized only with an Unreliable Datagram QP, and
the nature of this multicast is unreliable as well. It is possible that a subset
of the multicast group (created before the multicast on the IB switch) may
receive packets, but another subset may not receive any packets at all. At
the same time, this is an incredibly fast and scalable mechanism to perform a
multicast in a cluster. MPI component which utilizes this has to be intelligent
and implement several mechanism for creating multicast groups, and designing
an efficient reliability protocol over this Multicast interface [266]. MVAPICH
provides hardware level multicast in select conditions. This component is
utilized by the collective communication higher-level component.

Quality-of-Service (QoS). As indicated in Section 2.3.2.1, InfiniBand provides a set of mechanisms such as SLs, VLs and SL-to-VL mapping to provide QoS support. Using these mechanisms, an IB application can provide bandwidth provisioning by using specific VLs and SLs. During the initial deployment of IB products, the QoS mechanisms were not completely available in hardware and firmware. Gradually, these are becoming available in IB products and upper level protocols, and applications are trying to exploit these mechanisms to provide QoS support.

RDMA Operations. As discussed in Section 2.3.1.2, RDMA is one of the most powerful mechanisms provided by InfiniBand. It enables true zero-copy message passing, thus alleviating the CPU bottleneck. Using RDMA, the CPU is free to perform computation, whereas the network takes care of transmitting (via DMA engines) memory from the sending node to the receiving node. MVAPICH and MVAPICH2 software make heavy use of RDMA operations primarily to send and receive large messages. This enables MVAPICH software to attain very high bandwidths of up to 5GBytes/s when performing the bidirectional bandwidth test with 1MByte message size as observed in Figure 2.7d. The RDMA operations are also used by the protocol mapping and multi-rail components.

Atomic Operations. InfiniBand atomic operations allow remote nodes to atomically update 64-bit integer values in another node's memory. The area of memory being updated needs to be registered and should have appropriate access rights. It is important to note that this atomic operation is only atomic with respect to clients that are accessing the variable through the HCA. If the host (on whose memory the shared variable is hosted) modifies the variable by a direct load/store, then the process is not atomic at all. In this case, additional care needs to be taken to ensure atomicity. Regardless, this feature can be very useful in implementing the passive one-sided synchronization. As seen in Figure 2.8a, the use of this feature leads to significant application level benefit.

Multi-Path LID Mask Control. The Local Identifier (LID) is an address for the HCA at the InfiniBand subnet level, indicating a certain deterministic path through the network to reach it. Using a "mask," users of InfiniBand can choose multiple paths through the network for each QP connected through this HCA. Of course, using multiple paths means that congestion in the network can be alleviated by spreading the traffic around appropriately [443]. Figure 2.8b shows that utilizing this Hot-Spot Avoidance Mechanism (HSAM) leads to improved application performance. This feature can also be used by the multi-rail component.

End-to-end Flow Control. For reliably connected QPs, InfiniBand provides End-to-end flow control. This is enabled by default. This feature allows the users of the Send/Receive channel (Section 2.3.1.2), to post a send descriptor even though there may not be a receive buffer immediately available on the remote side. The HCA keeps track of how many are available on the remote side via these end-to-end credits. When the receive buffer becomes available at the remote side, the HCA sends a credit update message indicating that the send can proceed. In this manner, the network is not overloaded with continuous message send retries. This feature helps in congestion management and helps reduce the amount of buffer requirement per QP in send/receive channel.

Send/Receive. The Send/Receive channel is one of the methods to communicate using InfiniBand (Section 2.3.1.2). In this mode, the receiver needs to pre-post buffers on the QP where it expects messages to arrive. When the sender has a message to send, it sends it over this QP. After the network has received the message, it consumes one receive descriptor from the receive queue and generates a completion queue. The user application can poll the completion queue to look for incoming messages. This mode is available with both Reliable Connected (RC) and Unreliable Datagram (UD) transports. When used with the RC transport with normal receive queues, this mode is quite unscalable, as the user application (i.e., MPI library) needs to do some guesswork to determine which connections will send the most messages. As the number of remote peers grows, the requirement to make receive buffers available to each remote peer blows up the total memory requirement. This issue was fixed with the advent of Shared Receive Queues, where the MPI library can just post buffers to one receive queue instead of multiple. With the UD transport, again, one can just maintain one receive queue and there is no scalability problem.

Shared Receive Queues. As mentioned in the previous section, Shared Receive Queues (SRQs) were designed to overcome the scalability limitation in the ordinary send/receive mode. Using shared receive queues, a single receive queue can be maintained for multiple RC QPs. Thus, the MPI library does not have to do any guesswork as to which connections will be sending messages and which ones won't. As a result, it can drastically reduce on amount of buffer space required. This leads to enhanced scalability for up to large scale clusters (unreliable datagram is required to scale to ultra-scale clusters). The shared receive queue has a feature called "Low-Watermark" event, by which the network device can inform the MPI library that the amount of available buffer space for receiving incoming messages is running low. Using this mechanism, the MPI library can perform reactive flow control. This feature may be used by buffer management and flow control modules.

2.6.2   Case Study: Parallel File Systems

Modern parallel applications use Terabytes and Petabytes of data. Thus, parallel file systems are significant components in designing large-scale clusters. Multiple features (low latency, high bandwidth and low CPU utilization) of IB are quite attractive for designing such parallel file systems. Here we illustrate

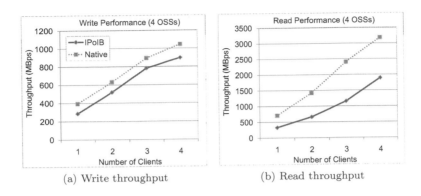

(a) Write throughput   (b) Read throughput

FIGURE 2.9:   Lustre performance over InfiniBand.

some of design challenges and performance numbers for Lustre [95], a popular parallel file system used in current generation high-end clusters.

Parallel file systems typically consist of a set of storage nodes, clients and a metadata server. Data from files are stripped across these nodes. A metadata server contains information regarding which files are located where. Clients first access the metadata server to get information about the location of the files on the storage nodes. Next, the clients access the data from the storage nodes in parallel.

Traditional parallel file systems have used TCP/IP stack with two-sided communication. Communication using these stacks do not deliver very high performance. On the other hand, RDMA features of InfiniBand (such as RDMA read and RDMA write) provide flexibility to use one-sided communication to achieve high throughput data transfer in parallel file systems [477]. The atomic operations of InfiniBand also help to design better lock management for parallel file systems.

Figure 2.9 shows the performance benefits achieved in the Lustre file system when RDMA protocols (identified as Native) are used instead of the TCP/IP protocols (IPoIB, emulation of TCP/IP over IB). Four storage servers are used in this experiment. As the number of clients increases, the implementation using the native protocol is able to deliver better performance compared to the IPoIB protocol. There is a substantial benefit for Read performance because of the caching effect of Read operations.

In addition to the basic performance improvement in latency and throughput, RDMA mechanisms of IB also consume less CPU while carrying out network communication. Figure 2.10 illustrates such benefits in CPU utilization in Lustre. When using the native RDMA protocol, the CPU utilization is considerably less compared to the IPoIB protocol. This trend is observed for both Read and Write operations in file systems. The low CPU utilization with

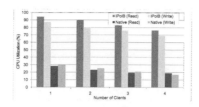

FIGURE 2.10: CPU utilization in Lustre with IPoIB and native (verb-level) protocols.

RDMA protocol in IB provides flexibility for higher overlap in computation and communication.

2.6.3 Case Study: Enterprise Data Centers

Together with scientific computing domains such as the MPI and parallel file-systems, IB is also making significant inroads into enterprise computing environments such as web data centers and financial markets [26]. In this section, we illustrate case studies where IB has shown substantial benefits in these areas. Specifically, we concentrate on two primary studies: (a) transparently accelerating web services such as Apache using high-performance sockets over IB and (b) efficient cache management to achieve strong cache coherency protocols designed over IB.

Transparently Accelerating Web Services. A large number of existing web services have traditionally been designed over TCP/IP. Accordingly, these applications have internally used TCP/IP sockets as their communication model. With the advent of InfiniBand, while a lot of new features are now available, it is difficult and impractical to migrate all these applications to use IB immediately because of the changes needed in the application to do so. Thus, researchers have been looking at solutions through which such applications can transparently utilize the performance and features provided by IB.

There are two ways in which sockets-based applications can execute on top of IB. In the first approach, IB provides a simple driver (known as IPoIB) that emulates the functionality of Ethernet on top of IB. This allows the kernel-based TCP/IP stack to communicate over IB. However, this approach does not utilize any of the advanced features of IB and thus is restricted in the performance it can achieve. The second approach utilizes the Sockets Direct Protocol (SDP) [389]. SDP is a byte-stream transport protocol that closely mimics TCP sockets' stream semantics. It is an industry-standard

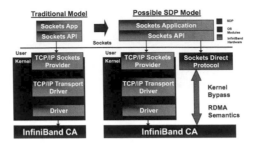

FIGURE 2.11:   SDP architecture (Courtesy InfiniBand Standard).

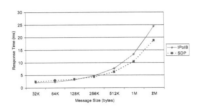

FIGURE 2.12:   Performance comparison of the Apache Web server: SDP vs. IPoIB.

specification for IB that utilizes advanced capabilities provided by the network stack (such as the hardware offloaded protocol stack and RDMA) to achieve high performance without requiring modifications to existing sockets-based applications. Figure 2.11 illustrates the architecture of the IPoIB and SDP protocol stacks.

Figure 2.12 illustrates the performance on Apache when layered on top of SDP, as compared to when it is layered on top of IPoIB. As shown in the figure, SDP's capability to utilize the advanced features of IB allows it to achieve a much better performance as compared to IPoIB, especially when users request large files.

Strong Cache Coherency over IB: While SDP internally utilizes a lot of advanced features provided by IB, it exposes only the sockets interface for applications to use. While this provides transparent performance for applications, in some cases, additional features can be very useful for additional performance improvements. Specifically, let us consider the example on strong cache coherency of dynamically generated content in enterprise data centers.

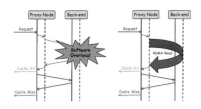

FIGURE 2.13:   Active polling to achieve strong cache coherency.

Dynamic content is becoming increasingly popular with user-personalized web content, transactional data (such as bank accounts, stock markets), ticketing web sites and other web content. However, for some of this content (especially those corresponding to financial aspects), it is critical that the data being served always be accurate. Specifically, each request for content needs to provide exactly the value stored by the back-end database storing this data.

A simple approach for handling such restrictions is to ensure that each request for content on such servers be directed to the back-end. However, this means that each request will need to be fully processed in order to provide the appropriate response. However, this is extremely time consuming and would thus lose performance. An alternative approach is to cache the processed data for different requests, and serve future requests from cache. For this approach, however, we need a mechanism to maintain cache coherency. That is, any response from cache should give the same value as if the request were responded to by the back-end itself. A popular approach to maintaining such cache coherency is **active polling**. Under the active polling approach, each cached object maintains a versioning scheme. Whenever the front-end gets a request, it checks the version of the cached object with that of the back-end database. If the version matches, it sends out the cached response. If not, the request is directed to the back-end [303].

While this approach is straightforward, in practice, even a simple check for version information is expensive on heavily load web servers as shown in Figure 2.13. For example, when the back-end is heavily loaded due to the processing of other requests, even a small request for a version check can take a long time, as the processing thread needs to be scheduled on one of the processors (or cores) for the version request to be processed and responded to. This is an inherent limitation of the two-sided nature of the sockets programming model (i.e., processes on both sides need to participate in the communication). On the other hand, if the apache web server could directly use IB one-sided RDMA operations, it can directly perform an RDMA read on the version information. Since the RDMA read operation happens completely

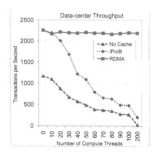

FIGURE 2.14:   Active polling performance

in hardware, it does not disturb the back-end processes at all and takes the same time as if the back-end was not loaded at all. Figure 2.14 illustrates the performance benefits of using IB RDMA communication as compared to using sockets.

2.7   Current and Future Trends of InfiniBand

Strong features and mechanisms of IB, together with the continuous advancements in IB hardware and software products, are leading many high-end clusters and parallel file systems. In less than nine years of its introduction, the open standard IB is already claiming 28.2% market share in TOP500 systems. This market share is expected to grow significantly during the next few years. Multiple parallel file systems, storage systems, database systems and visualization systems are also using InfiniBand. As indicated in this chapter, InfiniBand is also gradually moving into campus backbones and WANs. This is opening up multiple new applications and environments to utilize InfiniBand.

IB 4x QDR products are currently available in the market in a commodity manner. These are getting to be used in modern clusters with multi-core nodes (8 to 16 cores per node). In the near future, 8x QDR and 12x QDR products may be available. IB Roadmap also talks about the availability of EDR (Eight Data Rate) products during the 2011 time frame. Products with 4x EDR (80 Gigabits/sec line rate and 64 Gigabits/sec data rate) will be able to help designing balanced systems with next generation multi-core nodes (16 to 32 cores per node) and accelerate the field of high-end computing.

Chapter 3

Ethernet vs. EtherNOT

Wu-chun Feng
 Virginia Tech
Pavan Balaji
 Argonne National Laboratory

3.1 Overview

This chapter presents a perspective of how Ethernet has converged with EtherNOT, where EtherNOT is informally defined as any technology that is *not* Ethernet.

Over the past decade, many aspects of the convergence of Ethernet and EtherNOT can be attributed to high-end computing (HEC). While more than 85% of all installed network connections were Ethernet by the late 1990s,[1] Ethernet made up less than 2% of the networks used in the TOP500 supercomputers in the world.[2] In turn, more than 98% of the networks used in the TOP500 supercomputers were EtherNOT.

Back then, the HEC community relied on these exotic, speciality Ether-NOT networks to deliver the high-performance communication needed by supercomputers. However, achieving such high performance meant trading off generality and cost, e.g., eliminating interoperability with the ubiquitous Internet Protocol (IP) and significantly increasing the cost of hardware and the personnel to deploy it, respectively.

Much has changed since then. At a macroscopic perspective, Ethernet has gone from being practically non-existent on the TOP500 to occupying 56% of the TOP500, as shown in Figure 3.1. At a more microscopic perspective, different EtherNOT technologies continue to emerge and hold a significant minority stake in the TOP500 every few years.

---

[1]Source: International Data Corporation (IDC) and [120].
[2]Source: The TOP500 at http://www.top500.org/.

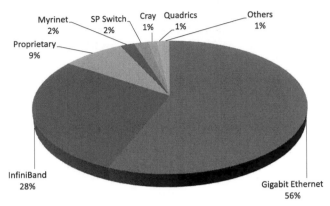

FIGURE 3.1:   Profile of network interconnects on the TOP500, November 2008. Data from TOP500 at http://www.top500.org/.

In late 1997, EtherNOT crossbar technology occupied 33% of the TOP500 while Ethernet occupied a mere 1%. Three years later in late 2000, the SP switch from IBM rose to supplant the crossbar by occupying 42% of the TOP500 as the crossbar ebbed to 22% and Ethernet rose to 3%. Another three years later in 2003, the EtherNOT of Myrinet catapulted from obscurity to nearly 40% of the TOP500 as both the EtherNOT SP switch and crossbar ebbed to 12% and 7%, respectively, while Ethernet rose more modestly to 22%.

By 2006, Ethernet surged to occupy more than 50% of the TOP500 with Myrinet falling back to 17% of the list and InfiniBand, the latest EtherNOT technology, rising to 7%. And now, three years later, InfiniBand has made a mercurial ascent, rising from the turmoil of surviving the withdrawal of major vendors from active InfiniBand development in the early 2000s to capturing nearly 30% of the TOP500, as shown in Figure 3.1. So, despite continued predictions that Ethernet will stamp out the existence of EtherNOT networks, the rapidly increasing market share of InfiniBand in the TOP500 indicates otherwise.

Before discussing the various aspects of technical convergence between Ethernet and EtherNOT in detail, we (1) present the early days of Ethernet and EtherNOT that helped to shape the current Ethernet vs. EtherNOT debate, (2) formally define what is meant by Ethernet and EtherNOT, and then (3) present additional background material on Ethernet and EtherNOT.

3.2 Introduction

On May 22, 1973, Robert Metcalfe of Xerox Corporation's Palo Alto Research Center (PARC) circulated a memo entitled "Alto Ethernet" that contained a rough schematic of how Ethernet would work. This memo marked the beginnings of what we know today as Ethernet. By November 11, 1973, Metcalfe and colleagues demonstrated the first functioning Ethernet system. Over the next few years, Ethernet continued to evolve, and by 1976, Robert Metcalfe and David Boggs had published "Ethernet: Distributed Packet-Switching for Local Computer Networks." And thus began the journey of Ethernet ...

How did this journey lead us to where we are today? That is, Ethernet vs. EtherNOT. To understand the context, we go back to the roots of Ethernet. Much like the many evolutions of EtherNOT technologies from the token ring of the past to the InfiniBand of today, the challenge back then for Ethernet was to build a network that would be *fast enough* (in this case, to drive the world's first laser printer and enable all of Xerox PARC's computers to be able to print to this printer). In addition to speed, two other major factors tipped the scales towards Ethernet in the early days of Ethernet vs. EtherNOT: (1) an open standard and (2) lower cost.

In 1978, Metcalfe left Xerox and formed 3Com, in order to push the commercialization and standardization of Ethernet. By 1980, DEC, Intel, and Xerox released the "blue blook" specification for Ethernet, and five years later, Ethernet became an official standard, specifically IEEE 802.3. The major competitor of Ethernet at the time was *token ring*, an EtherNOT technology pushed by IBM that was arguably an open standard that really was not that open but did eventually become IEEE 802.5. For instance, Metcalfe noted that IBM "played all sorts of screwy games with higher-level incompatibilities, so that when [one] tried to sell a token ring card into an IBM installation, it would never really work out" [291]. Furthermore, token ring consistently cost more than Ethernet. As a consequence, beyond IBM, the EtherNOT technology of token ring never garnered much support amongst vendors, who instead gravitated to Ethernet. Token ring began a rapid descent once 10Base-T Ethernet emerged. 10Base-T Ethernet relied on an open standard and cost less than token ring because 10Base-T was based on inexpensive, telephone-grade, unshielded twisted-pair, copper cabling. (The cost discrepancies were exacerbated when Olof Soderblom, the inventor of token ring, began demanding royalities from token-ring vendors and chip makers, thus driving costs even higher.)

In spite of all the past "battles" between Ethernet and EtherNOT, particularly in high-end computing, Ethernet has since become ubiquitous. As noted before, more than 85% of all installed network connections are Ethernet-based, with roughly a billion ports deployed worldwide [120]. Furthermore, more than 95% of all local-area networks (LANs) are Ethernet-based, according to the International Data Corporation (IDC). Thus, nearly all traffic on the Internet starts or ends on an Ethernet connection.

Do the above trends signal an end to the Ethernet vs. EtherNOT debate? This chapter suggests otherwise and outlines why recent trends promise to keep this debate alive despite the convergence of Ethernet and EtherNOT on multiple fronts.

As a hint of things to come, think about what made Ethernet so successful against EtherNOT technologies. Ethernet was a rapidly deployed and truly *open standard* that delivered *high performance* at a *low cost*. While Ethernet still arguably possesses the above characteristics, there are now "too many chiefs" bogging down the 40-Gigabit Ethernet and 100-Gigabit Ethernet standardization process, which started back in 2006 and has been drawn out by the multitude of vendors involved. In contrast, InfiniBand, the latest EtherNOT technology, has taken a page out of Ethernet's book by rapidly deploying an open standard *and* producing better performance than existing Ethernet and at a lower cost.

3.2.1   Defining Ethernet vs. EtherNOT

Strictly speaking, the term *Ethernet* refers to the data-link (or layer-2) functionality of the OSI/ISO model. Layers atop the Ethernet data-link layer implement additional functionality via protocols such as IP, UDP, TCP, SCTP, and iWARP, for example.

With the aforementioned strict definition of Ethernet, many would view Ethernet as lacking the robust performance and feature set of EtherNOT networks because EtherNOT networks *combine all the above layers together* into one name, i.e., InfiniBand, Myrinet, or Quadrics. However, the current generation of Ethernet also combines all these layers into a single unit, such as an iWARP or TCP offload engine, thus making Ethernet look much closer to EtherNOT networks in practice, even though Ethernet and its associated layers still maintain different names for each of the protocol stack layers.

So, technically, while EtherNOT networks far exceed the capabilities of the isolated layer-2 functionality of Ethernet, the focus of this chapter is on how EtherNOT compares to Ethernet *and* its associated layer-3 and layer-4 protocols, e.g., IP and TCP, respectively. That is, hereafter when we refer to Ethernet, we refer to the data-link layer (Ethernet) and its associated network layer (IP or equivalent) and transport layer (TCP or equivalent).

With the above as the definition for Ethernet, EtherNOT effectively refers to all technologies that are *not* Ethernet. EtherNOT technologies have traditionally been characterized as being exotic high-end solutions, where performance and functionality come at a higher cost. For instance, EtherNOT technologies generally offer the lowest latency and highest bandwidth in unicast performance, largely due to using non-TCP/IP stacks that provide ways to offload all or portions of the protocol stack to hardware. In addition, for multicast, they offer support for collective communication.

3.2.2   Forecast

The rest of this chapter is laid out as follows. In Section 3.3, we provide further background on Ethernet and EtherNOT in order to lay the foundation for discussing the convergence of Ethernet and EtherNOT in Section 3.4. Section 3.5 discusses how Ethernet and EtherNOT have affected the commercial market. Finally, Section 3.6 presents concluding remarks, including a prognostication of the future of Ethernet vs. EtherNOT.

3.3   Background

In tackling the challenge of building a networking system that would drive the world's first laser printer as well as interconnect all of PARC's hundreds of computers to be able to print to this printer, Metcalfe designed the way that the physical cabling would interconnect devices on the Ethernet, as shown in Figure 3.2, as well as the standards that govern communication on the cable. Ethernet has since become the most widely deployed network technology in the world and has withstood the advances of many EtherNOT technologies, including token ring, token bus, fiber-distributed data interface (FDDI) in the 1980s, asynchronous transfer mode (ATM) in the 1990s, and the crossbar, SP switch, and Myrinet in the early 2000s. Currently, the most visible EtherNOT technology comes from InfiniBand.

3.3.1   Ethernet Background

Based on the notion of a completely passive medium called *luminiferous ether* that was once thought to pervade the universe and carry light everywhere, Metcalfe coined the term *Ethernet* to describe the way that cabling, also a passive medium, could similarly carry data everywhere throughout the network.

Broadly, Ethernet refers to a family of frame-based computer networking technologies, originally for local-area networks (LANs), but now also for system-area networks, storage-area networks, metropolitan-area networks (MANs), and wide-area networks (WANs). At the physical layer, Ethernet has evolved from a thick coaxial cable bus to point-to-point links connected via switches. At the data-link layer, Ethernet stations communicate via unique 48-bit media access control (MAC) addresses.

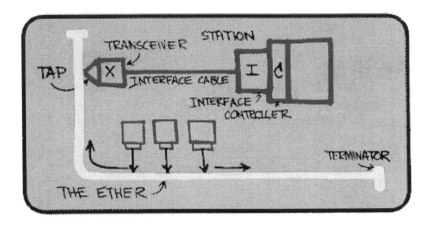

FIGURE 3.2:   Hand-drawn Ethernet diagram by Robert Metcalfe.

While the initial success of Ethernet over EtherNOT technologies may be attributed to the rapidity with which Ethernet was standardized by Digital Equipment Corporation (DEC), Intel, and Xerox and then further commercialized by 3Com; its continued success revolves around (1) providing a commodity solution to the masses at low cost, (2) having an open standard where all generations of Ethernet follow the same frame format, thus providing a consistent interface to upper layers and enabling interoperability across generations of Ethernet, (3) seamlessly working with the ubiquitous TCP/IP protocol suite, which sits atop Ethernet, to deliver critical protocol processing tasks such as data integrity, reliability, and routing, and (4) being just fast enough, e.g., 85% fast enough, for high-end computing at a low cost.

3.3.2   EtherNOT Background

Initially, EtherNOT solutions such as token ring, token bus, and FDDI competed directly with Ethernet with respect to speed, cost, interoperability, and openness. This has evolved, particularly since the early 1990s, from a direct competition to a complementary one where EtherNOT solutions targeted the high-end of the market, i.e., the last 15% of performance beyond the "85% solution of Ethernet," by sacrificing generality and cost.

For example, the arrival of gigabit-per-second (Gbps) networks in the early-to-mid 1990s shifted the burden of delivering high performance from the network hardware to network software. That is, network software had become the bottleneck to high-performance communication. This led to the development of OS-bypass protocols, which removed the OS kernel from the critical path, thus improving end-to-end performance with respect to both latency and throughput. An OS-bypass protocol moved substantial portions of a protocol from kernel space to user space, and later became known as a user-level network interface (ULNI) [44]. Examples of such OS-bypass protocols include FM [325], PM [420], U-Net [446], Scheduled Transfer [340], and VIA [448].

Coincident with the above development was the emergence of programmable network interface cards (NICs), e.g., [48]. Such NICs added programmable processors and memory, enabling them to share software processing of network traffic with the host. In turn, this allowed the host to free-up cycles for the application to use.

Both of the above developments marked the beginning of a new era of Ethernet vs. EtherNOT, one that would initially result in complementary competition rather than direct competition throughout the 1990s. However, by the early 2000s, the direct competition resumed, as did the converging aspects of Ethernet and EtherNOT.

3.4 Ethernet vs. EtherNOT?

As the Ethernet and EtherNOT technology families continue to expand in scope and the features they provide, these two families have converged in many aspects while staying divergent in others. Even for features that seem similar on cursory inspection, these two families have often chosen to utilize different underlying mechanisms to provide them. Thus, while a feature might be provided by both, each family might have a different restriction or performance implication depending on what underlying mechanism is used. In this section, we inspect the important features provided by these two families.

3.4.1 Hardware and Software Convergence

Here we discuss how Ethernet and EtherNOT have converged with respect to hardware and software.

Support for Large Data Streaming. Traditional Ethernet devices relied on kernel-level stacks such as TCP/IP to perform the required protocol processing. The TCP/IP stack would segment data to be communicated into packets that fit into the maximum transmission unit (MTU) of the device (typically 1500 bytes). Every time a packet is transmitted or received, the device interrupts the kernel to inform it about this event. As Ethernet speeds increased from 3 Mbps to 10 Mbps, 100 Mbps, 1 Gbps and 10 Gbps, researchers realized that software enhancements were not sufficient to meet the performance that the hardware was capable of. Several enhancements were proposed to reduce these overheads, including interrupt coalescing, jumbo frames and segmentation offload (also called virtual MTU).

Interrupt Coalescing. With interrupt coalescing, instead of giving an interrupt for every packet that is transmitted or received, the device can combine multiple events together and give one interrupt for all of them. This reduces the number of interrupts that the host has to process and can eventually help in better performance. While this approach is appropriate for large data streaming, it can hurt communication latency where the interrupt is delayed by the device "hoping" for more events to occur in the near future.

Jumbo Frames. Jumbo frames are another widely used enhancement in Ethernet where instead of using the standard 1500-byte MTU, devices can support larger MTUs of up to 9000 bytes. It has been shown that jumbo frames can significantly increase performance as well, as the number of interrupts is based on the number of packets, and with larger packets the number of packets is lesser. Jumbo frames also help in reducing the percentage of extra bytes transmitted for headers, because of their larger payload capacity. However, while jumbo frames are used in several local-area networks, they are typically not wide-area compatible as Internet-based routers generally do not support them.

Segmentation Offload. Modern Ethernet devices support what are called segmentation offload engines or virtual MTU devices. These devices allow hosts to achieve the benefits of jumbo frames without being restricted by wide-area incompatibility because of using a non-standard MTU size. With this mechanism, devices advertise a large MTU to the host kernel (sometimes even larger than jumbo frames as current devices support up to 16 KB). Once the kernel protocol stack hands over a large frame to the device, the device internally segments the data into small 1500-byte segments before transmitting them over the physical network. While this approach still suffers from the overhead of the percentage of bytes used for headers, the number of interrupts it generates is significantly lower than devices without this support, and hence, can achieve high performance.

EtherNOT devices, on the other hand, have typically dealt with data communication as messages rather than as a stream of bytes. Thus, most EtherNOT devices communicate data as complete messages. The user application hands over data to the device in the form of a *descriptor*, which is essentially a data structure that describes the virtual address (or in some rare cases, the physical address) of the buffer that needs to be used for data communication, length of the message, and others. The device breaks the message into smaller MTU-sized segments, transmits them, and lets the application know when the entire message has been transmitted. The application can either request an interrupt to be generated when the entire message has been communicated, or it can request a memory location to be marked with a completion event, in which case it can manually check this location at a later time for the completion.

As Ethernet has increased its networking speeds over the past decade, it has also adopted such message-based communication techniques from the EtherNOT family while still providing the aforementioned enhancements for regular streaming data. For example, Ethernet Message Passing (EMP) [397, 399] was one of the early protocols on Ethernet that provided such message-based communication capabilities. With the advent of the Internet Wide-Area RDMA Protocol (iWARP) [370], which we will discuss in more detail later in this chapter, such message-based communication is becoming more and more common in Ethernet.

Protocol Offload Engines. Traditionally, the processing of protocols such as TCP/IP is accomplished via software running on the host processor, i.e., CPU or central processing unit. As network speeds scale beyond a gigabit per second (Gbps), the CPU becomes overburdened with the large amount of protocol processing required. Resource-intensive memory copies, checksum computation, interrupts, and reassembly of out-of-order packets impose a heavy load on the host CPU. In high-speed networks, the CPU has to dedicate more cycles to handle the network traffic than to the application(s) it is running. Protocol-offload engines (POEs) are emerging as a solution to limit the processing required by CPUs for networking.

The basic idea of a POE is to offload the processing of protocols from the host CPU to the network adapter. A POE can be implemented with a network processor and firmware, specialized ASICs, or a combination of both. High-performance EtherNOT networks such as InfiniBand and Myrinet provide their own protocol stacks that are offloaded onto the network-adapter hardware. Many Ethernet vendors, on the other hand, have chosen to offload the ubiquitous TCP/IP protocol stack in order to maintain compatibility with legacy IP/Ethernet infrastructure, particularly over the wide-area network (WAN) [145]. Consequently, this offloading is more popularly known as a TCP offload engine (TOE) [25, 143, 471].

The basic idea of a TOE is to offload the processing of protocols from the host processor to the hardware on the adapter, which resides in the host system (Figure 3.3). A TOE can be implemented in different ways, depending on the end-user preference between various factors like deployment flexibility and performance. Traditionally, firmware-based solutions provided the flexibility to implement new features, while ASIC solutions provided performance but were not flexible enough to add new features. Today, there is a new breed of performance-optimized ASICs utilizing multiple processing engines to provide ASIC-like performance with more deployment flexibility. Most TOE implementations available in the market concentrate on offloading TCP/IP processing, while a few of them focus on other protocols such as UDP/IP.

The idea of POEs has been around in the EtherNOT family for many years, but it is relatively new to the Ethernet family. As the current state of these technologies stand, both provide similar capabilities with respect to hardware

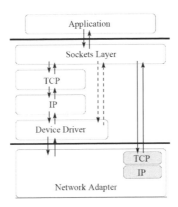

FIGURE 3.3:   TCP offload engines.

protocol offload. One primary difference, however, is that since EtherNOT networks have traditionally been optimized for high-end computing systems, their protocol stacks are typically not as elaborate as TCP/IP. Accordingly, the hardware complexity of these adapters is likely less than that of Ethernet-based TCP offload engines.

RDMA and Zero-Copy Communication-Capable Ethernet. Ether-NOT networks have provided advanced communication features such as zero-copy communication and remote direct-memory access (RDMA) [48, 207, 338, 353, 448] for many years. With zero-copy communication, the network hardware can directly DMA data from the host memory and send it out on the network; similarly the receiver side would directly DMA this data into the application memory without any intermediate copies. In RDMA, a node directly reads or writes into another remote process' memory address space completely through hardware. In the past, Ethernet relied on kernel-level stacks such as TCP/IP to perform communication, thus features such as zero-copy communication and RDMA were not provided. However, over the past few years, Ethernet has introduced the Internet Wide-Area RDMA Protocol (iWARP) that provides these features for Ethernet networks.

iWARP, as shown in Figure 3.4, is a new initiative by the Internet Engineering Task Force (IETF) [23] and RDMA Consortium (RDMAC) to provide capabilities such as remote direct-memory access (RDMA) and zero-copy data transfer. iWARP maintains compatibility with the existing TCP/IP infrastructure by stuffing iWARP frames within TCP/IP packets. The iWARP standard consists of up to three protocol layers on top of a reliable IP-based protocol such as TCP: (i) RDMA interface, denoted as RDMAP Verbs in Figure 3.4, (ii) Direct Data Placement (DDP), layer and (iii) Marker PDU Aligned (MPA)

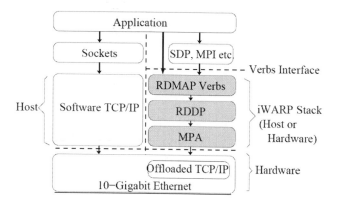

FIGURE 3.4: iWARP protocol stack.

layer. The RDMA layer is a thin interface that allows applications to interact with the DDP layer. The DDP layer uses an IP-based reliable protocol stack such as TCP to perform the actual data transmission. The MPA stack is an extension to the TCP/IP stack in order to maintain backward compatibility with the existing infrastructure.

While currently these capabilities are provided by both Ethernet and Ether-NOT families, the Ethernet family does so by maintaining backward compatibility with the existing TCP/IP infrastructure. Like protocol offload engines, this feature adds additional burden on the hardware and can potentially have performance implications as well.

Flow Control. Ethernet devices allow for stop-and-go kind of flow control based on PAUSE frames. Specifically, in order to prevent the sender from overrunning the switch or end receiver buffers, Ethernet allows the receiving device to ask the sender to temporarily pause transmission. This flow control is on a per-link basis and is implemented at the data-link layer (layer 2).

Several EtherNOT devices provide such flow control as well. In addition, some EtherNOT devices, such as InfiniBand, also provide end-to-end flow control between the sender and the receiver endpoints in conjunction with the per-link flow control, but at the transport layer (layer 4). However, the end-to-end flow control is with respect to the number of messages rather than the number of bytes transmitted. Thus, the end applications still need to

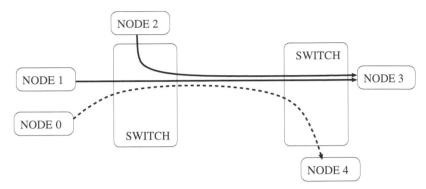

FIGURE 3.5:   Network congestion.

ensure that there is a large enough buffer allocated to receive the message, even with this flow control.

Like EtherNOT devices, Ethernet also provides end-to-end flow control that is implemented at the transport layer. When using kernel-based TCP/IP, such flow control is provided in software. However, when using hardware-offloaded TCP/IP, the flow control can also be provided in hardware. However, unlike EtherNOT devices, such end-to-end flow control in Ethernet is still based on the amount of bytes transmitted and not on the number of messages transmitted.

Congestion Control. While Ethernet provides flow control, as described above, congestion control is still needed in many circumstances. For example, consider the illustration in Figure 3.5. As shown in the figure, suppose there are two nodes (N1 and N2) sending data to one receiver node (N3), and one other node (N0) sending data to a completely separate node (N4). Now, let us assume the inter-switch link for both of these flows is the same. In this case, since N1 and N2 are both sending data to N3, the receiver node N3 cannot be expected to be able to receive at the same rate, thus causing a flow-control backlog on the network links. Thus, the flow-control credits between the two switches are completely exhausted by these flows. Now, when N0 is sending data to N4, since the inter-switch link has no flow-control credits, there is no data flow on that link either, despite the fact that the sender, the receiver, and all communication links connecting them are idle.

The Ethernet standard has traditonally relied on higher-level stacks such as TCP/IP to handle congestion control for it. When such congestion control is offloaded onto a TOE, the congestion-control policy that has been typically implemented is that of NewReno [146, 147] in light of its ubiquity in the early-to-mid 2000s. In recent years, however, software-based TCP protocol stacks have migrated away from NewReno to BIC [468] and CUBIC [178] in the Linux 2.6 kernel and Compound TCP [423] in Microsoft Vista with the primary intent to improve bulk data-transfer performance over the wide area, or more generally, any network with a large bandwidth-delay product, while still maintaining TCP friendliness.

EtherNOT networks, similarly, rely on their higher-level stacks (which are integrated on the network hardware as well) to perform congestion control. For example, the layer-4 stack in InfiniBand supports an elaborate congestion-control mechanism with forward explicit congestion notification (FECN), backward explicit congestion notification (BECN), and data throttling. In this aspect, Ethernet and EthetNOT networks seem to be converging towards a similar solution.

Multipathing. When systems are connected using a network fabric, it is possible that there are loops within the fabric. In such a case, packets sent onto the network could be forwarded around in circles, resulting in livelocks. Different networks use different solutions to deal with this problem. The Ethernet standard handles this problem by "reducing" the network to a spanning tree by disabling some of the links. Thus, between any two nodes, there is exactly one path that is enabled on which data can be sent.

This leads to two problems: (1) lack of robustness to network failures and (2) adverse impact to performance. With respect to network failures, researchers have proposed several techniques where, when a link fails, the remaining part of the network re-creates a different spanning tree using links that were previously disabled. The most prominent approach that is used is called the rapid spanning tree protocol (RSTP). With respect to performance, however, such tree topologies are not that helpful. That is, even when there are multiple paths available on the network fabric, using only one path can lead to network congestion and performance degradation.

Having recognized this issue, the Ethernet community proposed an extension to the Ethernet standard to logically divide a single physical network into multiple virtual local-area networks (VLANs), as illustrated in Figure 3.6. Each VLAN acts like an independent network, and hence, forms a spanning tree. However, multiple VLANs together can utilize all the links available in the system effectively. Specifically, Figure 3.6(a) shows the physical links available in the system, and Figure 3.6(b) shows how Ethernet would normally break it down to a spanning tree. Figure 3.6(c), on the other hand, shows how two separate VLANs (bold and dotted lines) can be used to form two spanning trees that collectively use up all the physical links in the system.

From an end node's perspective, it can pick which VLAN to send the packet on depending on what links it wants to use. Thus, by distributing packets sent over multiple VLANs, the nodes can effectively utilize all the available physical network links and improve performance.

VLANs are a powerful technique in Ethernet that allow it to perform many tasks including those of multicasting as well as many others, as we will see throughout this section.

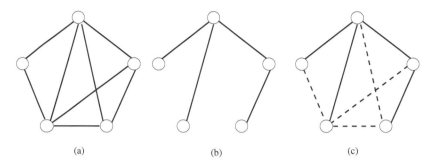

(a)   (b)   (c)

FIGURE 3.6:   VLAN-based multipath communication.

Quality of Service (QoS). QoS refers to the ability of the network to provide different classes of service to different communication flows. With the increasing importance of delivering QoS for voice, video, and data, Ethernet supports existing traffic management techniques, such as logical VLANs and IEEE 802.1p traffic prioritization at layer 2, and the Resource Reservation Protocol (RSVP) at layer 4, to deliver QoS over Ethernet.

Logical VLANs, as specified by IEEE 802.1Q, in tandem with IEEE 802.1p traffic prioritization allow end nodes to tag the priority of Ethernet packets in a way that can be communicated with network devices. In intelligent layer-3 and layer-4 devices, a network administrator can separate traffic by specific IP type and guarantee predictable delivery of this traffic.

On the other hand, EtherNOT networks such as InfiniBand, logically partition each link into multiple "virtual lanes," each of which can carry traffic at a different priority level in order to guarantee QoS.

Virtual Channels. Virtual channels are logical links dividing a single physical network link. Specifically, consider two nodes that are connected using a physical link; if both these nodes are a part of two different VLANs, data on one VLAN does not interfere with data on the other VLAN in the sense that head-of-line blocking on one VLAN will not stall communication on the other VLAN. Thus, the same physical link can be used to transmit two streams of non-interfering data. This concept is sometimes referred to as *virtual channels*.

An equivalent in the EtherNOT world can be drawn with InfiniBand's virtual lanes (VLs). InfiniBand allows its fabric to be broken down into up to 16 VLs. The number of VLANs in Ethernet, however, depends only on the number of bits that are allocated in its headers for separate VLAN IDs.

Multicast and Broadcast. Multicast refers to the capability of a sender to transmit a piece of data to multiple receivers. Broadcast, on the other hand, refers to the capability to send a piece of data to **all** receivers, making it a special case of multicast. Many Ethernet switches only support broadcast capabilities. However, multicast can be implemented in such infrastructures by combining broadcast with VLANs. Specifically, each VLAN can contain a subset of nodes in the system that are present in a multicast group. Thus, any broadcast on this VLAN would be received only by nodes in this subset. Such multicast groups can be set up in different ways, including (1) manual configuration by the system administrator, (2) packet snooping by switches looking for IGMP messages, and (3) synchronization with a multicast router in the system.

Many EtherNOT networks, on the other hand, do not have such multicast/broadcast capabilities. Some recent EtherNOT networks such as InfiniBand and Quadrics have provided these capabilities through different mechanisms. Quadrics provides hardware support to send messages to any set of contiguous nodes; thus, while it does not have any explicit group management, the sender can specify which contiguous nodes a message should go to. InfiniBand, on the other hand, has more explicit group management via the use of a "subnet manager" that allows processes to join and leave multicast groups by updating switch configuration tables dynamically.

Datacenter Support. While Ethernet's ability to maintain backward compatibility ensures that it continues to be the dominant network technology in the world, such compatibility also adds unnecessary overheads even in scenarios where they can be avoided. In order to avoid this, several users (mostly from the enterprise community) have conceived a new form of Ethernet, known as the *Datacenter Ethernet*. Datacenter Ethernet, also known as Converged Enhanced Ethernet (CEE) or Data Center Bridging (DCB), is essentially a collection of several advanced Ethernet features (e.g., QoS, multipathing), together with a few extensions to the Ethernet standard: (i) priority-based flow control, (ii) class-of-service (CoS) based bandwidth management, (iii) end-to-end congestion management, and (iv) lossless communication.

Lossless Communication. Datacenter Ethernet extends Ethernet flow control by using a credit-based mechanism. That is, the sender can send data as long as it has credits available. Whenever it sends a packet, it loses a credit. Once the switch has been able to forward this packet to the destination port, it sends an acknowledgement to the sender returning this credit. Thus, for each communication link, the sending end of the link never overflows the receiving end, thus avoiding unnecessary losses in communication. Note that this mechanism does not include explicit reliability, so it is not completely lossless; however, the loss probability is very low.

Priority-Based Flow Control. Priority-based flow control extends basic credit-based flow control, as described above, by combining it with virtual channels that divide a physical link into multiple logical channels (this is logically similar to VLANs). Specifically, instead of having a credit-based flow control and buffer allocations per physical link, this feature provides them per virtual channel. Thus, if one flow is throttled, another flow can still go ahead at its full capacity.

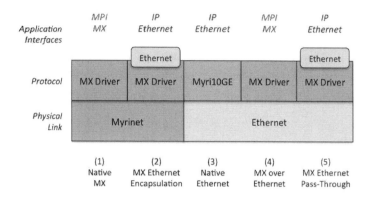

FIGURE 3.7:   Myrinet MX-over-Ethernet.

Class-of-Service (CoS) Based Bandwidth Management. CoS bandwidth management is very similar to QoS, except that it bundles a bunch of flows into each flow class and provides QoS to different classes. This is especially important when the network is dealing with different forms of traffic combined on the same physical network (e.g., LAN traffic, storage traffic, IPC traffic); the system can be configured to give a different CoS to each type of traffic. Also noteworthy is the fact that this CoS is not only with respect to prioritization of flows, but also with respect to bandwidth guarantees. That is, class A of flows can request 20% guaranteed bandwidth; the network fabric ensures that at least 20% of the data on the fabric belongs to this class at all times.

End-to-End Congestion Management. As described earlier in this section, Ethernet provides link-level flow control and relies on higher-level stacks such as TCP to provide congestion control. However, amongst the Datacenter Ethernet extensions is end-to-end congestion management. Specifically, when a node or switch sees congestion on the fabric, it sends a backward congestion notification (BCN) message to the sender on the previous network hop. Thus, for persistent congestion, the original data sender receives these BCN packets and can throttle its sending rate.

Switching Paradigm. Back in the 1990s, to assist in squeezing out the additional 15% of performance that the traditional 85% solution of Ethernet could not achieve, EtherNOT networks resorted to sacrificing generality and cost for performance. In the case of switching paradigm, EtherNOT technologies moved away from the store-and-forward switching that had become synonmous

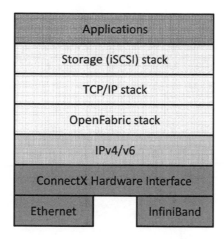

FIGURE 3.8: Mellanox ConnectX.

with Ethernet to the more efficient wormhole routing (actually, wormhole switching) and virtual cut-through switching [109].

Today, many Ethernet switches are also going the way of virtual cut-through switching to achieve better performance. Examples of such recent switches include the Fujitsu XG700 and Woven Systems' EFX Series of 10-Gb Ethernet switches.

Ethernet-EtherNOT Co-Existence. Nowhere is the convergence of Ethernet and EtherNOT more evident than with the pioneering Myrinet MX-over-Ethernet (Figure 3.7 [165]), and more recently, Mellanox ConnectX (Figure 3.8). In both of these cases, Myrinet and Mellanox provide support for both Ethernet and EtherNOT.

For example, as denoted by "(1) Native Myrinet" in Figure 3.7, Myricom originally provided a networking stack that was based on a specific link layer (Myrinet) and a specific software stack and interface (Myrinet Express or MX and MPICH-MX). However, because many end users wanted to run their IP- or sockets-based applications without modification, Myrinet provided Ethernet encapsulation, as captured by "(2) MX Ethernet Encapsulation," which, at the time, required specific hardware support or software bridging to enable the Myrinet-2000 Ethernet encapsulation to talk with a normal Ethernet network.

The above issue was addressed in 2006 when Myrinet launched its Myri-10G NIC, which was designed to be physically compatible with 10-Gigabit Ethernet.

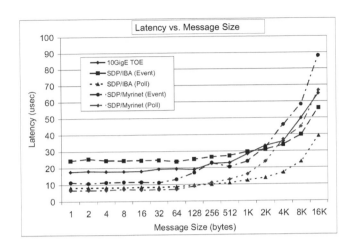

FIGURE 3.9:   Ethernet vs. EtherNOT: One-way latency.

The Myri-10G NIC could be plugged into either a Myri-10G switch, and talk the Myrinet protocol, or a 10G Ethernet switch, and talk native Ethernet, as denoted by "(3) Native Ethernet" in the figure.

While the "(2) MX Ethernet Encapsulation" implemented the Ethernet software interface atop the Myrinet link protocol, "(4) MX over Ethernet" does so over the Ethernet link protocol. So, while the software is the MX stack with support for OS-bypass and zero-copy communication, the NICs communicate with Ethernet packets underneath. Finally, because MX supports Ethernet encapsulation already, it could be run atop MX over Ethernet, as shown in "(5) MX Ethernet Pass-Through."

3.4.2   Overall Performance Convergence

In this section, we move from the qualitative descriptions of Ethernet-EtherNOT convergence of the previous section to a quantitative characterization of their performance convergence.

Micro-Benchmark Performance. Figures 3.9 and 3.10 show the comparison of two micro-benchmarks (ping-pong latency and uni-directional bandwidth) for three different network stacks — Ethernet (with TCP offload) and two EtherNOT networks (InfiniBand and Myrinet). IB and Myrinet provide two kinds of mechanisms to inform the user about the completion of data transmission or reception, namely polling and event-based. In the polling approach, the sockets implementation has to continuously poll on a predefined

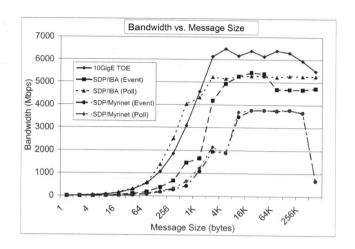

FIGURE 3.10:   Ethernet vs. EtherNOT: Unidirectional bandwidth.

location to check whether the data transmission or reception has completed. This approach is good for performance but requires the sockets implementation to continuously monitor the data-transfer completions, thus requiring a huge amount of CPU resources. In the event-based approach, the sockets implementation requests the network adapter to inform it on a completion and sleeps. On a completion event, the network adapter wakes this process up through an interrupt. While this approach is more efficient in terms of the CPU required since the application does not have to continuously monitor the data transfer completions, it incurs an additional cost of the interrupt. In general, for single-threaded applications the polling approach is the most efficient while for most multi-threaded applications the event-based approach turns out to perform better. Based on this, we show two implementations of the SDP/IB and SDP/Myrinet stacks, viz., event-based and polling-based; the Ethernet TOE network that we used supports only the event-based approach.

Apart from minor differences that are affected by the maturity of each specific network adapter version, and the associated software stack, the performance of all networks is in the same ballpark showing a general convergent trend in the performance.

Application Performance. Virtual Microscope (VM) is a data-intensive digitized microscopy application. The software support required to store, retrieve, and process digitized slides to provide interactive response times for the standard behavior of a physical microscope is a challenging issue [4, 82]. The main difficulty stems from the handling of large volumes of image data,

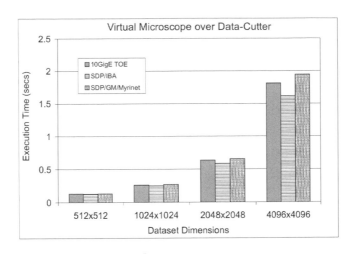

FIGURE 3.11:   Ethernet vs. EtherNOT: Virtual Microscope application.

which can range from a few hundreds of megabytes (MB) to several gigabytes (GB) per image. At a basic level, the software system should emulate the use of a physical microscope, including continuously moving the stage and changing magnification. The processing of client queries requires projecting high-resolution data onto a grid of suitable resolution and appropriately composing pixels mapping onto a single grid point. Figure 3.11 compares the performance of the VM application over each of the three networks (Ethernet, IB, Myrinet). Again, we see a comparable performance for all three networks.

MPI-tile-IO [371] is a tile-reading MPI-IO application. It tests the performance of tiled access to a two-dimensional dense dataset, simulating the type of workload that exists in some visualization applications and numerical applications. In our experiments, two nodes are used as server nodes and the other two as client nodes running MPI-tile-IO processes. Each process renders a 1×2 array of displays, each with 1024×768 pixels. The size of each element is 32 bytes, leading to a file size of 48 MB. We evaluated both the read and write performance of MPI-tile-IO over PVFS. As shown in Figure 3.12, once again the performance is quite comparable.

In summary, while there are some variations in performance between the Ethernet and EtherNOT networks, with one doing better in some cases and the other doing better in others, the performance of both families is similar. As these technologies mature, we expect their performances to converge even more.

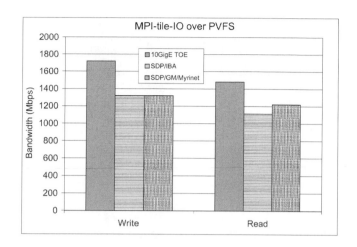

FIGURE 3.12: Ethernet vs. EtherNOT: MPI-tile-IO application.

3.5 Commercial Perspective

Thus far, we have examined the various features of Ethernet and EtherNOT and different aspects of their convergence from the perspectives of hardware, software, and performance. Here we take a more data-driven perspective on the "Ethernet vs. EtherNOT" debate.

In the datacenter, very few, if any, institutions want to cope with supporting multiple network technologies, e.g., Ethernet for connecting computers to corporate LANs, Fibre Channel for connecting to storage, and InfiniBand for connecting servers in a multi-tiered environment. Why? Each server would require as many as three different network adapters, three different networks using three different types of cables, and three management nodes. Thus, there seems to be a general consensus for a unified network fabric in light of this current hodgepodge of network adapters, switches, and cabling, typically consisting of some combination of Ethernet, InfiniBand, and Fibre Channel. Therefore, in this arena, the "Ethernet vs. EtherNOT" debate is also quite substantial. With the increasing commoditization of 10-Gigabit Ethernet (10GigE), coupled with the standardization process for Fibre Channel over Ethernet (FCoE), Ethernet proponents claim that victory will be theirs in their effort to unify cluster and storage connectivity. On the other hand, the more nimble InfiniBand community views their EtherNOT InfiniBand (IB) as being technologically superior to Ethernet, e.g., 40-Gbps IB (also known as quad data-rate or QDR IB) vs. the mere 10-Gbps of 10GigE.

In fact, one might argue that InfiniBand has turned the tables on Ethernet, the de facto standard that was quickly established as a "blue book" Ethernet specification in 1980 by DEC, Intel, and Xerox. If anything, the InfiniBand community is the more nimble one. Ethernet's current strength, i.e., its ubiquity, has also arguably become its biggest weakness, because modifying standards entails convincing the entire vendor community, which has been a slow and arduous process. For example, since 2006, an IEEE study group has been working on putting forth 40-Gbps and 100-Gbps standards; these standards are now not expected to be completed until late 2010. In the meantime, InfiniBand already has 40-Gbps networking technology available, leaving 10-Gigabit Ethernet in the dust, not only with respect to performance but also cost.

On the flip side, Mellanox, the only InfiniBand switch silicon vendor, is not all EtherNOT any more. They are hedging as well ... in March 2009, Mellanox announced the general availability of their ConnectX EN 10GigE network adapter. Similar to the Myri-10G network adapter that supports both Myrinet and Ethernet, the Mellanox ConnectX architecture supports both InfiniBand and Ethernet connectivity on the same adapter. In short, the ConnectX architecture unifies data traffic over a single I/O pipe instead of over a single protocol.

3.6   Concluding Remarks

This chapter presented a perspective on how Ethernet technology has been converging with EtherNOT technology, particularly over the last decade. The aspects covered in this chapter included support for large data streaming, protocol offload engines, RDMA and zero-copy communication-capable Ethernet, flow control, congestion control, multipathing, quality of service, virtual channels, multicast and broadcast, datacenter support, switching paradigm, and Ethernet-EtherNOT co-existence.

With Ethernet currently available only at 10 Gbps and the EtherNOT of InfiniBand already delivering 40 Gbps, InfiniBand should be "walking all over" Ethernet. Yet it is not. Despite InfiniBand EtherNOT vendors positioning InfiniBand as the datacenter network fabric of tomorrow, capable of carrying Ethernet and Fibre Channel, the economies of scale continue to point towards Ethernet. Recently, major network vendors, including Cisco Systems, Extreme Networks, Brocade/Foundry Networks, Force10 Networks, and Juniper Networks, have been pushing the economies of scale for Ethernet as well as the Datacenter Ethernet (DCE) angle. So, while the Ethernet standardization process for 40- and 100-Gbps Ethernet has been slow-going, the convergence of Ethernet and EtherNOT continues from both sides.

While the EtherNOT of InfiniBand may continue to lead Ethernet with respect to technological innovation and advances, Ethernet also continues to adapt and adopt "best practices" in a way that optimizes performance, cost, and interoperability. In short, what one might say about the Ethernet vs. EtherNOT debate is as follows: "I don't know exactly what the next networking technology will look like, but it will be called Ethernet."

Chapter 4

System Impact of Integrated Interconnects

Sudhakar Yalamanchili, Jeffrey Young
School of Electrical and Computer Engineering
Georgia Institute of Technology

4.1 Introduction

Multiprocessor interconnects have been the target of performance optimization from the early days of parallel computing. In the vast majority of these efforts, interprocessor communication and multiprocessor interconnection networks have been viewed primarily as part of the functionality of the I/O subsystem, and therefore network interfaces were treated as devices whose access was mediated and controlled by the operating system. It became clear early on that this was a significant impediment to performance scaling of parallel architectures, and this realization inspired the development of more efficient forms of interprocessor communication, including zero copy protocols, user space messaging, active messages, remote direct memory access (RDMA) mechanisms, and more recently the resurgence of interest in one-sided message communication. All of these approaches sought to make significant improvements in message latency. Now, as we progress into the deep submicron region of Moore's Law, multiprocessor architecture has given way to multi-core and many-core architecture, and we have seen the migration of high-speed communication protocols and attendant network interfaces onto the multi-core die. Such integration levels present new challenges to the chip architect as well as new opportunities for optimizing system performance.

This chapter presents an overview of the landscape of integrated interconnects and the technology trends that continue to motivate their development and evolution. In brief, we can broadly classify multi-core interconnects into three groups: i) on-chip interconnects, ii) inter-socket interconnects, and iii) board/rack scale interconnects. The last class is serviced by traditional local area interconnects and their derivatives led by Ethernet, Infiniband as well as by storage area networks. The first class of interconnects is an area of active research and is currently devoid of standards beyond industry standard buses which will not scale into many-core designs. It is the inter-socket interconnects that are playing a central role in the integration of high-performance multi-core die, memory systems, and I/O systems into modular blade infrastructures that are in turn driving the development of large-scale data centers and new high-performance computing platforms. This chapter focuses on the class of inter-socket interconnects. In particular, we do not cover the extensive literature on software communication stacks and network interface optimizations. Rather, we are interested in presenting the evolution of the hardware/architectural aspects of inter-socket interfaces and the impact that they can have on system design. This class of networks will have a significant impact on the design of future data centers as they i) will determine efficient patterns of sharing via the extent of coherence domains that they support, ii) affect the non-recurring engineering (NRE) costs of hardware via their impact on blade design, and iii) impact cost and power via low-latency access to remote resources, thereby promoting consolidation.

We first begin with an overview of technology trends that have motivated the evolution of integrated interconnects, followed by descriptions of three commercial inter-socket standards: AMD's HyperTransport (HT), Intel's QuickPath Interconnect (QPI), and PCI Express (PCIe). Strictly speaking, PCIe is not an inter-socket communication protocol but i) will impact on-chip networks, ii) is the major vehicle for many systems to access the local system interconnect such as Ethernet or Infiniband, and iii) shares many similarities with HT and QPI due to its board-level nature. Section 4.4 covers a system design inspired by the availability of low-latency integrated interconnects — the implementation of global address space architectures and the capabilities that can be derived from such designs. The chapter concludes with a description of some expectations that speculate on emergent functionalities and capabilities of future inter-socket interconnects.

4.2   Technology Trends

The industry is currently anticipating designs that should scale from thousands to millions of cores as exemplified by modern and emerging data centers and future supercomputing configurations at the Exascale level [234]. The end-to-end message latency in such systems is currently dominated by the end points rather than the node-to-node network switch and link latencies. In particular, the overall message latency is dominated by the intra-node delays at the end points (distance from the wire to host memory) and the software stack. For example, in a typical commodity implementation the network interface card may be sitting on a southbridge and must negotiate its way through the south and northbridges to the memory controllers via the assistance of some

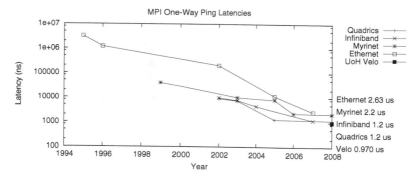

FIGURE 4.1:   Network latency scaling trends.

messaging layer stack. Thus, the physical location and access mode to the network interfaces play a key role in the performance of internode messaging. Left untouched, this overhead will in fact (relatively) grow with each new technology generation.

Consider the relationship between the physical properties of DRAM and networking components. In particular, note reported message latency trends as shown in Figure 4.1. Over the last decade we have seen the latencies of packet traversal across a link and the latency of packet cut-through routing in a switch drop faster than the latencies of DRAM accesses. Overall, as the figure indicates, we have seen dramatic reductions in network latency of a packet and only modest reductions in raw DRAM access time, which have been relatively constant in the 10-20 ns range. This is largely due to the fact that the raw memory cell access time in a DRAM does not dominate access latency and thus is not as sensitive to technology scaling. Rather, other factors such as bit line and word line length, and properties that are a function of the size of the memory cell array, dominate memory access. As a result, with increasing DRAM densities the positive impact of smaller cell size will continue to drop, leading to much smaller reductions in DRAM access latency relative to drops in network link and switch latencies. This view does not even consider the impact of the memory controller on the DRAM access path.

Now consider the integration of high-speed network interfaces on die. With integrated HT and QPI interfaces the distance from the wire to the memory controller has been significantly decreased. In particular, the raw congestion free memory-to-memory hardware latency between remote memory modules has been significantly reduced. An informed estimate, using 2008 technology and measurements from an HT implementation and analysis [263], points to approximately 600-800 ns hardware access time to a remote memory module over a single network link. Additional hops with state-of-the-art interconnection network technology should add no more than 20-50 ns per hop. The disparity between end-to-end network hardware delay and DRAM access times can

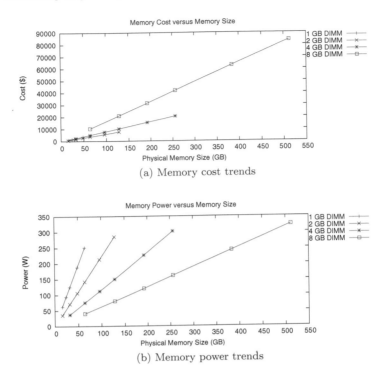

(a) Memory cost trends

(b) Memory power trends

FIGURE 4.2:   Memory cost and memory power trends for a commodity server system.

only be expected to grow in the future. *The overall impact of integrated interconnects is that the memory reach of a core can be extended beyond a single blade with relatively small hardware penalties.*

Now consider the trends in memory cost and power illustrated in Figure 4.2. With respect to cost, as the clock speed of DRAM subsystems rise, the number of connectors that can be supported on a single controller drop, due to physical signal integrity constraints. Further, keeping pace with memory volume demand requires the use of high-density memory chips. The overall impact on the high-end high-capacity blades is a high per gigabyte cost on the order of $1000/GByte [86,213]. For lower-speed memory subsystems, the costs are much lower, being on the order of $100/GByte. Thus, system design techniques and operation that can improve efficiency of memory usage will permit operating points lower down this non-linear cost curve with disproportionately higher returns in improved cost effectiveness. With respect to power, as Lefurgy [250] indicated, memory power can consume up to 40% of the overall system power, and energy consumption of computing is rising rapidly. For example, the energy consumption of data centers alone is expected to account for 3% of total US energy consumption by 2011 and will continue rising [439]. For every watt of power consumed, approximately 0.7W to 0.8W is expended in cooling [141]. For large data centers this can result in millions of dollars a year in energy costs alone! Figure 4.2 illustrates cost and power curves for memory derived from HP calculators [99, 190, 192] made available to size the infrastructure needs for installations of HP Proliant-based computing facilities.

The lesson learned from the preceding trends is the need to share physical memory more effectively to improve memory utilization and thereby improve both cost and energy consumption. Currently, server blades are typically provisioned with enough memory to support a maximum anticipated memory footprint across the applications. However, many large scale applications encountered in data center deployments tend to have time-varying memory footprints. The idea of provisioning blades for the peak memory footprint is inefficient in cost and power (which also has cost consequences) since peak memory demand is rarely, if ever, simultaneous across all applications. Thus memory utilization will be low since memory is provisioned for the worst case. Instead, the application characteristics suggest provisioning memory for the average case and using on-demand sharing of physical memory across blades to improve cost and power effectiveness. During periods of peak demand physical memory can be "borrowed" from an adjacent blade. By sharing we are not referring to sharing in the sense of a programming model such as coherent distributed shared memory. Rather, we are referring to simply making the number of DIMMS available to a blade temporarily larger via an interconnect such as HT or QPI, i.e., the physical memory accessible to the cores on a blade may change over time. How that additional physical memory is used is a matter for the blade operating system and blade applications. What is important is that the total memory statically provisioned per blade can be significantly reduced while having minimal practical impact on performance. Such sharing is feasible only if remote memory can be accessed without an overwhelming latency penalty. *The integration of high-performance network interfaces into the die and the consequent minimization of hardware memory-memory latency between blades makes the sharing of physical memory across blades practical.*

Finally, we can also examine overall trends from the perspective of available off-chip memory bandwidth. The memory demand bandwidth is growing faster than the available pin bandwidth. Jacobs [213] estimates that approximately 1 GByte/sec of memory bandwidth is required for each hardware thread. With core scaling replacing frequency scaling as the vehicle for sustaining Moore's Law, we are seeing a reduction in memory bandwidth per core, and in fact, the raw memory per core [213, 231]. Coupling this with increased consolidation via virtualization the memory pressure will continue to increase. Integrated interconnects will see increased pressure for bandwidth as the memory demand bandwidth overflows into interconnect bandwidth due to increased accesses to remote memory modules or increased page fault traffic. Note that the path to the memory controller from secondary storage may run through the on-chip integrated network interface. This is also true for operating system optimizations such as DRAM-based page caches and the use of solid state disks. The stress on the interconnect will grow, increasing pressure for reduced latency and increased bandwidth. *Integrated, very low-latency interconnect will have to serve as a pressure valve to release memory pressure by providing low-latency access to remote memory.* In effect, memory pressure demands will push for the translation of network bandwidth into additional memory bandwidth.

The preceding trends make a case for the importance of integrated inter-socket interconnects and how they might be used to address memory system challenges of future multi- and many-core architectures. The following section provides an overview of three commodity interconnects that currently form the bulk of this space.

4.3   Integrated Interconnects

In addition to the technology trends identified in the previous section, there are a number of pragmatic reasons for the integration of multiple high-speed, point-to-point interconnects onto the die. As cores, peripherals, I/O devices, and networks get faster the bottlenecks of the processor buses and I/O buses cause increasing degradation in overall performance. In the past, the presence of these slower buses was compensated for in part by the use of larger caches. However, that is no longer feasible since performance scaling now comes from increasing the number of cores on a die rather than the frequency of a core. As the number of transistors increases with each technology generation, more of them will have to be devoted to cores to scale performance and less will be available to mask the slow speed of the processor and I/O buses via the use of deeper cache hierarchies. Therefore, the interfaces will have to get faster. Second, consolidation and integration are major cost-reduction drivers. For example this reduces parts count as well as the number of connectors. Third, the pressure on the availability of pins is increasing and wide buses extract significant performance penalties in lost opportunity costs relative to the use of faster "thinner" off-chip communication channels. Hence, the evolution of high-speed point-to-point interconnects was inevitable, as is their migration on chip.

As the multi-core and many-core landscape evolves we can observe a further sharpening of the boundaries between various types of interconnects. The most visible are the LAN interconnects led by Ethernet. One can anticipate future designs where 10 GBit Ethernet media access control (MAC) components are integrated onto the chip. However, their primary role will be as a system interconnect between blades, racks, etc., rather than between cores or sockets. At the other end of the spectrum are on-chip networks — a vibrant research

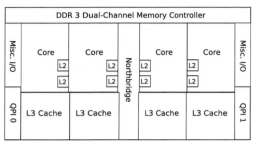

(a) Quad-core Nehalem floorplan

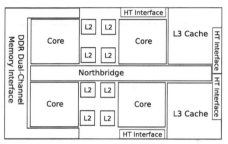

(b) Quad-core Barcelona floorplan

FIGURE 4.3: Approximate floorplans for quad-core processors from AMD and Intel. Adapted from die shot by Hans de Vries at www.chip-architect.com, 2008.

arena which at the time of this writing is in the formative and explorative stage of its evolution. Between these ends are the *inter-socket interconnects*. The leading commercial instantiations are AMD's HyperTransport and Intel's QuickPath Interconnect (QPI). While HyperTransport has been available since roughly 2001, QPI is a recent entry into the market. Both of these protocols support coherent memory address spaces, and as a consequence of being an inter-socket interconnect, rely heavily on implementations that can produce low latency and high bandwidth. In particular they evolved to overcome the bottleneck of bus based interconnections to I/O devices and co-processors. The evolution of both protocols has been accompanied by the integration of memory controllers on die and as we will see in Section 4.4 this can have a tremendous impact on system design performance. Finally, we include in this discussion the industry standard PCI Express (PCIe) I/O interconnect. Like HT and QPI, PCI Express is a packet switched, scalable width interconnect. As with HT and QPI, we can expect that PCI Express cores will inevitably migrate on chip.

The use of a point-to-point interconnect is common across all of the standards discussed here. The use of a point-to-point link implies limited electrical load

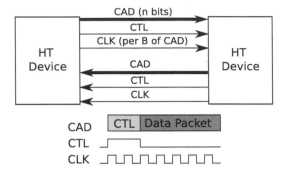

FIGURE 4.4:   Breakdown of a HyperTransport link.

and therefore high transmission speeds. The use of variable width or narrow links makes effective use of available (and limited) I/O pins where link speeds can be scaled. Further, in some instances it enables the response to certain classes of link failures where a new reduced link width can be renegotiated to switch out the failed bit lanes. All of the these standards support link-level error detection with retransmission. However, they also have clear and distinctive characteristics in their support for coherence domains, emphasis on latency vs. bandwidth, and support for levels of reliability, availability, and service (RAS).

Figures 4.3b and 4.3a show an approximate floor plan for AMD's quad-core Barcelona and Intel's quad-core Nehalem chips showing a number of integrated HT and QPI links. These are derived from die shots that can be found on the web [27, 306]. It is also of interest to note the integration of the memory controllers on die and to recognize the reduced physical latency between the memory controllers and the northbridge. In fact, multi-socket server blades interconnected by QPI and HT can be essentially viewed as networks of memory controllers transparently accessed by cores through the memory system. With a non-coherent mode, for example using non-coherent HT links, systems are less reliant on non-scalable (coherent) accesses to remote memory or devices.

4.3.1   HyperTransport (HT)

HT is a point-to-point packet switched interconnect standard [194] that defines features of message-based communication within a processor complex. HT was one of the first such public messaging standards to be developed and defined, being available in 2001. The specification for the non-coherent version of the protocol is public while the coherent specification is available under license. The HyperTransport Consortium (HTC) is the industry body that oversees the administration of the standard and shepherds the development of the industrywide HT ecosystem [193].

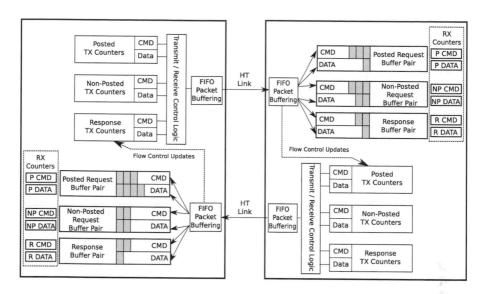

FIGURE 4.5: Organization of buffers/VCs on a HT link. Adapted from Don Anderson, Jay Trodden, and Mindshare. Hypertransport System Architecture. Addison-Wesley, 2003.

HT evolved in response to the bottleneck behavior of buses and was conceived as an intra-board interconnect. The goal was a standard that permitted scalable bandwidth across a range of design points. Each HT link consists of a pair of unidirectional channels using differential signaling. Each direction comprises two groups of signals. The link signals carry packets on the command, address, data (CAD) group of signals as shown in Figure 4.4. The CAD group is scalable in width from 2 bits to 32 bits. The link is source synchronous, and a separate CLK (clock) signal is provided for each byte lane in the CAD group. Finally, a control signal (CTL) is used to distinguish whether a control or data packet is being currently transmitted on the CAD lines. The links operate at double data rate. At 3.2 GHz up to 6.4 GTransactions/sec are possible in each direction. With 16-bit links that is 12.8 GBytes/sec in each direction and 25.6 GBytes/sec aggregate across a single link. If one considers the 32-bit links defined by the standard, the aggregate link bandwidth in both directions rises to 51.2 GBytes/sec. HT was designed for deployment as the interconnect fabric for the I/O subsystem and as such operates with both address-based semantics and messaging semantics. HT devices are configured with base address ranges in a CPU mapped I/O space and are compatible with the PCI view of devices and addressing.

The main HT packet types are control packets or data packets which are organized in multiples of four bytes. HT has a very simple and light packet format, considerably lighter than PCI Express (for example, see Section 4.3.3 which employs a traditional protocol stack approach leading to larger overheads per packet). The specific packet type being currently transmitted on a link is indicated by the CTL line. The advantage of this design is the ability to interleave packets — an optimization referred to as priority request interleaving. When a large data packet is being transmitted along a link, a control packet can be injected into the middle of the data transmission and distinguished by the state of the CTL line so that the packet is steered to the correct control buffer. In addition to maintaining separate buffers for control and data, there are separate buffers for the three message classes that traverse a HT link: posted messages, non-posted messages, and response messages. Posted messages do not require a response acknowledging completion of the requested operation, e.g., memory write. Non-posted transactions require a response. In the case of a non-posted write transaction this takes the form of a "target done" message. Response messages also occur in response to transactions such as memory read operations that return values. The messages are serviced by individual virtual channels for each message type. A group of virtual channels is referred to as a virtual channel group and HT defines support for multiple virtual channel groups. The buffering on a channel now appears as in Figure 4.5.

HT links employ credit-based flow control. The propagation of credit information is coupled with link management in the form of *NOP* packets that are continuously transmitted over an idle link. The packet contents include fields for both the propagation of credit information as necessary, as well as link diagnostic and management information. NOP packets themselves are not flow controlled and must be accepted. Flow control is indicated in units of the maximum packet size. In HT 3.1 link level errors are detected by use of a CRC and recovery is via retransmission by the transmitter. The NOP packets carry acknowledgment information specifying the last packet correctly received — erroneous packets are discarded. The retry sequence is initiated where the transmitter retransmits all packets from the last correctly received (acknowledged) packet.

Central to the HT specification is the concept of message ordering constraints. Transactions from a single source can be partitioned into groups and ordering constraints specified within and between groups. To assist in the maintenance of such ordering constraints and to provide control over memory consistency models, HT supports flush and fence primitives as well as atomic operations. The flush primitive flushes outstanding posted writes to memory from a device. A fence operation acts as a barrier to all posted writes across all devices and I/O streams and is only implemented at the host bridge which is at the end of a chain of HT devices. The specific rules concerning when certain posted writes on a specific virtual channel can pass a fence operation on another virtual channel can appear to be quite complex. The key idea underlying fence and flush commands is the ability to control and make assertions about the completion of posted writes (unacknowledged) to memory.

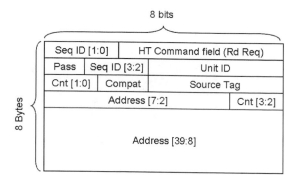

FIGURE 4.6: HT read request packet format.

A typical command packet is shown in Figure 4.6, where the fields specify options for the read transaction and preservation of ordering and deadlock freedom. The UnitID specifies the source or destination device and allows the local host bridge to direct requests/responses. The HT specification currently supports up to 32 devices. The Source Tag (SrcTag) and Sequence ID (SeqID) are used to specify ordering constraints between requests from a device, for example, ordering between outstanding, distinct transactions. Finally, the address field is used to access memory that is mapped to either main memory or HT-connected devices. An extended HT packet can be used that builds on this format to specify 64-bit addresses [194].

Distinctively, HT includes a coherent mode that utilizes a three-step coherence protocol. As with most snoop-based protocols, optimizations to limit the impact of snoop bandwidth are typically employed. However, HT as a whole has two additional and distinct advantages. The first is that HT has a non-coherent operating mode. This provides a low-latency communication substrate unburdened with the responsibility of maintaining coherence across the address space. An ecosystem for extending this non-coherent mode to a much larger node count via tightly coupled interfaces and a source-routed switched point-to-point network has been emerging in both compatible extensions to the base protocol [127] as well as a low-latency hardware infrastructure [263]. Collectively, these two extensions enable the support of a very low-latency systemwide global physical address space. In Section 4.4 we illustrate one consequence of this capability — namely physical memory consolidation leading to significant cost and power savings. The second distinct advantage is that the HT specification defines a physical layer for communication over longer distances referred to as the *AC operating mode*. This mode supports backplane, board-to-board, and chassis-to-chassis operation of the HT link. An AC mode link has (relatively) lower bandwidth and higher latency than the normal (or DC) mode. However, this capability permits the vendors to

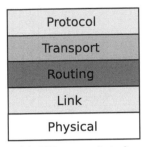

(a) QPI protocol stack

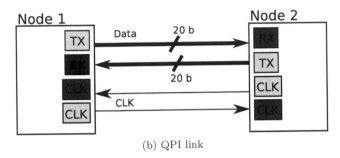

(b) QPI link

FIGURE 4.7: Structure of the QPI protocol stack and link.

use the same HT device for both physically short as well as (relatively) longer connections. In particular, low-latency interconnect can be extended to adjacent blades to facilitate sharing of memory or peripherals without introducing another interconnect or another layer of switching while the proposed switched extensions can achieve longer reach with minimal latency impact. Finally, and more recently, the HyperTransport Consortium has released a specification for connectors similar to PCIe, enabling the development of third-party HT cards with low-latency access to the multi-core host.

4.3.2　QuickPath Interconnect (QPI)

The Intel QuickPath Interconnect (QPI) is a high-speed, low-latency point-to-point interconnect designed as a replacement for the Intel Front Side Bus (FSB) that has served Intel well over many generations of products. The emergence of multiple QPI links on die coincided with the integration of memory controllers on die, creating Intel's next generation of high-end server processors.

Figure 4.7a illustrates the various layers of Intel's QPI protocol stack. Each QPI physical channel in Figure 4.7b is comprised of two unidirectional 20-bit channels — one in each direction. The interconnect supports multiple discrete link widths of 5, 10, and 20 bits where each 5-bit segment is referred to as

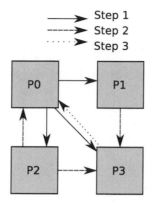

FIGURE 4.8: QPI's low-latency source snoop protocol.

a quadrant. On power-up, a link negotiation protocol determines the active link width. The 20-bit physical digits (or *phits*) correspond to 16-bit data and 4 bits of side band signals. The physical layer employs differential signaling and a number of novel techniques to support high-speed physical links over distances of up to 24 inches, with the ability to transfer data on each of the clocks, and (at the time of this writing) a 3.2 GHZ clock through which the link can support 6.4GT/sec in each direction for an aggregate throughput of 25.6 GBytes/sec per link. The link has one forwarded clock in each direction, and the 42 single-bit lanes correspond to 84 signals, that is, considerably less than the 150 signal count required of the Intel FSB [352].

The link layer is primarily concerned with management of the reliable transmission of 80-bit flow control digits or *flits* — each flit is comprised of four phits on a full width link. Credit-based flow control is used for managing the buffers on the receiver. Reliable transmission is provided at multiple levels. First, the link layer provides reliable transmission via the generation of an 8-bit CRC for each flit. When the receiver detects an error on a flit, a special control flit is transmitted to the sender causing the sender to roll back and retransmit flits that were not received. Such link-level retry mechanisms isolate link-level bit errors from higher layers of the protocol stack. If the error persists or unrecoverable hard errors are encountered, the link can enter a self-healing mode. In this mode the sender and receiver automatically initiate the renegotiation of the link width with data transmission automatically retried without loss of data. Finally, clock fail-over mechanisms enable recovery from failure on the clock signal by mapping the clock signal to a predefined data lane and operating the link at half width.

The link layer employs credit-based flow control in each of three virtual networks — VN0, VN1, and VNA. Within each virtual network independent virtual channels are available for each of up to 14 message classes, although only six message classes have been defined to date [352]. The virtual channels employ two-level buffering [221] with a minimum private buffer space allocated for each channel and a shared adaptive network. The message classes broadly fall into messages corresponding to coherence traffic and messages corresponding to non-coherent operations and accesses. These messages are managed at the level of packets and are handled by the packet layer. At the time of this writing, the routing layer and transport layer have not been publicly defined. Nominally the routing layer enables routing of packets among multiple QPI nodes while the transport layer has end-to-end reliability responsibilities to improve RAS properties (reliability, availability, serviceability), for example by providing end-to-end flow control for reliable transmission and buffer management.

In the QPI terminology, the coherence protocol is exercised between *caching agents* and *home agents*. The former operate on behalf of devices that may cache copies of the coherent memory space while the latter represents and manages part of the coherent memory space. An interesting feature of QPI is the support for two classes of coherence protocols. One protocol based on traditional directories designed for larger scale systems is a three-step protocol and is referred to as the *home snoop* version [352]. The second protocol, referred to as the *source snoop* version, defines a protocol where the source of a reference sends snoop messages directly to agents that may have cached copies. The latter coherence protocol is lower latency but consumes more network bandwidth and is not scalable to large systems. It is graphically illustrated in Figure 4.8. In step 1 the requesting node (P0) sends snoop messages to all nodes concurrently with a request to the home node of the requested cache line. In step 2 all snooped nodes respond to the home agent node (P3) with status information. If a node (P2) has a cached copy in the proper state as defined by the protocol, this node will return the cached line to the requesting node (P0). Thus, data can potentially be delivered to the requesting node in two steps and consequently this protocol is referred to as a two-step protocol. In step 3 the home node resolves any issues and communicates with the requester. The home snoop protocol is a three-step protocol, with higher snoop latency but better scaling properties with system size. One can anticipate that low core count systems would be configured with source snoop coherence while high core count systems would be configured with home snoop coherence.

4.3.3   PCI Express (PCIe)

Nowhere has the impact of bus limitations been felt more than the I/O subsystem. The previous generation of multi-drop I/O buses has now evolved into the packet-switched, serial, point-to-point high-bandwidth PCI Express (PCIe) standard as the dominant I/O interconnect in the desktop computing arena and enterprise arena (ignoring storage networks). The physical implementation is matched to modern implementation constraints while it maintains backward compatibility with PCI and PCI-X driver software by

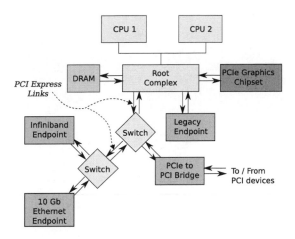

FIGURE 4.9: An example of the PCIe complex.

retaining the same memory, I/O, and configuration address spaces. Thus operating systems and drivers from PCI-based systems should be able to run unmodified. The I/O subsystem in many ways is more challenging. It is faced with backward compatibilities to I/O peripherals with vastly diverse interaction models, bandwidth requirements, and latency behaviors. Creating a protocol that can be configured across such a diverse performance spectrum while not penalizing high-performance peripherals is a daunting task.

A PCIe link is comprised of 1, 2, 4, 8, and 16 bit serial signal pairs in each direction, referred to as *lanes*. For example, 4 lanes in one direction is referred to as a 4x link (a by-four link). The link width is typically determined based on product target market with motherboards and blades being designed to provide various combinations of PCIe slots of various widths for different market segments. For example, the high-end gaming sector would have multiple 16x links to support multiple high-end graphics cards. Messages are byte striped across multiple lanes and transmitted using 8b/10b encoding — each byte is encoded in 10 bits leading to a 25% transmission overhead.

PCIe transactions comprise read and write transactions to the PCIe memory, I/O, and configuration address spaces. These transactions may be posted or non-posted. The former represents a unidirectional transaction in that no acknowledgment is returned to the sender, sacrificing knowledge of completion for lower transaction completion latency. The latter is implemented as a split transaction and requires a response. PCIe implementations prescribe a switch-based complex to connect a large number of devices as shown in Figure 4.9. Requesting devices can transmit packets (initiated from the transcation layer as described below) to end points where packets are routed through the switches. Specifics are best understood in the context of the overall architecture. The

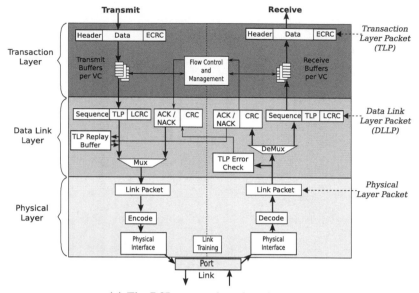

(a) The PCIe protocol stack architecture

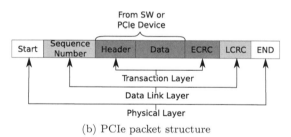

(b) PCIe packet structure

FIGURE 4.10:   Structure of the PCI Express packets and protocol processing. Adapted from Ravi Budruk, Don Anderson, Tom Shanley, and Mindshare. PCI Express System Architecture. Addison-Wesley, 2003.

protocol stack for PCIe and the attendant implementation view of each layer is illustrated in Figure 4.10a.

The transaction layer creates packets that are submitted to the data link layer which in turn relies on the physical layer for transmission. The transaction layer uses virtual channels of which there may be up to eight. As in HT and QPI, packets are mapped to message or traffic classes and so identified in the header. Each message class is mapped to a specific virtual channel and thereby inherits priorities in traversing the fabric as switches and devices are configured to prioritize traffic flowing through them. Buffer space is managed via a credit-based flow control mechanism across individual links.

The data link layer is responsible for reliable transmission. Like both HT

3.1 and QPI the data link layer protects packets with a link layer CRC. Packets are buffered by the data link layer and retransmitted when packets are received in error. Repeated errors result in reinitializion of the link while further errors can result in link shutdown. The transaction layer may choose to include an optional CRC. Like HT, for each virtual channel PCIe tracks and manages flow control for posted, non-posted, and response (completion in PCIe terminology) packets. Further, amongst each class, distinctions are made between control/header and data payloads. Hence there are six subsets of packet types that must be tracked and independently flow controlled. The PCIe specification prescribes the appropriate unit of management for each class of packets [69].

In addition to read/write transactions on memory, I/O, and configuration address spaces, PCIe supports a *locked* memory read operation. This operation locks the path from the requester to the destination and can only be issued by the root complex. Finally, PCIe supports a *message transaction* primitive. This one-sided message primitive can be used for posted transactions involving power management, interrupt signaling, unlocking locked paths, etc. Vendor-specific message transactions can also be defined.

4.3.4 Summary

The preceding three interconnect standards share many similar features. They are all point-to-point packet switched interconnects with variable-width links. Transparent reliability is provided at the data link layer with CRC support and error detection coupled with retransmission. They all emphasize the need for bandwidth, although HT and QPI have better latency properties. However, they also have significant differences. Both HT and QPI support coherent accesses and are optimized also for small transaction sizes and short distances, while PCIe compromises on latency in the interest of supporting communication with a range of device types and message sizes. PCIe was not designed as a protocol that would be integrated deep into the core and memory hierarchy like HT and QPI.

4.4 Case Study: Implementation of Global Address Spaces

This section describes an example of the use of integrated interconnects to realize a 64-bit global physical address space. Available memory bandwidth and interconnection bandwidth will have to track the increase in the number of cores to sustain Moore's Law performance growth with the scaling of cores.

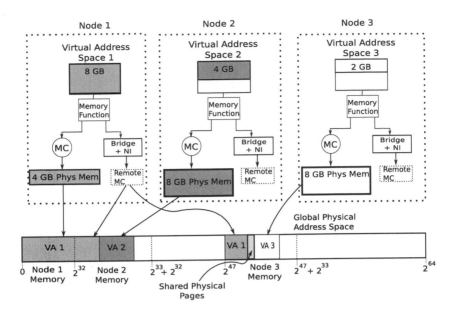

FIGURE 4.11:   The Dynamic Partitioned Global Address Space model (DPGAS).

The concept of a global (physical) system-wide address space can be used to improve memory sharing and increase memory utilization, thereby enabling significant reductions in power and cost. Global address spaces are possible due to the proximity of on-chip interfaces to the memory controllers and their low-latency characteristics.

To productively harness this raw capability for a global address space, it must be exercised in the context of a global system model that defines how the systemwide address space is deployed and utilized. Towards this end we have advocated and are exploring the implications and implementation of a global address space model that exploits the low-latency on-chip HT interfaces to provide a 64-bit global physical address space. In particular, we are concerned about the portability of the model and software implementations across future generations of HT-enabled multi-core and many-core processors with increasing physical address ranges.

Global address space models can be traced back to the supercomputer class machines of the early 1990s, notably the Cray T3D and T3E series machines which provided hardware support for global non-coherent physical memory [237, 472]. The distributed shared memory systems of the 1980s and 1990s provided a non-uniform access memory across a coherent physical global address space [245]. More recently we have seen the emergence of the partitioned global address space model in the supercomputing space. In this latest model, cores have access to a private memory as well as global shared memory. Additional language semantics capture the non-uniform access nature, for example the idea of places and remote activities in X10 [89] that enable global address space languages and associated compilers to reason about performance in generating efficient implementations.

Our approach differs in that we provide abstractions that enable the system to tailor the implementation to the specific distribution of memory controllers and their interconnection. The proposed memory model, a candidate implementation, and some representative performance evaluation results are presented in the remainder of this section with the goal of demonstrating capabilities that integrated interconnects now make feasible.

4.4.1 A Dynamic Partitioned Global Address Space Model (DPGAS)

The memory model is that of a 64-bit partitioned global physical address space. Each partition corresponds to a contiguous physical memory region controlled by a single memory controller. For the purposes of the discussion we assume that all partitions are of the same size. For example, 40-bit physical addresses produced by the Opteron processors correspond to a 1 Terabyte (TB) partition. Thus, a system can have 2^{24} partitions with a physical address space of 2^{40} bytes for each partition. Although large local partitions would be desirable for many applications such as databases, smaller partitions may occur due to packaging constraints; for example, the amount of memory attached to an FPGA or GPU accelerator via a single memory controller is typically far less than 1 TB. Thus we view the entire memory system as a network of memory controllers accessed from cores, accelerators, and I/O devices. This is graphically illustrated in Figure 4.11 which shows how virtual address spaces and physical address spaces at a node can be mapped into a global physical address space.

From a programming model perspective we are exploring one-sided message communication using a get/put model for managing data movement. The get/put operations are native to the hardware in our system model in the same sense as load/store operations. The execution of a get operation will trigger a read transaction on a remote partition and the transfer of data to a location in the local partition, while the execution of a put operation will trigger a write of local data to a remote partition. The get/put operations are typically visible to and optimized by the compiler while the global address space is non-coherent. At the same time, local partitions may have limited coherence maintained by local protocols. GASNet is a publicly available infrastructure for implementing global address space models [50].

An application (or process) now is allocated a physical address space that may span multiple partitions, i.e., local and remote partitions. Equivalently, the process's physical address space is mapped to multiple memory controllers, and thus an application's virtual address space may map to physical pages distributed across local and remote partitions. There are a number of reasons to dynamically distribute an application's physical address space. The most intuitive one is for sharing where processes from an application may share physical pages by having portions of their virtual address spaces mapped to the same physical pages, for example to share libraries. We also envision the sharing of peripherals and custom hardware through shared physical address spaces as these devices are often accessed and managed through memory mapped interfaces. Thus a remote device may be accessible through memory mapped I/O as if it were attached locally. There is a significant body of knowledge on managing remote storage that can be brought to bear in this context. Finally, we envision alternative messaging models that operate in non-coherent shared memory but are more easily subject to push vs. pull optimizations for message passing.

The set of physical pages allocated to a process can be static (compile-time) or dynamic (run-time). The nature of the page management changes in scalable systems due to the hierarchy of latencies necessitating optimizations that have little relevance in traditional operating systems in uniform access memory architectures, e.g., page placement. For example, when paging from disk, the choice of physical pages to place a page in becomes important. It may be necessary to maintain a list of remote partitions that can be accessed or information related to coherence/consistency management that may be maintained on a per page basis. Modern operating systems technology includes the ability to optimize NUMA systems, and some of those techniques are applicable here.

A set of cores on a chip will share one or more memory controllers and HT interfaces. All of the cores also will share access to a memory management function that will examine a physical address and route this request (read or write) to the correct memory controller — either local or remote — functionality that is typically encountered in the northbridge [97]. Several such multi-core chips can be directly connected via point-to-point HT or QPI links. Further, by using a custom network [263] or tunneling packets through a switched network, the memory or devices on remote nodes can be accessible.

To summarize, our model specifies get/put transactions for accessing physically distributed pages of a process and these transactions may have posted or non-posted semantics. The location of physical pages may be changed under compiler or operating system control, but the pages remain in a global 64-bit physical address space. All remote transactions are necessarily split phase. We now turn our attention to the process of implementing such an experimental DPGAS system and noting how the availability of low-latency interconnects is an enabling technology.

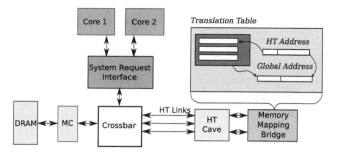

FIGURE 4.12: Memory bridge with Opteron memory subsystem.

4.4.2 The Implementation Path

In any system the equivalent of the northbridge will steer memory references to a local on-chip memory controller or to an HT interface that will create and manage a remote transaction. The HT interface must encapsulate a memory mapping unit that will i) map the referenced address to a remote physical address at a remote node, ii) encapsulate the reference into a remote message supported by some inter-blade communication fabric, and iii) manage the transmission of this request and any responses that result from this request. In our prior work, we used Ethernet as a demonstration fabric [475]. The memory mapping unit integrated into an HT interface is referred to as a *memory bridge* and the Ethernet version is referred to as HT-over-Ethernet (HToE). Its location in a typical node endpoint is illustrated in Figure 4.12 while a high-level breakdown of the microarchitecture components is shown in Figure 4.13.

Our demonstrator is based on the use of Ethernet as the commodity inter-blade interconnect primarily due to ready availability of hardware implementations. Furthermore, progress on data center Ethernet [136] [135] is addressing issues of flow control and error recovery that would enable (with some additional effort) the easier preservation of HT semantics across blades at low cost. The HToE bridge implementation uses the University of Heidelberg's Hyper-Transport Verilog implementation [406], which implements a non-coherent HT cave (endpoint) device. The detailed operation of the bridge can be found in [475] while a brief overview follows.

The three stages of encapsulation of a HT packet for a remote request are illustrated in Figure 4.13 for remapping a 48-bit local process address into a global address. An efficient implementation could pipeline the stages to minimize latency, but retaining the three stages has the following advantages: i) It separates the issues due to current processor core addressing limitations from the rest of the system, which will offer a clean, global shared address space, thus allowing implementations with other true 64-bit processors, and ii) it will be easy to port across switched networks. The receive behavior of

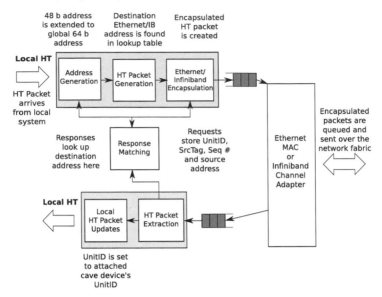

FIGURE 4.13:   Memory bridge stages.

the bridge will require a "response matching" table where it will store, for every request that requires a response, all the information required to route the response back to the source when it arrives. The management of this table has some complexities arising from the fact that HT was not designed as a switched interconnect, but by tunneling HT packets through an inter-blade interconnect such as Ethernet or Infiniband we are stressing the basic design of the HT protocol. Further, while the address mapping is a simple operation, it is on the critical path for remote accesses and therefore must be fast. Some of the results of our current implementation demonstrate that such mapping can indeed be performed quite fast as reported in the following section.

4.4.3   Bridge Implementation

Xilinx's ISE tool was used to synthesize, map, and place and route the HToE Verilog design for a Virtex 4 FX140 FPGA. The hardware details of the design can be found elsewhere [475], but we summarize relevant points here. The latency results of just the bridge are listed in Table 4.1 along with the associated latency of the Heidelberg cave device from [406]. Total read and write latencies are also listed that incorporate latency statistics discussed in relation to Table 4.2, which covers components on the complete hardware path. The key issue here is that the bridge and HT interface latencies are manageable and small relative to the Ethernet MAC latencies in contrast to the capabilities that are achieved. We note that there are several key system-level issues such as consistency protocols that were not addressed in this study and would have

TABLE 4.1:   Latency results for HToE bridge.

DPGAS operation	Latency (ns)
Heidelberg HT Core (input)	55
Heidelberg HT Core (output)	35
HToE Bridge Read (no data)	24
HToE Bridge Response (8 B data)	32
HToE Bridge Write (8 B data)	32
Total Read (64 B)	1692
Total Write (8 B)	944

TABLE 4.2:   Latency numbers used for evaluation of performance penalties.

Interconnect	Latency (ns)
AMD Northbridge	40
CPU to on-chip memory	60
Heidelberg HT Cave Device	35 – 55
HToE Bridge	24 – 72
10 Gbps Ethernet MAC	500
10 Gbps Ethernet Switch	200

to be accounted for in any detailed analysis. This study does, however, point out that the memory bridge by itself is not a limiting factor.

The latency statistics for the HToE bridge component and related Ethernet and memory subsystem components were obtained from statistics from other studies [97] [406] [210] and from our FPGA place and route timing statistics for our bridge implementation. 10 Gbps Ethernet numbers are shown in this table to demonstrate the expected performance with known latency numbers for newer Ethernet standards. We can construct a simple model for the overall latency of such implementations. Using the values from Table 4.1 for using the HToE bridge to send a request to remote memory, the performance penalty of remote memory access can be modeled by the following:

$$t_{rem\_req} = t_{northbridge} + t_{HToE} + t_{MAC} + t_{transmit}$$

where the remote request latency is equal to the time for an AMD northbridge request to DRAM, the memory bridge latency (including the Heidelberg HT interface core latency), and the Ethernet MAC encapsulation and transmission latency. This general form can be used to determine the latency of a read request that receives a response:

$$t_{rem\_read\_req} = 2^*t_{HToE\_req}\ 2^*t_{HToE\_resp} + 2^*t_{MAC} + 2^*t_{transmit} + t_{northbridge} + t_{rem\_mem\_access}$$

These latency penalties compare favorably to other technologies, including the 10 Gbps cut-through latency for a switch, which is currently 200 ns [354]; the fastest MPI latency, which is 1.2 μs [285]; and disk latency, which is on the order of 6 to 13 ms for hard drives such as those in one of the server configurations used for our evaluation of DPGAS memory sharing [418]. Additionally, this unoptimized version of the HToE bridge is fast enough to feed a 1 Gbps Ethernet MAC without any delay due to encapsulating packets. Likely improvements for a 10 Gbps-compatible version of the HToE bridge would include multiple pipelines to allow processing of packets from different virtual channels and the buffering of packets destined for the same destination in order to reduce the overhead of sending just one HT packet in each Ethernet packet in the current version.

4.4.4   Projected Impact of DPGAS

The latency numbers suggest that using the memory bridge to access physical memory from an adjacent blade would be feasible especially if coupled with an effective page migration/page swapping philosophy which can transparently update the memory bridge entries without requiring application-level interaction (remapping which physical addresses are remote). We expect this update cost is less expensive than updating page tables. However, the access to the memory bridge to perform remapping must be synchronized for correct operation. Assuming that physical memory can be "borrowed" from a neighbor, what sort of power and cost benefits can be expected? We performed such a study and found that benefits can be substantial [475].

Our analysis covered five benchmarks: Spec CPU 2006's MCF, MILC, and LBM [186]; the HPCS SSCA graph benchmark [22]; and the DIS Transitive Closure benchmark [122]. These benchmarks had maximum memory footprints ranging from 275 MB to 1600 MB. A 2.1 billion address trace (with 100 million addresses to warm the page table) was sampled from memory-intensive program regions of each benchmark. These traces were used in conjunction with application and system models to assess potential power and cost savings. We analyzed the impact of DPGAS by simulating a workload allocation across a multi-blade server configuration using a simple greedy bin packing algorithm and a "spill" model of memory allocation where if an application could not be allocated on a blade due to lack of space, additional memory is allocated from an adjacent node or the allocation fails. Several configurations with and without sharing were evaluated for system sizes of up to 255 blades (we used HP Proliant servers and HP provided configuration calculators as the data sources).

The detailed analysis can be found in our recent publication, "A HyperTransport-Enabled Global Memory Model For Improved Memory Efficiency" [475]. Our models predicted that in the base (250-server) case, DPGAS has the potential to save 15% to 26% in memory cost when the initial provisioning is high (64 GB/blade), which translates into a $30,736 savings for the low-end servers and $200,000 for the high-end servers. Blades with higher base memory provisioning guidelines should see higher savings. We should point out that this analysis was based on the analysis of equivalent memory footprints for shared and non-shared memory. However, it did not consider the performance impact of remote accesses. Further, the models were based on analytic models, not on software implementations servicing real applications executing on a cluster.

The models similarly discovered that the power savings using DPGAS allocation is substantial in the base case, with savings of 3,625 (25%) and 5,875 (22%) watts of input power for the low-end and high-end server configurations, respectively. When server consolidation onto 200 servers is used, power savings drops substantially to 800 and 500 watts for the same configurations. We found that typically in smaller installations and with workloads with many tasks with smaller memory footprints the comparative savings drops considerably when the workload is high since we can achieve good memory utilization. The scenarios where savings can be substantial are when the variance in memory demand across the applications is high and relative number of small footprint tasks is not high enough to fill memory holes.

4.5   Future Trends and Expectations

While trying to predict the future in any discipline is risky business at best, one can speculate about several evolutionary technology trends. We conclude this chapter with some thoughts on where these integrated interconnects might take us and some of the attendant research issues these expectations expose. First we exclude the exciting developments in optical interconnects from this discussion. That deserves separate treatment unto itself and promises to be a truly disruptive technology. On the I/O front we can anticipate that multiple PCIe interfaces will migrate on-chip reducing the raw physical latency between devices and the cores, and we can expect that product slices with varying I/O capacities for various market segments will emerge. One can also routinely expect standards evolution to faster, smaller, and more reliable instances. However, the requirements of the I/O space are quite distinct from the inter-socket space and it is this latter space that exposes some interesting trends.

First, with regards to interconnects such as HT and QPI, until scalable coherence mechanisms are realized we will continue to see large message passing-based systems evolve around islands of coherence within multi-socket blades. If one considers an 8 socket blade and the continuing trajectory of Moore's Law, then over the next two generations we will see 32 cores/socket and 256 cores/blade. A 256 core shared memory machine with coherent shared memory will likely suffice for applications pushing the memory and compute envelope today while the 1-2 socket blades will most likely account for the high volume segment. Thus, we can expect that standards such as HT and QPI might extend their coherence domains to remain within a blade but not cross blades, retaining the modularity and consequent lower non-recurring engineering costs (NRE) of blade manufacturing.

Second, we see an intermediate role for non-coherent, low-latency interconnects such as HyperTransport with extended (in terms of the number of nodes) addressing range [127]. Such networks can provide low-latency non-coherent access to remote memory and I/O devices and can be expected to be augmented with capabilities for supporting partitioned network operation. For example, a number of compute sockets/blades may be coupled with a number of DRAM memory blades, solid state disk (SSD) blades, and I/O devices into a logical platform. The communication between the platform constituents must be isolated from intra-platform communication in disjoint logical platforms to ensure compliance with promised service level agreements. Such partitioning capability is central to cloud computing infrastructures today, and systems will be under pressure to produce better hardware utilizations in future data centers and clouds to reign in escalating power consumption and cooling costs. The on-chip integration and low-latency characteristics of these inter-socket interconnects effectively extends the physical reach of a multi-core processor socket. This capability is central to resource virtualization where resources can be freely composed and shared across multiple virtual machines. The flexibility and extent of such composition (and consequently the benefits of resulting consolidation of hardware resources) are a function of the low-latency reach of these integrated networks. Extensions across blades will impact blade cost, perhaps spurring innovations in backplanes to retain modularity and consequent cost benefits. An example of one such set of innovations includes extensions to HT [127, 263].

At the outer end of the coherence domains and the non-coherent partitions of resources lay the interfaces to commodity fabrics such as Infiniband and Ethernet. It is unclear whether HPC machines will see benefit in integrating Ethernet and Infiniband interfaces on die. Cost and power concerns suggest that sharing of bandwidth on a blade basis will be the norm, leaving the inter-socket communication gap to be filled by the likes of HT and QPI. Further, in comparison Ethernet and Infiniband have a unique intellectual history of addressing a wider range of communication issues and not being as focused on low latency. Thus, while we can expect to see integrated Ethernet MACs, it appears unlikely that they or Infiniband will be the main means of low-latency inter-socket communication. Commercial standards such as HT and QPI will fill this gap.

The integration of networks on die also introduces some interesting problems at the interfaces with the inevitable on-chip networks. Today Barcelona and Nehalem have multiple HT and QPI interfaces on chip respectively. The number of such interfaces will not grow as fast as the number of cores. Consequently these interfaces may become congestion points for on-chip traffic leaving the chip. Such non-uniformities will affect the design of the on-chip network and likely the floorplanning in the placement of these network interfaces in future designs. For example, will we need non-uniform on-chip network designs to deal with these potential choke points? A related constraint is the path from the network interface to the memory controllers. This must be a fast path for remote accesses and must retain its low-latency properties while at the same time dealing with increased stress on the DRAM memory bandwidth and capabilities of the DRAM memory controller. It is most likely that as a consequence we will see some integration of network and memory controller functionality. Today there is a tremendous amount of interest in memory controller design, particularly in scheduling disciplines. The presence of remote accesses can have disruptive impact on the locality properties of the local cores. Scheduling and optimization heuristics that worked well with single threaded workloads are now being subsumed by local multithreaded and multi-core access patterns. This disruption will intensify with the addition of remote traffic into the mix, and we can anticipate an impact on the scheduling disciplines and controller design to distinguish between and manage remote and local traffic.

The growth of low-latency non-coherent network links has a singular focus on minimizing the latency from the wire to the DRAM controller. Implementations concentrate on the lowest memory-to-memory latency, with other functionality layered on top of this layer. For example, one could pursue region-based memory semantics (following the lead of recent work on region tracking [76] for example) where pages/regions of memory may have shared memory or transactional semantics. These semantics are a property of the application rather than the hardware implementation (the fast path). The prevalent shared memory coherence protocols have inherent barriers to scalability, primarily because they possess two characteristics: i) uniform treatment of accesses to any memory address, and ii) application agnostic behavior in that all memory references are treated equivalently. Mechanisms such as snooping, flooding, and the use of directories are global mechanisms for synchronizing updates to shared variables. By enabling non-uniform treatment of memory references or memory regions, we feel one can develop approaches to coherence that are tailored to the reference behavior of the applications and the sharing pattern amongst cores (equivalently memory controllers). Such limited, dynamic, or on-demand coherence can be effectively supported via the flexible global address space mapping mechanisms described here, supported by optimized page management policies. For example, the address space mappings in our case study in the previous section implicitly define the set of memory controllers that must participate in coherence operations at any given point in time. We can consider support within the network for dynamically created coherence domains (partitions of the system) whereby coherence is maintained only within the current constructed coherence domain. Required functionality includes flexible data structures that can be tailored to the current set of pages that are to be coherent and managing traffic accordingly without penalizing non-coherent traffic. These are considerable challenges but may represent a technology-enabled approach to providing new avenues for pursuing scalable shared memory.

In summary, this chapter has focused on the class of inter-socket interconnects that fills the gap between on-chip and traditional high-speed networks. By virtue of their integration on die, these networks provide new capabilities as well as opportunities and challenges, and remain an active area of research and development as they mature and evolve to meet new and emerging HPC challenges.

Chapter 5

Network Interfaces for High Performance Computing

Keith Underwood
Intel
Ron Brightwell, Scott Hemmert
Sandia National Laboratories [1]

5.1   Introduction

The network interface provides the hardware that supports the implementation of the features exposed by the software APIs. Modern network interfaces leverage various approaches to implementing the network API. Most network interfaces accelerate some portion of their native low-level programming interface (see Section 6.3); however, how much of that interface is implemented on the NIC varies from almost all [353] to almost none [13]. To complement the work done on the NIC, various networks leverage everything from threads [301] to interrupts [63] to simply application level libraries [206]. In this chapter, we explore the issues and trade-offs in modern network interface design and place them in the context of several modern network interfaces. We conclude with a look at several research efforts with implications for future network interfaces.

5.2   Network Interface Design Issues

Numerous issues face the designer of a modern network interface. One must target an existing network application programming interface (API) (Chapter 6), or design a new one. While the API limits the features that must be (or can be) exposed to the user, it also imposes a set of requirements on

---

[1]Sandia is a multiprogram laboratory operated by Sandia Corporation, a Lockheed Martin Company, for the United States Department of Energy's National Nuclear Security Administration under contract DE-AC04-94AL85000.

the network interface in terms of ordering and the interaction model between the NIC and the host processor. In this section, we attempt to capture some of the major issues facing modern network interface designs.

5.2.1   Offload vs. Onload

Whether a network *can* offload MPI functionality or other higher-level API functionality is determined by the choice of network API. After choosing an API, one of the first decisions that must be made is whether or not to provide hardware on the network interface to offload the functionality provided by the API. The offload-versus-onload debate is a long-standing one. Proponents of onload argue that the performance of the message passing libraries can be scaled with the number of processors. They point to the low performance of embedded microprocessors as a bottleneck in NIC implementations that offload the message processing. In turn, proponents of offload point to the ability to more easily provide independent progress by offloading operations to the NIC. Offloading also reduces or eliminates the contention for host processing resources and the inevitable cache pollution that accompanies that contention.

Of course, the choice is not a black and white issue. Rather than asking "Should I use offload or onload?" the network interface designer is faced with the question "How much of the API should I offload?" The network interface is always going to offload something. For example, most network interfaces have a DMA engine, and CRC generation and checking is always offloaded in modern networks, because the performance of hardware is dramatically better than the performance of software. Beyond that, however, the choice is much less obvious. At one extreme, some Ethernet interfaces require the host software to place the data into a ring of packet-sized buffers. Most HPC networks, however, allow the software to interact with the network at a higher level with the network interface providing the packetization, depacketization, and basic header processing. At the opposite extreme from Ethernet is the Quadrics Elan family of network interfaces [353]. These network interfaces offload virtually all of the functionality provided by the API.

The decision about what to offload or onload is heavily influenced by the semantics of the API. For example, a two-sided interface like the standard MPI send/receive calls can require a substantial amount of processing to offload the interface. Since MPI does not directly require independent progress, many implementers choose not to provide offload for these functions at either the API or hardware level. Or, they may choose to use a software thread to provide the independent progress allowed by their API. In contrast, truly one-sided operations like those provided by SHMEM are much less processing intensive. In addition, one-sided operations can impose a requirement for independent progress. Due to the expectation that one-sided operations are extremely lightweight, providing these semantics through a software thread based implementation is unlikely to meet the performance expectations.

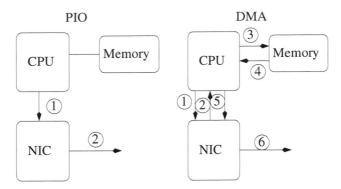

FIGURE 5.1: Comparing programmed I/O and DMA transactions to the network interface.

5.2.2 Short vs. Long Message Handling

A high performance network must handle messages ranging from 8 bytes to several megabytes in an efficient manner. The performance of 8 byte messages is dominated by the metrics of latency and message rate. In turn, messages of more than a few kilobytes are dominated by the bandwidth of the network. Unfortunately, the mechanisms to achieve the minimum message latency and the mechanisms to achieve the maximum message bandwidth directly conflict. Figure 5.1 illustrates the difference between programmed I/O and DMA transactions.

The interface between the processor and the network interface has always been a bottleneck for both latency and bandwidth. In the case of latency, programmed I/O transactions offer the lowest latency because they only require one transaction across the bus at the source ((1) in Figure 5.1) followed by placing the data on the network (2). Unfortunately, they also rely on the host processor to copy the data directly to the network interface. Due to the typical interactions between the CPU and network interface, this operation is almost always bandwidth limited to something significantly below the network bandwidth. In contrast, DMA transactions only require the processor core to copy a simple command to the network interface. The network interface can then have a large number of outstanding requests to the host memory to cover the bandwidth-delay product of the host interface. This yields the optimal bandwidth for network transfers, but there is a an extra startup overhead that prevents it from achieving the minimum latency. As shown in Figure 5.1, the command must propagate from the core to the NIC (1), and then the NIC incurs the latency to read host memory ((2) through (5)) before the message can be sent out on the wire (6). A network interface designer must decide where a network interface crosses over between using programmed I/O and DMA.

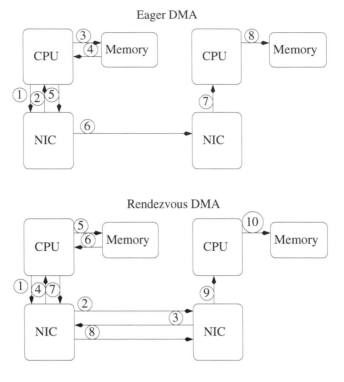

FIGURE 5.2:   Comparing messages using an eager protocol to messages using a rendezvous protocol.

A similar switch between short and long messages typically occurs in how the hardware and software manage buffer space and flow control for messages going over the network. Many network APIs have the concept of an *unexpected* message. Unexpected messages are those that arrive at the target before the corresponding buffer has been provided. For example, an `MPI_Isend()` can reach the target before the corresponding `MPI_Irecv()` has been called. With no corresponding buffer for the message, what should the target do with the data? One approach is to send messages eagerly, and then to buffer unexpected messages in state provided by the communications library for unexpected messages. Unfortunately, for long messages, it is easy to overwhelm any state that has been set aside by the communications library. Thus, when messages are sent eagerly, small messages are typically buffered with long messages being dropped and retransmitted when the user posts the matching buffer. This approach can waste a significant amount of network bandwidth when a large number of unexpected large messages are encountered.

In contrast, a message can be sent using rendezvous, where the initiator sends only a message header and then sends the body of the message when the

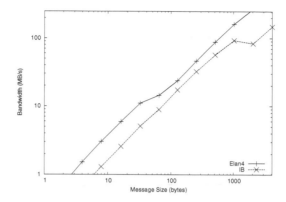

FIGURE 5.3:   Ping-pong bandwidth for Quadrics Elan4 and 4X SDR InfiniBand.

target has indicated that a matching user buffer has been posted. This prevents wasting bandwidth for large messages, but imposes an extra round-trip across the network that can carry an extreme latency penalty for small messages. The contrast between eager and rendezvous protocols is illustrated in Figure 5.2. Note that the rendezvous approach has two extra steps across the potentially high latency network between the two NICs.

To prevent having to drop data sent over the network, most networks use an eager protocol for short messages and a rendezvous protocol for long messages. The network must then define the cross-over between short and long messages and ensure that adequate buffer space is available for all of the unexpected short messages that an application might receive. The correct crossover point is a balance of the extra latency imposed by the rendezvous transaction and the buffer space required for the small messages. The penalty of such a rendezvous transaction can be mitigated by implementing it in hardware (in the network interface), which can allow the total amount of buffer space needed to be reduced.

The graph of ping-pong bandwidth in Figure 5.3 (from a previous study [59]) illustrates the impact of the trade-offs associated with short vs. long messages. Inevitably, there is a transition between short and long messages that appears in the performance graphs. The transition always appears as a flattening in (or even drop in) the bandwidth curve when going from a smaller message size to a larger message size. A key transition for Elan4 is between 32 and 64 bytes, while a more dramatic change occurs for InfiniBand between 1KB and 2KB. In each case, this is the added latency of the "long protocol" for the network.

5.2.3   Interactions between Host and NIC

The interactions between the host — a processing core — and the NIC have a dramatic impact on a variety of system performance issues. However, the two metrics that are most heavily impacted are the latency and the small message rate. Four issues dominate the interaction between the CPU core and the network interface:

- Host Interface — A network interface interacts with the processor through an I/O bus of some sort.

- Cache Interaction — A network interface may participate in the processor's coherency protocol, or not.

- The OS and OS-bypass — A network interface can require OS and/or application intervention for the various tasks it must perform.

- Command queue management — The host places commands in a location where the network interface can find and process them. The placement of commands into command queues must be flow controlled between the host and network interface.

- Event posting/completion notification — A network interface must notify the host processor when a message completes.

- Interrupts — Interrupts are the primary mechanism for asynchronous interaction between a network interface and the processor core.

5.2.3.1   Host Interface and Cache Interaction

A network interface typically attaches to a single type of host interface. Examples of typical host interfaces include PCI-X, PCI Express, HyperTransport (HT), and the Intel QuickPath Interconnect (QPI). The variants of PCI offer the most ubiquitous connectivity to platforms with the strongest guarantee of long term compatibility. In contrast, HT and QPI are much less portable between platforms, but offer the highest level of performance in a given generation.

If the network interface is attached to a processor front-side bus style interface, like HT or QPI, it must then decide whether or not it is going to participate in the coherency protocol. If a network interface chooses to participate in the processor's coherency protocol, it can reap several significant advantages. For example, by participating in the caching protocol, the NIC can perform atomic operations that are truly atomic relative to the host processor. It can also avoid some of the performance issues associated with having the network interface's command queues in uncacheable space (see Section 5.2.3.3).

5.2.3.2 The OS and OS-bypass

Another interesting choice for the network interface is how much to involve the operating system (OS) in communications. The OS can provide most of the protection needed for communications processing — as long as it is involved in every network transaction. The network needs to identify the user and process ID in a trusted fashion. In addition, the network needs to provide isolation between the network interface level resources available to the various processes. All of this can be achieved by trapping to the kernel on every interaction between the application and the network interface. Unfortunately, kernel traps on modern systems still add *at least* 60 ns to every transaction. Such overheads can have a dramatic impact on message latency and message rate.

Most network interfaces targeted towards HPC provide a feature known as OS-bypass. OS-bypass requires two features on the network interface. First, the network interface must have resources to create a command queue per process (see Section 5.2.3.3). This command queue is then memory mapped in the address space of one, and only one, process using the processor's virtual memory system. Second, the network interface must maintain *trusted state* for each command queue (i.e. for each process with direct access to the network interface). Trusted state implies that the state can only be modified by the OS itself on behalf of the process. This state is used to identify the user ID and process ID associated the application that is allowed to access a given command queue.

5.2.3.3 Command Queue Management

Modern network interfaces always provide a mechanism for the processor to enqueue multiple requests, so that the processing of the requests can be pipelined through the network interface. These command queues can reside in either host memory or network interface memory. While older Ethernet adapters typically created ring buffers of commands in host memory, most modern HPC network interfaces now put the command queues directly on the network interface. The command queues are then mapped directly into the address space of either the OS or the application (see Section 5.2.3.2). This creates a range of issues to deal with in the management of the command queues.

Cached vs. Uncached Accesses. A command queue in the network interface can either be in a cacheable or an uncacheable address range. Obviously, to place the command queue in cacheable space, the network interface must participate in the host processors coherency protocol. While this is a significant added complexity, it can provide performance advantages. Uncached accesses are unnatural for the processor pipeline. They tend to carry memory ordering and access semantics that create stalls in the pipeline (see the next section on write-combining). In contrast, cacheable accesses would allow the processor to build commands in a command queue that was in the local cache. Those commands would automatically flush to the network interface when the cache line was flushed *or* when the network interface was ready to read that cached location. This would enable a substantially higher message rate than can typically be achieved with command queues mapped into uncacheable space.

Unfortunately, in addition to the complexity of placing the command queue into cacheable space, there are some negative performance implications as well. Most cache coherency protocols are much higher overhead for data transfers. For example, a store to non-coherent space can be as simple as a single transaction that carries exactly the data that was written. In contrast, most modern CPU caches do *write allocation*, which means that they read a full cache line from the memory location being written to. This would double the traffic required for commands between the host processor core and the network interface. Whether the command queue is in cached or uncached space significantly impacts the correct approaches to things like managing flow-control and credits for the command queue.

Interacting with Write-Combining. When the command queue is placed in an uncacheable address space, stores to the command queue can become extremely slow. For example, many processors define uncacheable space (or I/O space) as only allowing one outstanding store at a time. This preserves very strong memory ordering semantics, but it imposes the round-trip latency on every store to the command queue. Even in the best case, commands often require multiple stores to complete, and writes across the I/O interface can easily have a several hundred nanosecond round-trip time.

To deal with this issue, the concept of a *write-combining* space was created. Write-combining space is a space with a much weaker memory ordering semantic. Stores to write-combining space are not guaranteed to be ordered, and they can be aggregated at the processor before being sent over the I/O bus. In fact, stores to write-combining space can be aggregated *until the next serializing event*. These properties allow the implementation of write-combining to significantly accelerate stores to uncacheable I/O space, but come with significant penalties as well.

Because stores to write-combining space can be delayed until the next serializing event, it is necessary to place a serializing event in the instruction stream. An sfence() is usually placed after a full command is written to the command queue. Stores that occur *before* an sfence() must complete before stores that occur after the sfence(). In fact, depending on how the command queue is managed (see Section 5.2.3.3), it can be necessary to place multiple sfence() operations into the instruction stream. For example, if the host knows it must store to the tail pointer after the command queue entry is filled in, it may need an extra sfence(). Unfortunately, an sfence() when stores into write-combining space are outstanding often takes as long as a single store to uncacheable space.

Flow-Control and Credit Management. Any time there is a command queue for an I/O device, the host processor must know when it is safe to write a new command. This means that some form of flow-control and credit management must be provided. Several options exist for how to manage this flow control. In one common variant, the network interface maintains a head and tail pointer, with commands written at the tail and read from the head. In this case the tail pointer is updated by the host and the head pointer is updated by the network interface. The host then reads, and potentially polls, the head pointer when it needs more command queue slots. Uncached reads of the head pointer from the network interface can be extremely high latency and yield significant performance variability depending on how often the head pointer must be read and/or polled over the host interface. In addition, the write combining semantics (Section 5.2.3.3) of the processor can make explicitly updating the tail pointer an expensive operation.

A variant of the head/tail pointer scheme that performs better when the command queue is placed into uncacheable space reduces the *read traffic* over the host interface. In this case, the tail pointer is updated implicitly by network interface hardware when new commands are written. The head pointer is kept in host memory, where it is updated by the network interface when new items become free. This allows the processor to *only* read and poll on host memory and *never* issue a read in the critical path that crosses the network interface.

Properly sizing the command queue on the network interface depends on the bandwidth and latency of the host interface. It also depends on the target message rate for the network interface. If the command queue is going to be used as the place to put data for a programmed I/O (PIO) message, the size of the command queue (in bytes) must be greater than the bandwidth delay product of the interface, as shown in Equation 5.1; however, the latency that is relevant is not only the latency of writing a command, but also the latency of returning a credit to the host processor.

$$CQ_{size} \geq BW_{PIO} \times (Latency_{CQWrite} + Latency_{CreditReturn}) \qquad (5.1)$$

$$CQ_{entries} \geq \frac{Messages}{s} \times (Latency_{CQWrite} + Latency_{CreditReturn}) \qquad (5.2)$$

Another key factor in sizing the command queue is determining how many command queue entries there will be. It is often easier for commands to end on a fixed boundary in the command queue; thus, the bytes in the command queue CQ_{size} are often divided among some fixed number of command queue entries. To achieve a target message rate (messages per second), the number of entries in the command queue must also be sized to cover the latency of the interface — as shown in Equation 5.2.

5.2.3.4   Event Posting / Completion Notification

Because the network interface allows the processor to enqueue multiple commands, it must also have a way to notify the host processor that the command has completed. There are two types of commands that can complete: transmit and receive commands. For onload types of interfaces, these two commands can complete in very similar ways; however, offloading implementations need to deliver much more information (e.g., the *MPI envelope*) when it notifies the host of a receive completion.

Transmit completion can be accomplished in numerous ways. It can be coupled to the command queue management such that a command is not removed from the command queue until it has completed. In this case, the application library knows that a command has completed, because the command queue has been drained. Other transmit completion mechanisms include posting an event [62, 64, 65], clearing a flag in the command (common in older Ethernet adapters), and incrementing a counter either in the NIC or in host memory. Transmits may complete out of order, because they may require an acknowledgment from the remote node to complete. This could discourage a scheme where a simple counter or consumption from the command queue was used to signal completion.

Receive completion can be very similar to transmit completion, but it can also be very different. In an "onload" implementation, messages are often received into a queue of untagged, identical buffers. The host can be notified of the receipt of a message into these buffers in any of the ways noted for transmit completion, but the exact mechanism chosen may be different from the mechanism chosen for the transmit completion because the optimization space is slightly different. For example, the receive queue is frequently consumed *in order*, whereas the transmits may complete *out of order*. This may encourage the receive completion mechanism to use a simpler mechanism like a pointer or counter update to indicate that a buffer has been consumed.

In striking contrast, an offloading implementation must pass a significant amount of information to the host when a message completes. As an example, an interface that offloads MPI semantics will match the incoming message to a posted receive and must then provide the MPI implementation with information about the source of the message, the communicator it was on, and the tag that it used. One option is to place the header in a separate space specified by the posted receive and use a light weight notification of completion, though that notification must include which posted receive was targeted. Another option is to post an event that includes the MPI header information and an indication (e.g., a pointer to) which posted receive completed.

5.2.3.5 Interrupts

Interrupts are the only mechanism available to the network interface to immediately (asynchronously) obtain the assistance of the processor. Unfortunately, interrupts can easily take over $1\mu s$ to service [63]. A network interface designer must determine how often to use interrupts and for what purpose. A network interface can use interrupts to process every message [63], or for exception conditions only [37]. As an intermediate, a network could use an interrupt to process unexpected messages, while handling expected messages on the network interface.

5.2.4 Collectives

Collective operations are a construct that is unique to the world of parallel computing, because a collective operation requires that all participating processes (or MPI ranks, SHMEM PEs, etc.) contribute to a global operation. Network interface designers have two choices for collective operations. They can expect the upper layer software library to build them using standard point-to-point operations, or they can provide hardware support for them on the network interface.

Collective operations have typically been implemented using point-to-point communications with the computation occurring on the host processor. Unfortunately, such implementations face a major challenge in obtaining high performance. As noted earlier, the latency of the network interface to host link is typically higher than the latency from one NIC to another. A large collective operation will typically have several communication and computation steps. At each step, the message must cross the network interface to processor boundary, obtain the attention of the processor to have it do the work, and cross the processor to network interface boundary again. This adds a substantial amount of latency at each step.

In contrast, a collective operation can be implemented almost entirely in the network. The network interface can implement the collective algorithm and the routers can even assist in the computation. This approach greatly reduces the latency of a collective operation; however, it also faces significant challenges. Foremost, the *best* algorithm for a given collective operation is heavily dependent on network characteristics; thus, the algorithms are constantly evolving in a way that makes them less amenable to hardware implementation. In addition, collective operations come in a range of sizes, and supporting that range can strain the resources available on a network interface. Finally, many collective operations require floating-point arithmetic, which can be expensive to implement on a network interface.

5.3   Current Approaches to Network Interface Design Issues

The approaches to solving the issues associated with network interface design are as myriad as the issues themselves. This section examines how modern high-performance networks address the issues raised in the previous section.

5.3.1   Quadrics QsNet

The Quadrics QsNet network evolved from the interconnect developed for the transputer-based Meiko Computing Surface platform, which was deployed in the mid-1980s. There have been several generations of QsNet, all based on successive generations of the Elan network processor and Elite switching fabric. The current-generation is QsNet-III, which is based on the Elan5 [37] network processor.

The Quadrics Elan family of network processors are the epitome of fully offloaded network interfaces specifically targeted at massively parallel high-performance computing systems. Quadrics provides a software development environment that allows a user-level process to create threads that run on the network interface. A low-level network interface programming library provides basic constructs, such as local and remote DMA queues and event management. This approach provides the ultimate flexibility for upper-level network protocol designers. Each upper-level protocol can have a dedicated NIC-level thread that handles all network transfers. Quadrics also provides several mid-level network programming interfaces that provide portability and abstract the differences in functionality between the different Elan processor versions.

The Elan5 network interface supports a PCI Express host interface and contains seven packet processing engines that perform high-speed data streaming and realignment. Each packet engine has a 500 MHz dual-issue proprietary RISC processing core and a 9 KB DMA buffer. The Elan5 NIC also has 512 KB of on-chip SDRAM and contains 1024 separate hardware queues that serve as communication channels between the individual packet engines. The NIC also has a memory management unit that allows it to replicate the virtual address memory translations of the host processor(s).

To initiate a short message put operation, the user process writes a command to the Elan5 command queue using PIO. The command contains the desired operation, the destination virtual address, the destination rank, and the data. One of the packet processing engines builds a packet and sends it to the network. For longer put messages, the process writes a DMA descriptor into the command rather than the data.

Short MPI messages are similar to short put messages, only part of the data represents the MPI header information. A packet processing engine at the destination is used to match the MPI header in the incoming message against a list of receive requests. Long MPI messages leverage a rendezvous protocol implemented on the NIC. These are handled as remote get operations. Rather than sending the data immediately, the remote NIC issues a get operation when a matching receive request is found. The Elan5 adapter also contains support for offloading several of the MPI collective operations, such as barrier and broadcast [373].

5.3.2   Myrinet

As with Quadrics, Myrinet evolved from technologies originally developed for massively parallel processing systems. When Myrinet was initially released in the early 1990s, it ignited a flurry of research because it was fully programmable. Each Myrinet network interface contained a LANai processor, and Myricom provided a LANai software development environment that provided the ability to create a custom Myrinet Control Program (MCP). Many in the high-performance networking research community used Myrinet and custom MCPs to explore issues such as onload vs. offload.

The latest Myrinet PCI Express network adapter contains a LANai-Z8ES chip, which is a low-power programmable RISC processor. Each NIC also typically has between 2 and 4 MB of high-speed SDRAM memory.

The default MCP that currently ships with Myrinet exports the Myrinet eXpress (MX) programming interface, as discussed in Chapter 6. The implementation of MX uses a combination of onload and offload. MX uses a two-level protocol. Short messages are sent eagerly and long messages use a rendezvous protocol. Message matching is a coordinated activity between the network interface and the user-level library. The MX library has a message progression thread that is awakened upon the arrival of a rendezvous message to insure independent progress of MPI long messages. Myrinet does not have any native support for offloading MPI collective operations.

5.3.3   InfiniBand

Unlike Quadrics and Myrinet, which are proprietary, InfiniBand was designed as an industry standard from the physical layer up. While there are very few InfiniBand silicon producers, there are several vendors that design and manufacture InfiniBand network adapters, switches, and routers.

InfiniBand is generally viewed as an offload-capable network by the general-purpose computing community. It supports a two-sided send/receive model of message passing, as well as the remote DMA (RDMA) model of data transfer, and also enables a limited set of remote atomic memory operations. These models all enable an OS-bypass implementation, and network-based interrupts can be enabled or disabled on a per-connection basis.

Despite the ability to do offload, InfiniBand does not have the level of offload required for preserving independent progress for MPI messages. Messages are received in FIFO order with no capability for message selection. It also does not have any native support for implementing MPI collective operations.

Current-generation InfiniBand network adapters support a PCI-Express interface. They are capable of performing PIO to optimize latency for very short messages and switch to using DMA for longer messages. In general, the message passing functions are used to handle short message transfers and RDMA is used for long messages. The RDMA read capability allows for a rendezvous protocol optimization where the initial request-to-send is sent using message passing and the data is pulled by the receiver when the ultimate destination of the message is determined.

5.3.4   Seastar

The Seastar interconnect [63] was developed by Cray for the Red Storm supercomputer [13] that later became the Cray XT3 [187, 440] product line. The Seastar chip unified both a 3D torus router and a high performance network interface into a single device. The network interface is directly attached to HyperTransport (HT), which is the equivalent of a front-side bus on an AMD Opteron processor. Although Seastar sits on the processor to processor interconnect, it does not participate in the coherency protocol and only leverages the non-coherent portion of the HyperTransport standard.

Seastar provides independent progress for MPI through the Portals 3.3 network programming API. Originally, independent progress was achieved exclusively by using an *eager* protocol for all network messages. Short unexpected messages (below a size threshold) are buffered at the target in memory provide by MPI through the Portals API, and long messages are dropped if they arrive before the corresponding posted receive and later retrieved using a `PtlGet()` that is issued by the recipient. More recent versions of the MPI library default to a rendezvous implementation of MPI over Portals that does not provide independent progress, but still provides the option to use exclusively eager messages. Seastar can provide independent progress through either interrupt driven processing (an interrupt for every message received) of the Portals semantics, or through offload to the PowerPC440 on the network interface [63, 334]. The PowerPC on the Seastar network interface does *not* have a floating-point unit; thus, no attempt is made to offload collective functions.

Command queues exist in a scratch memory on the network interface that is directly mapped into the kernel's address space as a write-combining address space. Processes trap to the kernel to provide authentication when sending (or receiving) a message by writing a command to the command queue. Each command queue has several entries, and credits in the command queue are obtained by the host processor reading the head pointer directly from the network interface. When a command completes, an event is posted directly into a queue in host memory.

5.3.5   PathScale InfiniPath and Qlogic TrueScale

The PathScale InfiniPath [121] adapter originally made a very similar choice to the Seastar interconnect: it attached directly to the HT interface and did not participate in the processors cache coherency protocol. In more recent generations, however, the product has been renamed to Qlogic TrueScale and now connects to PCI-Express interfaces. This allows the adapter to connect to a broader range of processors.

In contrast to the Quadrics Elan approach, the PathScale InfiniPath approach is the epitome of a fully onloaded network interface. Even the protocol and end-to-end error handling are placed on the host processor. In fact, the original Infinipath adapter even used the processor core to copy data to the network interface rather than providing a DMA engine.

The programming API for InfiniPath does not provide a mechanism for the offload of MPI matching semantics or a mechanism to provide independent progress (other than software thread based independent progress in an MPI library itself). It uses an eager protocol for short messages and a rendezvous protocol for long messages.

The command queues for InfiniPath are mapped directly into the application's address space with write-combining semantics. The application even builds the packet headers in the command buffer, and then increments a pointer to indicate that the buffer is ready for transmit.

5.3.6   BlueGene/L and BlueGene/P

The BlueGene/L [2] and BlueGene/P systems from IBM chose to integrate the network interfaces directly onto the die; thus, the network interacts directly with the processor's memory hierarchy. BlueGene/L actually used the processors to push or pull every word of data in to / out of the network. In BlueGene/P, the system was augmented with a DMA engine that could interact with the coherency logic in the socket. The approach is still predominantly an "onload" based approach, much like the InfiniPath/TrueScale interconnect.

Little is known about the specifics of the command queue interface or completion semantics within BlueGene/L and BlueGene/P systems. One can reasonably assume that in BlueGene/L, the network is represented as a simple FIFO where the processor can write data. The relatively low data rate of the BlueGene/L network combined with the tight coupling of that network to the processors would suggest that it can be mapped into a standard uncached space using any of the variety of mechanisms discussed in Section 5.2.3.3.

One novel feature of the BlueGene/L and BlueGene/P network interfaces is that they interact with a collectives network as well. This collectives network can provide some types of integer operations that can be used to build up some floating-point collectives [11] without the intervention of the host microprocessor.

5.4   Research Directions

The debate between "onload" and "offload" largely centers around issues of cost and performance, because traditional systems have used embedded processors for offload [13,48,353], and this has sometimes impacted performance. Similarly, the offload of collective operations is often pointed to as complex, high silicon area, and subject to continuous optimization at the MPI layer. Here, we discuss research topics that seek to improve the performance and lower the cost of offloading communications functions.

5.4.1   Offload of Message Processing

The offload of message processing, including the MPI matching operations, to the NIC can provide an array of advantages. It provides a straightforward architectural mechanism for providing independent progress without interrupts or host processor threads. In doing so, it minimizes the contention with user applications for host processor resources, including such issues as cache and TLB pollution. In the near future, perhaps the most important advantage of offload will be the ability to improve the energy efficiency of the system: hardware customized to solve the MPI matching problem can provide the functionality at substantially lower power than performing the same operation on the host cores.

Nonetheless, there are many challenges to providing MPI offload. There is typically a limited number of processing resources on the NIC for a growing number of processor cores on the host. The processing on the NIC has traditionally been provided using embedded processors [13, 48, 353] that are often not as high performance as host processor cores.

Recent research has focused on how to provide high performance message processing in the resource constrained environment of modern HPC NICs. These have ranged from augmenting the processor with associative structures to accelerate list traversal to dedicated hardware to manage the list state. These techniques have shown significant potential for providing *better* MPI processing performance with hardware that is far simpler than even a single host processor core.

5.4.1.1   Associative List Processing to Accelerate Matching

The first step in accelerating matching in the network interface is to eliminate the need for a linked list traversal in the *common* case. Most applications build an MPI *posted receive queue* that has a few, but not an extreme number, of entries [61]. To accelerate the processing of the first few items in a list, it is possible to build an associative structure that examines all of the list items in parallel [436]. This is called an *associative list processing unit* (ALPU).

Architecture of an ALPU. An associative list processing unit (ALPU) can be built up from the ternary cell shown in Figure 5.4(a). The cell contains storage for both the match bits (the MPI matching information) as well as corresponding mask bits (for wildcard bits within MPI). The set of mask bits can range from a pair of bits (one each for the two fields in an MPI_Irecv that can be wildcarded) to a full width mask as is needed by the Portals interface [64–66]. In addition, a valid bit, indicating whether or not the entry contains valid data, and a tag field, used at the discretion of the software, are stored. Stored data is passed from one cell to the next. Compare logic (factoring in a set of mask bits that indicate "don't care" locations) produces a single match bit. The basic cell then has three additional outputs that feed into the higher level block. The first is a single bit that is the logical AND of the match bit and valid bit (invalid data cannot produce a valid match). The second is the tag which is muxed through priority logic to select the right match. The final output is a valid bit to allow the higher level block to manage flow control.

At the next higher level (Figure 5.4(b)), a group of cells is combined into a cell block. In addition to the match cells, the cell block contains a registered version of the incoming request (to facilitate timing), logic to correctly prioritize the tags, and logic to generate a "match location." Priority in the unit is determined based on position. Data enters the unit in the lowest priority cell and moves up in priority with each subsequent insert. The cell block manages the flow of data through the unit using the insert and delete signals. Insert is a single bit that tells the block that data is being inserted into the unit. The delete signal includes the address of the entry to be deleted and a bit to indicate the delete is active.

An ALPU can chain several cell blocks together along with control logic to interface to the rest of the network interface. The cell block outputs are combined and prioritized (data enters an ALPU through the lowest priority block) in the same manner as cell outputs are combined in the cell block. Effectively, several cell blocks are combined to create one large, virtual array of cells.

One of the key challenges is to efficiently manage the flow of data through the ALPU. Items must be inserted atomically relative to searches of the list. Further, items must be deleted when they match, and the resulting free space must be managed efficiently. The atomicity requirement leads to overhead in

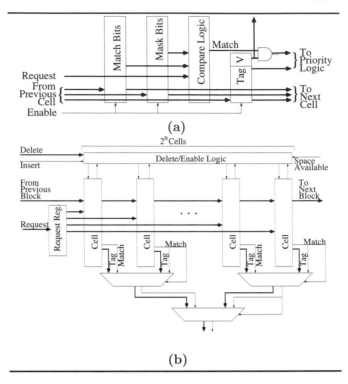

FIGURE 5.4:   (a) A cell with a single match unit for the posted receive
queue; **(b)** A block of cells. (Reprinted with permission from K. Underwood,
K. S. Hemmert, A. Rodriques, R. Murphy, R. Brightwell, A Hardware
Acceleration Unit for MPI Queue Processing, In *19th IEEE International
Parallel and Distributed Processing Symposium (IPDPS'05)*, ©[2005] IEEE.)

managing the ALPU because the ALPU may need to stop matching operations
when data is being inserted. In general, the ALPU must be put into "insert
mode" before data can be inserted into the unit. When the ALPU is in insert
mode, it will continue servicing match requests, until a request fails to match.
At this point, the ALPU must wait until insert mode is terminated because
items being inserted into the ALPU may match the request.

The goal of the free space management is to minimize data movement while
maximizing the availability to insert operations. In addition, it is desirable to
localize the control logic to the cell blocks to facilitate timing. This is done by
having cell blocks maintain information about the global state of empty space
in the unit. In particular, each cell block maintains a counter (referred to as
the free counter) which tracks the number of entries which are free in lower
priority blocks.

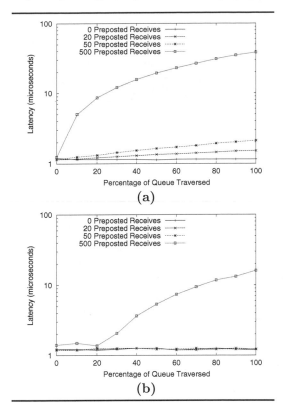

FIGURE 5.5:   (a) Growth of latency with standard posted receive queue; (b) Growth of latency using a 128-entry ALPU.

The free counter is maintained using only the global insert and delete signals. On power-up or reset, the free counter in each block is initialized to the total number of cells in blocks with lower priority (e.g., if the block size is 8, then the free counter in block 0 is set to zero, block 1 to 8, block 2 to 16, etc.). Thereafter, the counter is decremented for each insert and incremented for each delete which occurs in a lower priority block. The counter "saturates" at zero so that it will never go negative, as you cannot have a negative amount of free space.

The entry management control is localized to the cell block and uses the block's free counter as well as the valid signals from all cells in the block. On a delete, the valid bit for the deleted entry is set to zero and the free counter is modified as explained above; thus, there is no data movement on a delete. On an insert, all entries below the lowest priority free cell are moved up one entry to make room for the insert. The blocks handle this by dealing with three cases:

Case 1: When the free counter is greater than 0, there is room in lower priority block to "absorb" the insert, therefore the current block does nothing.

Case 2: When the free counter is equal to 0 and at least one of the cells in the block is empty, the block "absorbs" the insert by moving up all entries below the empty cell (including accepting the highest priority entry from the previous block to the lowest priority cell in the current block). Entries above the first empty cell in the block are not affected.

Case 3: When the free counter is equal to 0 and no cell in the block is empty, the insert must be "absorbed" in a higher priority cell. The current block simply moves all entries up one cell. This includes moving the top entry to the next higher block and accepting the top entry from the previous block into the lowest cell in the current block. There is guaranteed space in a higher priority cell because the ALPU maintains a count of the total number of entries in the unit and disallows inserts when the unit is full.

Performance Advantages of an ALPU. Including an ALPU in the processing pipeline on a NIC can provide a dramatic improvement in list processing time. Figure 5.5 compares the performance of a baseline NIC to the same NIC enhanced with a 128-entry ALPU. The graphs have some interesting traits. For the baseline NIC (5.5(a)), the low end of the graph shows each entry traversed adding an average of 19 ns of latency. By comparison, for a Quadrics Elan4 NIC, each entry traversed adds 150 ns of latency. The almost 10× performance improvement is not surprising because the NIC being modeled has a significantly faster clock (2.5×), is dual issue (for integers, floating-point does not get used), and has separate 32 KB instruction and data caches. When the queue is too long to fit in cache, the average time per entry traversed grows to 75 ns. This overhead shows up even when the entire list is not traversed. For example, the time to traverse an entire 400-entry list is $18\mu s$ and the time to traverse 80% of a 500-entry list is $30\mu s$.

Incorporating an ALPU yields two significant advantages. The most dramatic advantage is a flat latency curve until the length of the queue traversal crosses the size of the ALPU. The penalty is a 40 ns increase in the baseline latency (zero-length posted receive queues) as the processor incurs overhead from interacting with the ALPU. With 4 entries in the posted receive queue, the ALPU breaks even. Thus, an MPI library could be optimized to not use the ALPU until the list is at least 4 entries long. The second advantage provided by the ALPU is the reduction in the usage of the cache. By using the ALPU, the processor is not required to traverse the first N entries of the queue, even if the ALPU does not find a match. The storage required by the ALPU is relatively small (the entire queue entry does not have to be stored). Each entry in the ALPU contains matching data only, but the processor stores several other pieces of data in the queue entry. Thus, the number of cache lines the processor must retrieve from memory is dramatically reduced if it does not have to search the first several entries.

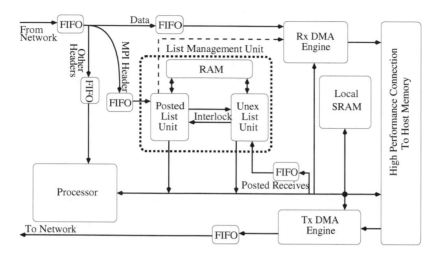

FIGURE 5.6: NIC architecture enhanced with a list management unit.

While an ALPU like structure is by far the fastest mechanism for searching a list, it does pay a penalty in terms of power and area. Associative structures are known to be power hungry and area intensive. To support a typical 2009 era computing node of two quad-core processors (8 cores total) with one process per core and 16 entries per process would require a total of 128 entries in the ALPU of roughly 128 bits each for a Portals 4.0 based API.

5.4.1.2 List Management in Hardware

The next opportunity for optimizing the processing of MPI lists is to build a hardware list management unit [438]. The latency to a local memory is typically much higher for an embedded processor than a hardware unit directly coupled to an SRAM. This advantage can be combined with specialized structures for issuing list lookups and pipelining list entry state to implement a much more efficient list management engine.

NIC Architecture Impacts. Figure 5.6 shows the architecture of a NIC enhanced with a list management unit. MPI headers go directly from the network to the list management unit, with the NIC processor only seeing headers for such things as rendezvous responses and out of band exchanges. Similarly, new posted receives go directly to the list management unit. Thus, the list management unit handles the traversal of the MPI posted receive queue and the MPI unexpected message queue. On a successful match in either queue, the match result is output to the NIC processor. An optional path (shown as a dashed line) allows the posted receive unit to generate DMA commands directly.

Each list unit within the list management unit consists of a *list walking component* and a *matching unit*. List walking components traverse a linked list in a local SRAM. This SRAM is equivalent in size to what would have been statically allocated in the processor's scratch RAM for maintaining lists. In turn, the *matching unit* is specific to the higher level protocols being implemented.

The posted list unit and unexpected list unit can operate concurrently, but passing data between them requires an exclusive region. An interlock protocol was designed to create the appropriate exclusive region. The interlock protocol requires that a unit obtain a lock from the partner unit by passing a "STOP" message and waiting for a "STOPPED" response. After a "STOPPED" response, a unit can pass an unlimited stream of "misses" (insertions) to the partner unit. When all of the insertions have been passed, a "START" message is sent. Sending a "STOP" and receiving a "STOPPED" can occur in as few as 5 clock cycles, but can require much longer as the partner unit must first enter a safe state.

One race still remains in the protocol described above. If both units simultaneously have a "miss," they will both pass "STOP" messages and begin waiting on a "STOPPED" message from the partner. To break this deadlock, one unit must always defer to the other. Thus, one unit has priority over the other. This is most easily implemented by giving priority to the posted list unit. In the case where both units have simultaneous misses, the lower priority unit must take the insertions it receives and pass them by each of the misses waiting in its outgoing queue to insure that no misses pass without attempting to match.

Performance of the List Management Unit. List management hardware achieves its performance advantage by decreasing the time to process each list item. Whereas a baseline NIC incurs a 20ns penalty, the hardware penalty is only 16ns. This delta seems small, but adds up to 10% of the latency at a posted receive queue length of 30 and completely dominates the latency at 300 posted receives. The hardware approach can also acheive a significant performance increase by allowing the list management unit to directly program the DMA engine. The offloads work from the embedded CPU, allowing it to service other tasks. Figure 5.7 shows latency (5.7a) and streaming bandwidth (5.7b) results for the 3 configurations (baseline, list manager and list manager directly pogramming DMA).

5.4.1.3   A Programmable Unit for Header Matching

While hardware to perform list management is an attractive solution, it comes with a certain degree of inflexibility; thus, the natural question that arises is how to retain most of the programmability of an embedded processor while obtaining most of the performance, area, and power advantages of pure hardware. One solution is a programmable unit which is specialized to efficiently perform the matching function [185].

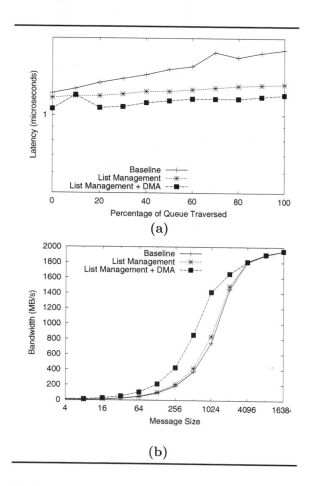

FIGURE 5.7: (a) Latency of 8-byte messages with 30 elements in posted receive queue. (b) Streaming bandwidth results for 30 element posted receive queue. (Reprinted with permission from K. S. Hemmert, K. Underwood, A. Rodriques, An Architecture to Perform NIC-based MPI Matching, In *IEEE International Conference on Cluster Computin (CLUSTER'07)*, ©[2007] IEEE.)

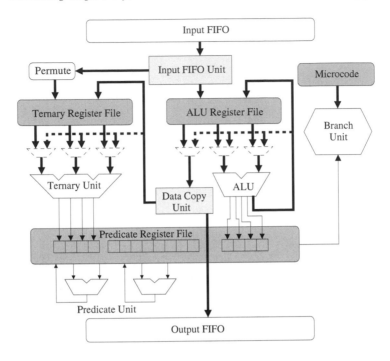

FIGURE 5.8:   Microcoded match unit. (Reprinted with permission from K. S. Hemmert, K. Underwood, A. Rodriques, An Architecture to Perform NIC-based MPI Matching, In *IEEE International Conference on Cluster Computin (CLUSTER'07)*, ©[2007] IEEE.)

Architecture of a Programmable Unit for Header Matching.   The architecture of a specialized match unit is driven by three main considerations: high throughput, irregular alignment of input data, and program consistency. Ideally, the unit would be capable of processing data as quickly as it arrives. However, there is a trade-off between circuit complexity and throughput. This trade-off led to an architecture with a small number of computational units operating in parallel and a three stage pipeline. In general, the first cycle reads operands from the appropriate register file, the second cycle does the operation, and the third writes the result to the register file. The architecture of the match unit is shown in Figure 5.8 and is capable of executing six simultaneous operations: 1) input an item, 2) output or copy an item, 3) perform an ALU operation, 4) perform a ternary operation, 5) perform two predicate merge operations, and 6) resolve a branch. Due to the streaming nature of the processing, input and output are accomplished through the use of FIFOs, allowing easy integration with other system components.

Inherent in header and list item processing is the need to process data of varying bit widths packed into native size words (64 bits). This led to

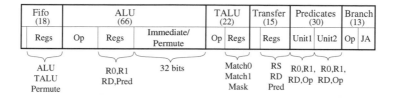

FIGURE 5.9:   Match unit's wide instruction word.

specialized functions that combine and reorder data, as well as SIMD-like functionality in the ALU and ternary unit. Finally, given the streaming nature of the processing, we require strict ordering semantics for the program: all operations in the same instruction word are independent and their results are available for use by the next instruction. If the input FIFO is empty and an element of the wide instruction requires input data, the issue of the entire wide instruction is delayed until the FIFO has data. This makes it necessary to include result forwarding (forwarding paths shown as dashed lines in Figure 5.8) and to require some of the register file write ports to be write before read. It is also necessary to modify the pipelining of the predicate unit.

To accomplish the six parallel operations, the match unit includes 4 computational/control units, 4 memories and 2 data transfer units. The computational units include a 64-bit arithmetic logic unit (ALU), a 64-bit ternary unit, a predicate unit and a branch unit. The four memories consist of the microcode memory and three register files: the ALU registers, the ternary registers and the predicate registers. The data transfer units perform data copies: 1) from the input FIFO to the ALU and ternary register files and 2) from the ALU register file to the output FIFO or the ternary register file. Each unit has dedicated ports into the register files and is controlled independently.

Each of the 6 functional units has an instruction slot in the wide instruction word format of the microcode (shown in Figure 5.9). The bit widths of the major unit instruction fields are shown below the label for each field (the overall instruction word is 164 bits). The minor fields for each instruction are also shown. As a by-product of the extensive concurrency, the highest level of programming for the match unit uses assembly language, which is translated directly into microcode. The level and specificity of concurrency within the core would be difficult to exploit with a high level language.

To better support header matching, the match unit includes functions which are not typically found in general purpose CPUs. First, all comparison operations write results to a dedicated predicate register file and are then combined using two predicate units. This frees the other units for more complex operations. Second, the ALU includes a permute operator, which can arbitrarily combine two registers on 8-bit boundaries. This means that each byte of the output can be chosen from any of the 8-bytes from either of the two inputs (or can be set to zero).

Third, because headers have several fields ranging from 16 to 64 bits, both of the 64-bit functional units include SIMD-like functionality, allowing then to process data smaller than 64-bits. The units are divided into four 16-bit sections that can be aggregated into larger operations. This function is controlled by 3 SIMD bits that indicate which of the internal 16-bit boundaries the operation will cross. In addition, each of 16-bit sections has a bit in the control word that indicates whether a comparison result for that segment will be written (these are valid only for comparison operations). The SIMD functionality allows the match unit to work on multiple, non-natively aligned fields simultaneously.

Finally, the match unit includes a dedicated ternary unit. Unlike the ALU, the ternary unit performs only a single type of operation (although it can be used for multiple functions). There are three inputs to the ternary unit: match0, match1 and mask. The mask is used to perform equal comparisons — only the bits which are specified in the mask are used in the comparison, all other bits are ignored. Like the ALU, the ternary unit has SIMD functionality. The ternary unit serves two purposes for the matching functionality. The first is to determine if the match bits in the header match the mask and match bits in the posted receive. In addition, the ternary unit can do equal comparisons at smaller granularity than 16 bits, which is used to pull out single bit flags from packed control words. This feature also allows for limited boolean functions (wide *and* functions with optional negation of inputs) to be performed on flags which are found in the same word. For example, if a, b and c are one bit values packed into a single control word, the ternary unit could perform !a && b && !c.

The ternary unit also includes a simple permute unit on the input to the register file from the input FIFO. This permute divides the input into four 16-bit fields (corresponding to the four fields in a SIMD operation) and allows each of the 4 output fields to arbitrarily select any of the 4 input fields. This is useful for allowing the ternary unit to pull multiple flags out of a single 16-bit word. If this simplified permute is not sufficient, the more complex permute of the ALU can be used to properly stage the data.

Matching Code Characteristics. The match unit is programmed entirely in an assembly language that translates one-to-one into microcode instructions. Part of the design target of the architecture was to minimize the number of instructions (and, therefore, cycles) required to implement the matching code. Table 5.1 gives a breakdown of the 44 instructions required to implement the matching operation. Initialization is needed to establish constants that will be used in the primary loop and is only executed at boot time. The primary loop for matching includes one execution of the header code, one or more executions of the list item code (typically shared code plus the fast path), and

TABLE 5.1: Breakdown of the assembly code.

Code segment	Instructions
Initialization	4
Header	9
List Item (Shared)	6
List Item (Fast Path)	7
List Item (Slow Path)	10
Flush	8
Total	44

one execution of the flush code for a total of at least 31 cycles (the extra cycle arises from a pipelining impact). Each additional list item traversed adds at least 8 cycles (assuming the common case).

The header code must read an 8 item header (one per cycle) from the input FIFO. Because header fields and list item fields are not typically identical, this code must reformat the data to match the list items to improve matching speed. The list item code then reads list items and compares them to the header. This code is split into "fast" and "slow" paths that share a common preamble, where the "slow" path supports a less common Portals semantic on a per list item basis. The "fast" path is optimized to complete a list item as soon as a match fails, but this cannot be less than 8 total cycles, as it takes 8 cycles to read the list item from the FIFO. The most common match failure is the test of the Portals match bits. This failure requires 8 cycles per list item, where a full match will execute 12 instructions in 14 cycles (on a match, only 5 of the 6 "shared" instructions are executed). Finally, in most cases, the list manager will have sent an extra list item to be matched — not knowing that a match will be found. This requires that the flush code execute (8 cycles) to drain the extra list item from the input.

Performance Relative to Processors and Hardware. As shown in Figure 5.10 the performance of a programmable unit for header matching can approach the performance of a dedicated hardware solution on a per item basis. The graphs show simulated results for the traversal of posted receive queues of length 10 and 50 (5.10(a) and reffig:match-data(b), respectively). The graphs include data for the match unit, a dedicated hardware unit, and an embedded CPU, all operating at the same frequency. The x-axis shows the percentage of the queue traversed to find a match and the y-axis shows the average time per list item traversed (log scale). For long list traversals, there is almost no difference in performance for the match unit compared to a dedicated hardware solution. For short list traversals the primary difference is found in the one-time start-up costs and final match cost.

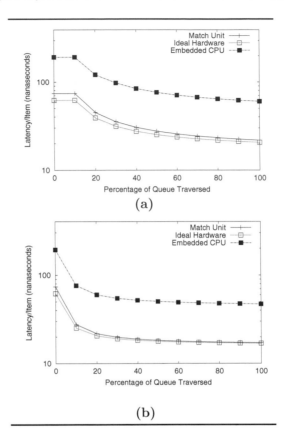

FIGURE 5.10:   Comparison of average time per item traversed for posted receive list of **(a)** 10 items and **(b)** 50 items.

5.4.2   Offloading Collective Operations

The offload of collective operations poses several substantial challenges. First, the network interface needs a way to perform the appropriate arithmetic operations. Ultimately, this means that a floating-point unit and an integer arithmetic unit are needed. Second, the complexity and constant evolution of collective algorithms poses a significant challenge. If these challenges can be addressed, research indicates that offloading of collective operations can offer a significant advantage [70, 297, 373, 449].

To address these issues, it is important to provide building blocks for the user level message passing library (e.g., MPI) to use to offload collective operations. An interesting building block introduced in the Portals 4.0 API is the notion of triggered atomic operations (see Section 5.4.2.2). When implemented in the

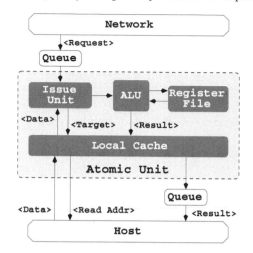

FIGURE 5.11: NIC-based atomic unit. (Reprinted with permission from K. Underwood, M. Levenhagen, K. S. Hemmert, R. Brightwell, High Message Rate, NIC-based Atomics: Design and Performance Considerations, In *IEEE International Conference on Cluster Computin (CLUSTER'08)*, ©[2008] IEEE.)

network interface, triggered atomic operations enable the application to specify a complete collective operation that is then performed within the network without further intervention from the application. Not only does this enable the offload of collective operations, but it also enables the implementation of non-blocking collective operations with independent progress.

5.4.2.1 NIC-Based Atomics

Atomic operations implemented in a network interface can provide a good basis for offloading collective operations; however, several design issues must be considered to effectively implement atomic operations. For example, should a cache be used on the NIC? How large should that cache be? What should the associativity be? On many host interfaces (e.g., PCI Express), it is not possible to keep such a cache coherent with the host. In such cases, how often should results from the cache be propagated to the host to mitigate the usage of host interface bandwidth?

Figure 5.11 illustrates an architecture that was used to study many of these issues [437]. Atomic operations received by the NIC are directed to the atomic unit (routing path not shown in the figure), which makes a request of the local cache for the data value to update. If the request misses the cache, the cache requests the data from the host. The ALU then updates the value and writes the result back to the local cache. If requested, the updated value can be returned over the network to the requester.

The local cache is configured in write-through mode: every value written to the cache from the functional unit is immediately propagated to host memory. While most caches operate in write-back mode, write-through mode simplifies the NIC-to-host interaction. For example, a local read of a local variable that is updated by the atomic unit should only go to local memory. A write-through cache provides immediate updates to the host and eliminates the need to have something like a timer per cache line. For some cases, this configuration has no performance penalty (e.g., HPCC RandomAccess); however, atomic operations from the network frequently target only a small number of memory locations that can be cached. In these cases, it is still desirable to reduce the bandwidth requirements on the host link.

Reducing Host Bandwidth Usage. The atomic unit separates cache requests from the actual operations (see Figure 5.11). The issue unit translates operations into a stream of cache requests that cause the cache to generate a queue of values back to the issue unit. Once the data is available, the request can be issued to the ALU. This arrangement can be used to limit the rate at which data is written back to the cache, and therefore to the host. Specifically, host writes are limited by executing multiple operations that target the same address and arrive in close temporal proximity before writing the result to the cache. Three resources are needed to accomplish this. First, the issue unit is provisioned with logic to look at a window of requests that have backed up at the input to the atomic unit. Second, a small register file is needed to hold intermediate values. Third, a method for tracking busy cache entries is needed. A cache entry is busy if it has been read and not yet written back (either to the cache or the register file).

To reduce host traffic, the issue unit may rewrite the source or destination of an atomic request. Data is written to the register file if an operation in the look-ahead window targets the same address as the current operation. A later instruction targeting the same address would then use the temporary register as a source and write the result back to the cache, thus causing a write to the host. Once an instruction is issued, it cannot be changed; therefore, the tracking of busy cache entries becomes necessary. If the data is unavailable, the issue unit blocks.

Using a register file has the same effect as a write-back cache if a long stream of accesses to a single variable arrives in temporal proximity. To guarantee regular updates to the host, the issue unit monitors the status of the queue to the host. If the queue to the host is *empty*, then the result will be written back to the cache to generate a write to the host. If the queue to the host is full, then the host link is being overutilized, and it is better to suppress the write by targeting the register file.

Resolving Read-after-Write (RAW) Hazards. A cache for the atomic unit on the network interface can create a read-after-write (RAW) hazard. Specifically, the sequence: 1) perform an atomic operation on address A, 2) write the result to host memory at address A, 3) evict address A from the cache, and 4) perform a second atomic operation on address A. Step 4 in this sequence will cause a host memory read after the host memory write from step 2. If those operations are temporally close, it is possible (even likely) for the read access to pass the previous write. This situation can easily occur with pipelined I/O interfaces that do not guarantee ordering of requests.

This problem can be solved with a unit to buffer host writes until they have completed to host memory. This buffer acts as a secondary cache structure until writes to host memory complete to a level of the memory hierarchy where ordering is guaranteed. Items are held in this buffer until the host bus interface returns an acknowledgment that the write request has completed. Host interfaces that do not return write acknowledgments must implement a *flush* function that does not complete until *all* outstanding requests are completed. In this case, it is not possible to track when individual writes complete; thus, when the buffer is full, a *flush* request is issued and the buffer is emptied.

Pipelining.   To sustain high performance under certain workloads (those that primarily miss in the cache, such as the HPCC RandomAccess benchmark), it is critical that the function unit and cache be pipelined to maintain a sufficient number of outstanding accesses to host memory to cover the round-trip latency. Pipelining of accesses to the host begins with the issue unit providing a stream of address requests to the cache. When a cache miss is encountered, the cache must forward the request to the host interface and attempt to service the next request. The issue unit associates a tag with each cache request, allowing the results to be returned out-of-order, and making it possible for processing to continue on operands that hit in the cache. This is key to allowing enough host requests to be in-flight in the case where cache hits and misses are interspersed. Although the cache results can be returned out-of-order, instructions are issued to the function unit in-order. Depending on the ordering constraints imposed on the atomic operations, this could be relaxed and instructions could be issued out-of-order, but this would also increase the complexity of the atomic unit.

Performance of an Atomic Unit on the NIC.   Simulations of the architecture in Figure 5.11 have provided insight into the answers [437] for many of the questions above. For example, a cache can be critical as a buffer for covering the latency of the host interface, as illustrated in Figure 5.12. The bottom line shows the performance of multiple remote PEs (processing elements) targeting a single memory location. The performance is poor, because the atomic unit must wait for the round-trip to host memory for *every* *access*. In contrast, adding any cache to the atomic unit moves the atomic operation bandwidth up to the point where it saturates the simulated network. At the other end of the spectrum, if each PE accesses several random memory locations ($16 \times P$ locations targeted at a given NIC), a cache is not needed. In fact, a small, direct mapped cache performs *much* worse than no cache at all. In this case, the access stream has sufficient parallelism to cover the access

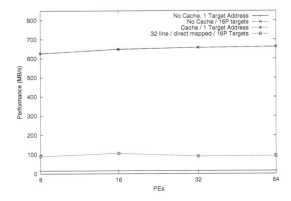

FIGURE 5.12:   Comparing the performance with and without cache.

latencies to the host; however, an access stream with high concurrency does require a substantial amount of memory buffer somewhere on the NIC and it is straightforward to use a cache for this purpose. Thus, even these workloads can benefit from a cache.

The question, then, is how large should the cache be and at what level of associativity? Figure 5.13 shows the performance of targeting a large number of memory locations (typical of operations like HPCC RandomAccess [271, 343]) for various cache sizes and associativities. A relatively high associativity is needed to prevent performance degradation attributable to conflicts in the cache.

5.4.2.2   Triggered Operations

Triggered operations were introduced in Portals 4.0 and discussed in Section 6.7. When combined with network interface based atomic operations, triggered operation semantics can provide a useful building block for creating collective operations. This is illustrated in the example of a double precision, floating-point Allreduce in Figure 5.14.

To implement atomic operations and triggered operations on the network interface, a few subtleties must be considered. For example, a network interface pipeline tends to be such that operations are not particularly strongly ordered. In the example of the Seastar network interface, an embedded microprocessor would program a receive DMA engine and then later post an event [63]. If atomic operations and triggered operation semantics were both implemented in this architecture, it would be critical for the counting event update to *only* happen after the result of the atomic operation was *globally visible*. In this case, such a guarantee is straightforward, but consider the architecture in Figure 5.11.

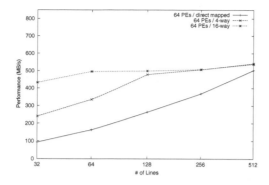

FIGURE 5.13:   Assessing the impact of size and associativity on atomic unit cache performance.

If the network interface implemented the atomic unit as a separate block with a cache, the atomic operation may issue, and even complete, in the atomic unit. This would allow the counting event to be incremented. In theory, this would allow the triggered operation to issue; however, if the data had not propagated through the cache and into the host memory, that triggered operation could DMA from a memory location that has not yet been updated. This would result in sending old data and would not be acceptable. Such issues must be accounted for in the design of triggered operation support.

5.4.2.3   Operation Associativity... or Not

A fundamental issue that creates tension between application programmer needs and the performance of collective operations is the issue of operation associativity. Some operations are truly associative, like integer adds and integer or floating-point minimums, but some floating-point operations — particularly add and multiply — can yield slightly different results based on the order in which they are applied. Because this is a known issue, message passing libraries do not *require* that a collective operation yield the *exact* same results every time it is called with the same inputs. For example, MPI [290] only suggests that a *high quality* implementation will return the same results each time it is called with the same arguments. The highest performing offloading implementations *will not* guarantee the ordering of floating-point operations and, thus, *may* return slightly different results on each invocation — even when the inputs are the same. This may contrast with the expectations of a user.

```
// Example uses a radix-4 tree for simplicity.
// Complexities of unexpected messages not shown.

// Allocate one handle for reduction and one for broadcast
PtlCTAlloc(mynihandle, PTL_CT_OPERATION, cthandle1);
PtlCTAlloc(mynihandle, PTL_CT_OPERATION, cthandle2);

// ME1 and MD1 describe location where reduction is done.  This
// example assumes that is a separate buffer from the user buffer.
// ME and MD setup are not shown.  ME Handles are a  reference
// to the ME that is returned.
PtlMEAppend(mynihandle, collectivePT, ME1, PTL_PRIORITY_LIST,
    reduce_usr_ptr, MEhandle1);

// ME2 and MD2 describe broadcast buffer. ME1 references cthandle1
// and ME2 references  cthandle2.
if (local_node != root)
    PtlMEAppend(mynihandle, collectivePT, ME2, PTL_PRIORITY_LIST,
    bcast_usr_ptr, MEhandle2);

// Not all arguments to PtlAtomic, PtlTriggeredAtomic, and
// PtlTriggeredPut are shown

PtlAtomic(UserBufferMDhandle, local_target, collectivePT, PTL_SUM,
    PTL_DOUBLE);

if (local_node != root) {
    // TriggeredAtomic triggers when all 4 children and local
    //operation complete locally.
    PtlTriggeredAtomic(MDhandle1, parent, collectivePT, PTL_SUM,
        PTL_DOUBLE, cthandle1, 5);

    if (local_node != leaf) {
        for (j = 0; j < num_children; j++){
            // At intermediate node, when parent sends response,
            // send a put to child
            PtlTriggeredPut(MDhandle2, child, collectivePT,
                cthandle2, 1);
        }
    } else {
        // Wait for parent to complete
        PtlCTWait(cthandle2, 1);
    }
} else {
    for (j = 0; j < num_children; j++)
        // At root, when all children complete, send responses to
        // children
        PtlTriggeredPut(MDhandle1, child, collectivePT, PTL_SUM,
            PTL_DOUBLE, cthandle1, 5);
}
```

FIGURE 5.14:   Pseudo code for an Allreduce using triggered operations

5.4.3 Cache Injection

Cache injection is broadly defined as the network interface placing data directly into the cache of a host processor. This can apply to anything from placing the full data of the message into the processor's cache, to just the header, to a completion notification. Cache injection also comes in two forms. It can refer to updating data *only* if it is already in cache, or can refer to *evicting* other data that is in cache to make room for the network data [199].

Evicting data that was in the cache to make room for the full data payload of network messages is only a good idea for network-dominated work loads. It is most meaningful for a network stack that is interrupt driven and that processes a significant portion of the data payload when a message is received. An interrupt-driven stack like TCP/IP over Ethernet can get significant advantages by placing the header directly into the cache before the interrupt, or even by placing the entire body of the packet in the cache if the processor handles the TCP/IP checksums. This prevents the network stack from incurring the latency and bandwidth impacts of cache misses. Similarly, in some circumstances, injecting data for a completion notification can significantly improve message passing performance as long as the network stack is executed before another cache miss causes an eviction of that data item.

Blind injection of the entire data from a network packet can actually hurt performance [199, 259]; however, simply replacing the data that is already in the cache is always beneficial. Many network APIs poll a memory location to determine when a message has completed. That polling operation typically happens to a cached location that reduces overall load on the memory system; however, in most memory hierarchies, the network interface writing to a memory location will cause the cached value to be invalidated and re-read from memory. This is a significant latency penalty for the software and uses a significant amount of memory bandwidth needlessly. In these cases, simply updating the memory hierarchy to replace data that is already in the caches is a significant advantage.

5.5 Summary

In this chapter, we explore the issues and trade-offs in modern network interface design along several dimensions, including the issue of onload vs. offload of the network API, support for long vs. short messages and collective communications, and the types on host-NIC interactions the NIC enables. We place the discussion in the context of several modern network interfaces, and present results for experimental analysis which illustrates the performance impact of various design decisions. In addition, we point out several research directions aimed at improved performance which address various costs associated with offloaded communication functions.

Chapter 6

Network Programming Interfaces for High Performance Computing

Ron Brightwell

Sandia National Laboratories [1]

Keith Underwood

Intel

This chapter is dedicated to the memory of Jarek Nieplocha, who made many significant contributions to the field of high-performance networking for parallel systems.

6.1 Introduction

The network application programming interface (API) is an important layer in the networking hierarchy. It is the place where the capabilities of the underlying network hardware combine with the resource management mechanisms of the operating system to fulfill the functional requirements and performance demands of high-performance computing applications. In this chapter, we present the state of the art and discuss technology trends surrounding low-level network programming interfaces. We discuss several of the important issues that differentiate modern-day HPC network programming interfaces and describe our particular research path.

---

[1]Sandia is a multiprogram laboratory operated by Sandia Corporation, a Lockheed Martin Company, for the United States Department of Energy's National Nuclear Security Administration under contract DE-AC04-94AL85000.

6.2   The Evolution of HPC Network Programming Interfaces

In the late 1980s, distributed memory massively parallel processing machines (MPPs) emerged as viable platforms for high-performance computing. Several vendors, such as nCUBE and Intel, offered systems that contained several hundred to a few thousand processors connected by a proprietary high-speed networking fabric. Network bandwidth and latency were recognized as important factors in delivering application performance and scalability, and platform vendors viewed the network programming interface as a differentiator. As such, most of the low-level network programming interfaces offered on these machines were specific to their underlying network. Application developers were expected to use these APIs directly, so ease-of-use was an important consideration.

The research community was exploring alternative approaches to kernel-based protocol stacks, where the network adapter delivers data into operating system buffers, the OS does protocol processing and copies data into user-space. In an environment where the network bandwidth is at least as great as memory bandwidth, the extra data copies severely limit message passing performance. Using system calls to initiate data transfers was also expensive, on the order of hundreds of microseconds.

One of the fundamental communication models to appear during this time was Active Messages (AM) [447]. The AM model associated a small amount of user-level computation (a handler) that would be invoked upon the initial arrival of a message. This approach eliminated the need for system buffers and allowed an application extreme flexibility in determining the destination of messages and how to respond to them.

Virtual Memory Mapped Communication (VMMC) was another model of communication that addressed several of the limitations of high-performance networks in the early- to mid-1990s. In VMMC, remote regions of memory are mapped into a local address space. Accesses to these local memory regions would then initiate network transfers to/from the remote region. Once the remote region was mapped, these transfers could avoid any interaction with either the local or remote operating system. (VMMC is load-store-based.)

The U-Net model [446] captured several fundamental mechanisms for providing direct user-level access to a network adapter. It established a queue pair model of communication where requests for network transfers could be given directly to the network adapter and all data transfers could bypass the operating system.

About this same time, network technologies that had been initially developed for MPP systems were being developed for broader use. Networks like Myrinet [48] and Quadrics [353] were becoming popular for supporting networks and clusters of workstations as a way to create a more cost effective MPP-like platform. The programmability of these network interfaces fostered even more research into new APIs and models for high-performance network programming interfaces.

MPP vendors continued to explore enhancements to their proprietary network programming interfaces, while the systems research community continued to explore more effective models and interfaces for network communication. In response, the application community struggled to develop higher-level message passing interfaces that would provide the required application-level semantics and provide a portability layer. Initially, there were several domain-specific APIs, like TCGMSG. This lack of portability led to the development of the MPI Standard, which provided a portability layer for application developers and a higher-level interface that systems researchers and vendors could target.

In the late 1990s, as the success of cluster-based MPP systems was becoming apparent, the bottleneck shifted from the network itself to the I/O bus. This led to an effort to provide higher-speed I/O buses that could keep up with or even exceed the raw performance of the network. VIA [96] emerged as a standard user-level network interface, but failed to provide hardware interoperability. InfiniBand [207] overcame this limitation and has become a commodity HPC network.

More recently, there have been several efforts to bring high-performance networking techniques, like remote DMA (RDMA) capability, to a larger commodity computing community. An example of this is the Internet Wide Area RDMA Protocol, or iWARP. iWarp is an Internet Engineering Task Force (IETF) update of the RDMA Consortiums RDMA over TCP standard. This later standard is zero-copy transmission over legacy TCP/IP protocol stack.

6.3   Low-Level Network Programming Interfaces

In this section, we provide a brief overview of the current low-level network programming interfaces that are in active use on HPC systems, more specifically, those running on machines in the current (November 2008) list of the five hundred fastest computers. Historically, the area of network programming interfaces for HPC has been very active, so it is impossible to completely capture all of the ongoing research projects.

6.3.1   InfiniBand Verbs

InfiniBand (IB) is a commodity high-performance networking fabric that is growing in popularity among HPC cluster systems. IB has its roots in several different technology areas. From the network programming interface perspective, the IB specification [206] does not include the definition of a standard API, but rather specifies functionality through an abstraction called *Verbs*.

Unfortunately, this lack of standardization initially led to numerous vendor-specific APIs as well as several efforts to establish common APIs for IB hardware. At roughly the same time, there were similar standardization efforts [363, 364] occurring in the ethernet community, where gigabit ethernet (GigE) was just becoming available and efforts were already underway to explore 10-gigabit ethernet capabilities. Eventually, several of these standards groups were combined and/or disbanded, resulting in two common APIs for IB networks.

The first common IB API was established by the Open Fabrics Alliance, which was founded to unify an IB API and to provide a set of drivers for the Linux operating system and encourage hardware interoperability. The organization has since expanded to include other transports, including ethernet, and encompass other operating systems. This API is generally referred to as "IB Verbs" or "OFED Verbs."

As IB was initially aimed at serving the needs of high-performance storage, there were efforts in that community to develop an industry standard IB API for storage applications. The DAT Collaborative was formed to standardize a set of transport- and platform-independent APIs that exploits Remote Direct Memory Access (RMDA) capabilities (described in more detail below). This process resulted in the establishment of a second command IB API called the User Direct Access Programming Library (uDAPL) [114].

Both of these APIs have their roots in the U-Net [446] user-level networking API developed at Cornell in the mid-1990s. One of the distinguishing character-istics of the U-Net interface is the *queue pair* abstraction for two-side message passing. A queue pair consists of a queue of send and receive requests that are established for each connection. The sending process enqueues sends for the network interface to deliver, while receive buffers that have been enqueued are filled up with incoming messages directly by the network interface. Unlike kernel-based networking protocols, where all data transfers pass through the operating system, the U-Net model allowed the network interface to transfer data directly to and from user-level processes.

The Virtual Interface Architecture (VIA) [96] was a predecessor to IB that attempted to standardize the U-Net programming interface without providing hardware interoperability. VIA also extended the U-Net programming interface to include remote direct memory access (RDMA) operations, which allow one-sided remote data transfers after an initial setup phase. The lack of hardware interoperability in VIA was one of the primary motivators of the InfiniBand specification.

6.3.2   Deep Computing Messaging Fabric

The Deep Computing Messaging Fabric (DCMF) [240] is a recently developed network programming interface developed by IBM for their Blue Gene/P (BG/P) massively parallel processing machine. DCMF was designed to support large-scale tightly-coupled distributed memory parallel computing systems. DCMF is based on the active message [447] paradigm and provides abstractions that allow for simultaneously using the multiple network interfaces on a BG/P system.

6.3.3   Portals

Portals [65] is an evolution of the user-level network programming interface developed in early generations of the lightweight kernel operating systems [272, 400] created by Sandia National Laboratories and the University of New Mexico (UNM) for large-scale massively parallel distributed memory parallel computers. Early versions of Portals did not have a functional programming interface, which severely hampered implementing the API for intelligent or programmable networking hardware. In 1999, Sandia and UNM developed a functional programming interface for Portals that retained much of the desired functionality but removed the dependence on a lightweight kernel environment.

In 2002, Cray chose to use Portals version 3.2 as the lowest-level programming interface for their custom designed SeaStar [13] network interface on the XT3 massively parallel system. Portals continues to be the lowest-level network programming interface on the current-generation Cray XT5, the largest of which has broken the petaflops barrier. The continued evolution of Portals is discussed in detail in Section 6.7 below.

6.3.4   Myrinet Express (MX)

The Myrinet Express MX [301] interface is the latest supported programming interface from Myricom for their Myrinet [48] network hardware. MX was designed to address the shortcomings of the previous Myrinet programming interface, GM [299], with respect to MPI performance. The MX API was designed specifically to support important features of MPI, such as the ability to more efficiently overlap computation and communication. Recently, the MX programming interface has been extended to work on ethernet network interfaces, resulting in the Open MX [166] API.

6.3.5   Tagged Ports (Tports)

The Tagged Ports (Tports) network programming interface was initially developed by Meiko for their supercomputing products circa 1994. The Meiko networking technology as well as this API was carried forward by Quadrics into their QSNet [353] line of products. The Tports interface was created prior to the development of the MPI Standard, and has since been enhanced to better support the features and requirements of an MPI implementation. Currently Tports is supported on three generations of Elan network hardware from Quadrics.

In addition to Tports, Quadrics also provides a lower-level functional interface for accessing the network. The Elan library provides DMA queues and an efficient event mechanism that can be combined with a NIC-level thread to implement almost any type of network operation. All functional APIs, including Tports, are built using the primitives provided by the Elan library. However, it appears as though this interface is hardware-specific, so the API may not be consistent across all generations of Elan product. The Tports API is consistent and is used to abstract the differences in hardware across generations.

6.3.6 LAPI

The low-level API, or LAPI [392], was developed by IBM in collaboration with Pacific Northwest National Laboratory for the IBM RS/6000 SP line of parallel computers. Similar to DCMF, LAPI is also based on the active messages concept. The initial design goal of LAPI was to provide primitives for optimized one-sided communication operations on the IBM SP series of machines.

6.3.7 Sockets

Sockets are likely the longest-living and most popular low-level network programming interface for general-purpose computing. Sockets began as the fundamental interprocess communication mechanism in the UNIX operating system and were extended to support network communication using various types of network protocols, such as TCP/IP. Because sockets were not designed specifically for HPC, they have several limitations as compared to other HPC-specific communication interfaces. There have been several research projects associated with extending sockets to meet the need of HPC, but as of yet, none of them have been standardized or are in widespread use on commodity HPC systems.

6.4 Distinguishing Characteristics

In this section, we discuss the important characteristics that distinguish these networking programming interfaces from one another. Understanding these traits provides insight into their semantics, objects, and mechanisms that each provides. HPC network programming interfaces are driven first and foremost by performance. All strive to deliver the lowest latency and highest bandwidth provided by the hardware. However, there are also other important performance characteristics to consider that we will discuss. Beyond performance, these interfaces also try to provide the appropriate data movement semantics that higher-level network programming interfaces desire. These capabilities help shape the objects and functions that each interface exposes to the network programmer.

6.4.1 Endpoint Addressing

One important differentiator in network programming interfaces is how a process discovers a peer process with which to communicate. Some interfaces make this an explicit operation and provide services for discovering the address of peers, while for others this is an implicit operation. General-purpose networks, like IB and Myrinet, have more general mechanisms for performing address lookup and provide an address that is independent of any other component in the system. For example, MX assigns each network interface a unique 64-bit identifier and provides a function for discovering an ID based on the name of the host. Other interfaces, like Portals and DMCF, that are designed specifically for the internal network on a massively parallel system, provide more specific endpoint addressing, such as numbering each network node sequentially. Some APIs depend on an external run-time system that provides an implicit address translation and prevents processes from communicating outside an established group of endpoints. Tports is one such example, where each peer process is assigned a rank identifier and translation to a hardware destination address is handled by mechanisms further down in the network stack.

6.4.2 Independent Processes

Similarly, some network programming interfaces provide the ability for arbitrary independent processes to communicate, while others are more restrictive. For example, the endpoint addressing restriction for Tports does not allow any process to establish communication with any other process in the system. In contrast, the endpoint addressing scheme used in Portals allows any two processes on the network to communicate. There is no external mechanism required to initiate network transfers. The same is true for Myrinet and IB even though they employ an more general endpoint addressing scheme. Support for independent processes allows one to construct more general-purpose networking applications beyond the traditional fixed-size network domain of an MPI application. On a system like a Cray XT, several components in the system require network connectivity and the dynamic nature of the system mandates that processes be able to come and go independently.

6.4.3   Connections

Upon discovering a method of identifying a peer process with which to communicate, some programming interfaces require an explicit connection establishment step, while others do not. The queue pair abstraction in IB mandates that the peer processes explicitly establish a connection before any data transfer can take place. For interfaces with logical endpoint addressing, like Tports, no explicit connection must be established, since presumably some external entity, like the runtime system, has already established the pathways for communication. The Portals API is connectionless as well. No explicit connection establishment is needed in order for a process to send or receive data with another process.

There are several concerns related to connection-based programming interfaces, especially with respect to scalability. Of primary concern is the amount of state needed for connections. Most HPC network programming interfaces are aimed at supporting MPI, which is a fully connected model where any and all processes can communicate. If the amount of state needed for connections scales linearly with the number of connections, then this can potentially lead to resource exhaustion issues on machines with an extremely large number of endpoints. Even connectionless networks typically maintain a small amount of state per endpoint for reliability mechanism, like sequence numbers, but generally this state is relatively small and is fixed for the number of endpoints.

6.4.4   Privacy

One of the motivations for a connection-oriented model is that it provides a process with explicit control in determining with whom it communicates. In some cases, the ability to refuse communication is just as important as the ability to enable communication. In cases where enabling communication is not explicit or implicit in the model, it is necessary to provide mechanisms to disable communication between peers. For example, even though Portals provides the ability for any two processes in the system to communicate without explicitly establishing a connection, a process must explicitly construct local objects in a way that enables communication to occur.

6.4.5   Operating System Interaction

Low-level network programming interfaces also tend to reflect the requirements of the underlying operating system. This is evident in several places, such as using operating system assigned process identifiers in endpoint addressing (ala Portals) or providing functions for registering regions of memory used by the network interface with the operating system. General-purpose operating systems with demand-paged memory typically require that any buffers used for network transfers be resident in physical memory. Some APIs, such as IB Verbs, make this memory registration step explicit, while others, like Tports, provide this capability behind the scenes.

6.4.6   Data Movement Semantics

Once all of the setup for communication is complete, the transfer of data can begin. The semantics of data movement is largely where network programming interfaces differ, both in terms of what capabilities they provide and how they are provided. We discuss several important characteristics of data movement below.

6.4.6.1   Data Transfer Unit

Low-level network programming interfaces expose different units of data transfer. Sockets, for example, can transfer byte streams or datagrams. Other APIs are message based and some expose the concept of a message header that contains context information about the data being transferred. The transfer unit not only has implications on the API itself, but it can also have secondary impacts on the semantics associated with data delivery and completion, which we discuss in more detail below.

6.4.6.2   Reliability

Different APIs offer different levels of reliability, and some, like IB, offer the user the ability to choose between several. In general, message-based programming interfaces provide a best-effort level of reliability, largely because this is what MPI requires.

6.4.6.3   Ordering

Another important distinction is the ability to guarantee ordering between subsequent transfers to the same peer. For message-based communications, it may be desirable that messages be non-overtaking. That is, the messages arrive in the order in which they were sent. However, in some cases, the meaning of *arrive* can be different. For example, the pairwise ordering constraint in MPI only mandates that messages be *matched* in the order in which they were sent. This distinction does not guarantee that a large message sent before a small message will be completely delivered before the small message. Rather, the messages will arrive at the proper buffer but the small message may complete before the large message does.

6.4.6.4   Non-Contiguous Transfers

Another property that some APIs provide is the ability to transfer non-contiguous blocks of data using scatter/gather mechanisms. The APIs in this analysis vary from providing no support (Tports) to providing very flexible support using address/length pairs to describe disjoint regions of memory. Support for arbitrary blocks of data can be inefficient when supporting higher-level programming interfaces like Cray SHMEM that support only fixed-stride non-contiguous transfers.

6.4.6.5   Peer Communication

We further classify peer communications into two categories: one-sided and two-sided. One-sided operations are data transfer operations where the initiator specifies both the source and destination of the data transfer and no explicit API call must be made to enable successive transfers of data beyond an initial connection or setup phase. We consider the RDMA model to be a subset of one-side communications that is distinguished by an initial exchange of information between the initiator and the target. In contrast, for two-sided operations, also known as the send/receive operations, the target must make explicit API calls to enable each data transfer and each peer only specifies their local part of the data to be exchanged. Two-sided operations also provide the ability to tag individual transfers in order to provide some context along with each message.

An important distinction between one-sided and two-sided data movement is that one-sided operations do not require any extra buffer resources. Since the initiator determines the destination of the data transfer, the data can be placed directly into its final location. With two-sided communication, the sender must speculate that there are sufficient resources at the receiver to consume the incoming message. This model can lead to the concept of unexpected or early arrival messages — messages for which there is no matching receive operation at the target. Determining the amount of resources to allocate for unexpected messages and the flow control protocols required to ensure that these resources are not exhausted is one of the most significant challenges in designing a highly scalable low-level network programming interface. One-sided operations eliminate the need for this complexity.

6.4.6.6   Group Communications

Group communication operations, sometimes called collective operations, are also an important component of network programming APIs for high-performance computing. Group communications are those that involve the participation of more than two processes to complete the operation. In some cases this operation can be a simple synchronization operation, such as a barrier, or involve the transfer of data, such as a broadcast operation. The majority of communication interfaces focus on providing peer communication operations upon which collective operations can be layered. However, some network programming interfaces, like DCMF, provide collective operations directly because the underlying hardware has native support for them.

6.4.7   Data Transfer Completion

An important distinction between communication programming interfaces that may not be obvious is how the completion of data transfers is communicated between the network hardware and the user process. The Active Message model [447] provides the ability to run handlers or callback functions that are invoked upon message header arrival and when the data transfer has completed. Handlers are generally designed to be small amounts of computation that perform a single operation in response to a network event. For example in LAPI, a header handler is invoked when a message header arrives. The role of the header handler is to return an address into which the network is supposed to place the data body of the message. Another handler is invoked when the data transfer is complete. This model provides a significant amount of flexibility for message processing. The drawback of the AM model is that the mechanisms for invoking user-level activity from the network are subject to many other parts of the system, such as the operating system scheduler. In contrast to AM, other APIs simply deliver event notifications to the user to signal that data has arrived either partially or wholly. Several APIs also have the ability to consolidate event completion notifications to minimize the overhead of checking for completion on multiple communication paths.

6.4.7.1   Multi-Protocol Support

Another important feature of a network programming interface is the ability to support several simultaneous uses by different higher-layer protocols. This is especially true for the internal network of a massive parallel computer where applications rely on data transfers for more than just message passing with MPI. Within a process, the network will also be used for handling file and storage operations, remote procedure calls, and potentially mid-level network programming interfaces. The underlying network programming layer needs to provide the ability to multiplex different higher-layer protocols efficiently. The AM model does this by providing a handler specific to each protocol. The Portals API provides a message selection capability that can be used to differentiate structures on a per-protocol basis.

6.4.7.2   Failures

Exposing network failures can be an important capability for higher-layer protocols. For some higher-level protocols, network failures need not be exposed at the application level. For example, the semantics of MPI allow an implementation to capture certain network failures and handle them within the context of MPI. Any catastrophic network failure is catastrophic to the process or the entire parallel application. In contrast, persistent system-level services, such as those used in I/O servers, should not allow a catastrophic network failure to result in denial of service. Therefore, the low-level network programming interface needs to provide a mechanism to expose failures and allow the higher-level protocol to determine what action needs to be taken.

6.4.8   Portability

Portability is a consideration of some low-level networking programming interfaces. Some vendors see the network API as a way to emphasize the capabilities of their specific hardware, while some are designed to allow software to be developed for multiple different hardware platforms.

6.5   Supporting MPI

MPI is the dominant parallel programming model and application-level networking programming interface in high performance computing. In many cases, low-level network programming interfaces are evaluated and judged on their ability to support a high-performing implementation of MPI. In this section, we discuss several important factors that impact the design and implementation of MPI.

6.5.1   Copy Blocks

In order to remove this high overhead operation from the latency path, a strategy that uses copy blocks (sometimes called bounce buffers) is usually employed. Copy blocks are buffers that are allocated and registered when the functional interface to the network is initialized. A short message is then copied into one of these regions when it is sent. On the receive side, the incoming message is deposited into one of these regions, and copied into the actual destination buffer when the process polls for completion of the transfer. In a sense, polling for message completion actually performs data delivery. The distinction between polling for data delivery and polling for delivery completion is important, especially considering that the MPI Standard only supports polling as the method for determining the completion of a message transfer.

6.5.2   Progress

Low-level network programming interfaces generally insure the progress of data transfers. Once a communication operation has begun, it will complete. However, progress at the network level does not always guarantee progress for MPI.

It is desirable to have a network interface that will allow the MPI implementation to make progress independent of calls to the MPI library. The MPI Standard mandates a Progress Rule for asynchronous communication operations. This rule states that once a non-blocking communication operation has been posted, a matching operation will make progress regardless of whether the application makes further library calls.

For example, if rank 0 posts a non-blocking receive and performs an infinite loop (or a significantly long computation) and rank 1 performs a matching blocking send operation, this operation will complete successfully on rank 1 regardless of whether or not rank 0 makes another MPI call. This desire can be viewed as a more succinct statement of the polling requirement discussed above. In this case, not only must polling not be used to complete local data transfers, but it must not be required to complete non-local data transfers either. That is, rank 0 must not be required to poll the network so that the send operation on rank 1 can complete.

6.5.3   Overlap

It is also desirable to have a network programming interface that will support overlap of computation and communication as well as support the overlap of communication with communication, where several data transfer requests can be outstanding. The benefit of overlap is that the host processor need not directly be involved in the transfer of data to its final destination, allowing the CPU to be dedicated to computation.

It is possible to support overlap without supporting independent MPI progress. Networks capable of performing remote DMA read and write operations can fully overlap communication with computation. However, the target address of these operations must be known. If the transfer of the target address depends on the user making an MPI library call (after the initial operation has begun) then progress is not independent. If the transfer of the target address is handled directly by the network interface, or by a user-level thread, then independent progress can be made.

Conversely, it is possible to have independent progress without overlap. An example of this is the implementation of MPI for ASCI Red [67], where the interrupt-driven nature of the network interface insures that progress is made, but the host processor is dedicated to moving data to the network (at least in the default mode of operation where the communication co-processor is not used).

6.5.4   Unexpected Messages

One of the most important issues that a low-level network programming API needs to address to support MPI is handling unexpected messages. The semantics of MPI allow for messages that arrive without a matching receive request. In order to optimize latency, small messages are typically sent immediately and buffered at the receiver if unexpected. In a fully-connected model such as MPI, it is possible for a process to receive a message from any other process in the parallel job. In scaling MPI to tens or hundreds of thousands of processes, the resources needed to buffer unexpected messages can be significant. Flow control can be used to limit the number of outstanding messages per destination rank, but the overhead of flow control must be carefully weighed against the need to have determinism, low latency, and high message rate for small messages. The ability of the low-level API to provide the features necessary to scalably and efficiently process unexpected MPI messages is critical.

6.6   Supporting SHMEM and Partitioned Global Address Space (PGAS)

The SHMEM [98] programming interface was developed at Cray in the early 1990s for the Cray T3 series of machines. SHMEM represents a second common parallel programming model and application-level networking programming interface in high-performance computing. It is closely related to the emerging PGAS programming models, such as UPC [77] and Co-Array Fortran [311]. Unlike MPI message passing, SHMEM is characterized by one-sided operations where the target of an operation is a remote virtual address. Support for SHMEM and PGAS is an emerging criteria by which low-level network programming interfaces are evaluated. In this section, we discuss several important factors that impact the design and implementation of SHMEM and PGAS communication libraries.

6.6.1   Fence and Quiet

Two interesting routines in SHMEM provide key pieces of the memory semantics needed for many PGAS languages: `shmem_fence()` and `shmem_quiet()`. The semantics of `shmem_fence()` require that operations to a specific target node that occur before the `shmem_fence()` call are completed before operations that occur after the call. In contrast, `shmem_quiet()` requires that *all* outstanding operations complete before the call returns. Providing support for `shmem_fence()` is very simple on an ordered network — in fact, `shmem_fence()` becomes a no-op in those cases. Implementing `shmem_quiet()` is most easily accomplished by providing lightweight remote completion tracking of all commands issued at the initiating processor.

6.6.2   Synchronization and Atomics

Because PGAS languages are based on a globally visible memory model, they require significantly more support for synchronization and remote atomic operations than MPI based programming models typically do. Supporting PGAS well typically involves strong support for global barrier operations. In addition, many atomic operations, such as fetch-and-add and compare-

and-swap can be useful in building remote lock constructs. Even a simple atomic add or atomic logical operation can be useful for building lockless data structures for various operations.

6.6.3   Progress

SHMEM has a very different view of progress than MPI. While MPI technically allows a weak interpretation of progress, a SHMEM PE can update a location in the memory of another process without the involvement of that target process. This requires that the network operations complete *without* the intervention of the target process. However, achieving progress for SHMEM is in some ways easier than it is for MPI. While progress for MPI requires that matching occurs at the target to complete a message, a SHMEM message uses a virtual address directly. This translation can be done at the initiator or target.

6.6.4   Scalable Addressing

SHMEM also requires a scalable mechanism to address remote virtual memory. Because SHMEM uses a target PE and target virtual address at the initiator of an operation, there must be a way to translate these virtual addresses to physical addresses at the target. Unlike the send/receive model of MPI, which requires certain types of information to be placed in the header of the message for matching at the target, SHMEM *only* has the target PE and virtual address to work with. While this *sounds* easier, it still does not map well to many modern networks. For example, the symmetric heap requirements for SHMEM require significant work in the run-time system. In addition, for an network with RDMA capabilities that are similar to those provided by InfiniBand, the remote virtual address is translated into an *RKEY*, so that an RDMA operation can be issued. In a large system, tracking all of the RKEYs for all of the peer processes can require enormous storage resources.

6.7   Portals 4.0

In this section, we discuss the continued evolution of the Portals network programming interface. Implementing and evaluating Portals 3.3 for the SeaStar network on the Cray XT series of machines was a valuable experience that helped reaffirm the ability of Portals to support the requirements of a massively parallel processing system. However, it also exposed some of the limitations of Portals, which we have tried to address in the next generation of the API. We discuss the motivating factors and outline the major changes below.

6.7.1   Small Message Rate

In previous generations of Portals, the major performance considerations were latency, bandwidth, and the ability to overlap computation and communication. More recently, we've begun to target small message rate as another key performance metric because of the important role it plays in achieving full bandwidth utilization of the network. Rather than only considering the latency path that an upper-layer protocol like MPI might follow, we also considered the impediments to achieving a high small message rate.

6.7.1.1   Unexpected Messages

One of the significant barriers to achieving a high message rate is operations that require round-trip communication with the network interface across the I/O bus. The majority of Portals functions are non-blocking. A request can be given the network interface and response is not required in order for the function to return control to the caller. However, there was one key function for managing MPI unexpected messages that required a round-trip operation.

The semantics of MPI allow for messages to be received before the user has posted a matching receive request. Before posting a receive request, the unexpected messages must be searched to see if a matching message has already arrived. If there is no match, the receive is posted. The search-and-post operation for MPI must be atomic to avoid a race condition where a matching message arrives after the search completes.

In Portals 3.3, there is no concept of unexpected messages. Therefore, the MPI implementation must set aside appropriate resources for unexpected message when the MPI library is initialized. To the Portals layer, MPI unexpected messages do not appear to be any different than any other message. When posting a receive, the MPI library must search through the messages that have arrived, consuming all of the available message arrival events. The atomic search-and-post operation is then called to post a receive. This operation will fail if another message event arrives before the receive can be posted, at which point the MPI library must consume this event and check for a match. This strategy requires that the MPI library wait for a response from Portals to know whether the receive has been posted successfully or whether there are still outstanding events that need to be processed. For the SeaStar network, the time for a round-trip operation was more than a microsecond, which severely limited the achievable small message rate.

The only way to avoid the round trip required for the search-and-post operation is to unify this operation so that the alternative paths are known and taken in the same context where the search is performed. For Portals, this meant introducing the concept of unexpected messages into the API. To accomplish this, we introduced the concept of an optional overflow list that is searched whenever receive request is posted. This way, the MPI library can simply post a receive request, and Portals will generate one of two types of events that indicate that the message has been received into the posted buffer, either directly via the network or through the implementation copying a previously received message, or that a matching message had previously arrived but only part of the message was delivered. This ability to receive either a whole message or a partial messages allows for implementing a two-level protocol where short unexpected messages are buffered at the receiver and completed via a memory copy, and long unexpected messages are buffered in place at the sender and completed via a remote read operation.

6.7.1.2   Resource Isolation

In the previous generation of Portals, we tried to enhance the usability of the API by embracing symmetry in objects and operations. In some cases, the generalization that accompanied symmetry created complexities that degraded performance. For example, Portals 3.3 had a single memory descriptor object that could be used for describing memory to be used for a put operation and/or for a get operation. Situations where a buffer is used for both sending and receiving are common. An example is a long message protocol in MPI where the buffer being sent could potentially be read by the receiver if it is unexpected. Creating a single object that can be used for both put and get operations simplifies the API and is convenient. However, this flexibility creates a situation where extra coordination between the host and the network may be required to manage resources. For example, if a use-once memory descriptor is created and used to initiate a put operation, the network interface must maintain the state of the memory descriptor to insure that it has not already been consumed by an incoming message. Separating objects used for host-side operations (sending) from those used for network-side operations (receiving) allows for the state of the object to be maintained wherever it is most appropriate and reduces the amount of coordination that must occur between the host and the network.

Separating these objects also led to other positive side effects, such as reducing the space needed to contain initiator events. A unified event structure accompanied a unified memory descriptor. For initiator operations, a large portion of the event structure contained either unneeded or redundant information. Reducing the size of initiator events can potentially increase small message rate as well.

6.7.1.3   Logical Endpoint Translation

Another change targeted at increasing small message rate performance was to provide the ability for the Portals implementation to cache a translation from a logical endpoint (rank) to a physical endpoint (node identifier and process identifier). This feature was available in a previous generation of Portals, but it was provided in a way that required a reverse lookup from physical to logical each time a message was received. The current specification allows for translating from a logical to a physical address at the initiator without requiring the reverse lookup at the target.

6.7.2   PGAS Optimizations

Previous generations of Portals supported the operations necessary to implement the data movement operations needed for Partitioned Global Address Space (PGAS) models, although they were not highly optimized for this purpose. In particular, the matching overhead required to support MPI semantics is unnecessary for PGAS models. In an effort to provide more optimal support for PGAS operations, we extended the specification to allow for bypassing message matching, to support lightweight counting events, and to provide a broader set of remote atomic memory operations.

PGAS models are distinguished from message-passing models by requiring ultra-low-latency one-sided communication operations where the virtual address of the target data is known at the initiator. For these types of operations, the message selection capability that Portals provides results in an unneeded overhead, in terms of extra computation required for match processing, larger amounts of memory used to signify message events, and more unused bits transfered along the wire. In an environment where remote reads and writes of 8-byte quantities are typical, 64-byte message headers and 128-byte message events are extremely inefficient.

We have extended the Portals API to support a non-matching network interface that allows an implementation to use a minimal network header and avoid most of the Portals address translation computation at the target. Additionally, we have added lightweight counting events that can be used in place of more heavyweight message events. Counting events are simple mechanisms that can be used to keep track of the number of transfers that have completed or the total number of bytes that have been received. The API has also been extended to include a larger set of remote atomic memory operations, where the initiator can pass in data and an operation to be performed on remote memory.

6.7.3   Hardware Friendliness

In some instances, Portals was modified to better accommodate a direct hardware implementation of the specification. This was a different approach than we had taken with the previous version of the specification where we had aimed to design an API that could be implemented by a programmable network interface containing a general-purpose embedded processor, like the SeaStar. For instance, Portals 3.3 allowed the user to do arbitrary list inserts when adding to a list of posted receive requests. In practice, arbitrary list inserts are uncommon — upper-layer protocols only append requests. Arbitrary inserts are straightforward to implement as a linked list on a general-purpose processor, but are extremely more complex to implement in hardware, especially when considering that a FIFO structure is sufficient.

6.7.4   New Functionality

The Portals API has also been extended to support some new capabilities that we believe will be more important for future generations of massively parallel systems.

We have added the concept of triggered operations in order to support non-blocking collective operations more efficiently. Triggered operations allow for associating a data movement operation with counting event. When the counting event reaches a specified threshold, the posted operation is initiated. These kinds of event-driven operations are not new, but they are a necessary component for insuring the progress of non-blocking collective operations. For example, a non-blocking tree-based broadcast operation can be constructed by issuing triggered put operations to children that fire when a message from the parent is received.

Finally, we have extended Portals to better support the flow control needed for handling large numbers of unexpected messages in a scalable way. Rather than impose a specific flow control protocol, Portals provides the infrastructure necessary to efficiently communicate resource exhaustion failures and allow an application to gracefully recover.

Chapter 7

High Performance IP-Based Transports

Ada Gavrilovska

Georgia Institute of Technology

7.1 Introduction

Most massively parallel systems and many cluster systems based on inter-connection technologies such as InfiniBand rely on custom protocol stacks to provide end-to-end transport services for their applications' data movement needs. However, more and more we are experiencing the need for IP-based high performance communications services, "outside-of-the-box," across distributed infrastructures and across the public Internet.

The motivations for this are multifold. First, the bandwidth capabilities of the underlying physical infrastructure are creating opportunities for high-speed data paths to be established and utilized by distributed end-user applications. This allows traditional HPC applications to be decoupled across multiple distinct high-end platforms or deployed on distributed computational grids. Numerous collaborative research efforts require large and timely data exchanges across globally distributed laboratories, data sources such as astronomic ob-servatories, particle accelerators, or other advanced instruments on one side, high-end machines running complex data-intensive simulations on another side, and end-user scientists interacting with both the data and the ongoing com-putation. These types of applications rely on the communication capabilities of the underlying IP-based network (including high-speed networks such as Internet2 and the National Lambda Rail (NLR)) and protocols to carry huge amounts of data at acceptable data rates. Next, beyond the high-end scientific applications, more classes of enterprise applications exhibit increase in data volumes and stricter response time guarantees, ranging from financial trading and information systems and global parcel delivery and tracking tools, to distributed file sharing services, to high-end video-conferencing, immersive and tele-presence applications. Finally, the multi-core nature of modern platforms, and the trends to consolidate computation into Compute Clouds, raises the networking requirements for these types of environments too, both due to the aggregate communication needs of the workloads hosted on such cloud platforms, as well as due to the core cloud services used to enable dynamic coordination with and migration across potentially globally distributed clouds.

In order for these types of applications to best take advantage of the increased capacity of the global networking infrastructure, they must rely on a network protocol stack which can be supported in the public IP-based Internet. While there are non-IP based alternatives for long distance/wide-area high-speed data transfers, e.g., based on InfiniBand over Wide-Area (IBoWA) solutions such as those supported by Longbow XR from Obsidian Strategics [361], these require substantial infrastructure investment, which is not generally available.

In this chapter, we focus on issues related to attaining high performance IP-based transport services. Particularly, we discuss the Transmission Control Protocol (TCP), which jointly with IP, has been a core component of today's Internet since its inception. Addressing the performance issues in the TCP/IP stack opens opportunities for a large class of TCP-based applications to benefit from the high data rates sustainable in current and future networks. In fact, this is part of the reason why a series of challenges have been organized over recent years to push the limits of the performance of TCP (or TCP compliant) transports over the public Internet infrastructure [173, 451, 474]. In the following sections we provide a survey of some of the key TCP variants and the main mechanisms used to improve the data rates they can deliver. In addition, we discuss certain implementation techniques aimed at eliminating various sources of bottleneck in the execution of the TCP protocol processing engine. Finally, we also include a brief discussion of other IP-based transport alternatives.

7.2   Transmission Control Protocol — TCP

7.2.1   TCP Origins and Future

In the 1970s the Department of Defense's Advanced Research Project Agency (DARPA) initiated an effort to redesign the protocol technology used by its locally supported research network - ARPAnet, so as to enable its growth beyond its existing boundaries. The first design was proposed by Vint Cerf and Bob Kahn [85], and by 1974 the first official specification for the then called Internet Transmission Control Program was written [84]. The original protocol went through several revisions, and by 1978, Vint Cerf and Jon Postel split it into two protocols: Transmission Control Protocol (as opposed to Program) — TCP, responsible for the end-to-end communications behavior (e.g., flow control and reliability management), and Internet Protocol — IP, responsible for the hop-by-hop routing of packets in the distributed network. This was the birth of TCP/IP, still the most popular protocol in the Internet today. In 1983, the Department of Defense placed a requirement that every network connected to the ARPAnet must switch over to TCP/IP, and soon afterwards ARPAnet was renamed Internet. While there were other commercial attempts to introduce alternative internetworking protocols, the Internet and the TCP/IP prevailed, and, at least conceptually, have remained unchanged until today.

The creation of Internet2 in 1996, and the continuous community committment to upgrade its infrastructure to the most advanced networking technologies available, has created a testbed for experimentation with the limits of TCP, and other IP-based network stacks, among other ways, through challenges such as the Internet2 Land Speed Record, and the Bandwidth Challenge, organized by the SuperComputing community. One goal of these challenges is to demonstrate the possibility of eliminating the "wizard gap," the mismatch between the theoretical end-to-end performance levels and those attained in practice via existing protocol choices and configurations. It is exactly these challenges which demonstrate that eliminating this gap requires significant skills and wizardry!

More importantly, this testbed has opened unique possibilities to investigate other methods needed to support emerging communication needs, including the need for end-to-end resource reservation, guaranteed performance levels for high-volume, and highly interactive applications, etc. One important class of applications driving this level of innovation are *3DInternet* applications such as tele-presence and large-scale immersive distributed games and collaborations.

More recently the networking community has started to raise the question of the future of the Internet and the continuous viability of the TCP/IP Internet stack and the core Internet architecture design as the networking solution of choice. The scale of the Internet, the challenges raised by connectivity of billions of mobile devices, and the performance, security and QoS requirements of emerging classes of applications have raised the need for many ad-hoc, one-off solutions, "barnacles" [336] on the otherwise elegant stack of well defined protocols with clean interfaces. The direction of the future Internet architecture and the networking protocols it will support remain an open question, and several efforts led by the National Science Foundation, the Network Science and Engineering Council, and the GENI program [161] are creating a platform to foster this type of innovation.

The reasons for the success of the TCP protocol lie in the functionality it provides. TCP is a connection-oriented protocol, which provides reliable, ordered data transfers with end-to-end guarantees, and ensures fair utilization of the aggregate shared networking resources. It has been a fundamental part of today's Internet for several decades, and as such it has proven that it is suitable for many types of applications with diverse communication requirements. Irrespective of what the future Internet will look like, and what other types of protocols it will support, TCP is expected to remain one of its major components for a long time.

7.2.2   TCP in High Speed Networks

While conceptually TCP has retained its original design, several variants have emerged over the years. The primary reason for the emergence of new versions of the protocol was to address issues with the protocol's behavior during congestion, and were first triggered by the "congestion collapse" first observed in 1986. During this event, the throughput rates between Lawrence Berkeley Laboratory and UC Berkeley, sites merely 400 yards apart, dropped to one thousandth of its original bandwidth — from 32kbps to 40bps [214]. The cause of this phenomena was the lack of adequate congestion detection and avoidance mechanisms in the protocol. Namely, under congestion, packets or acknowledgments for their receipt would be dropped, data sources would determine that the transmission was unsuccessful, and would re-initiate it, thereby further increasing the load in the network, and worsening its congestion level.

Congestion results in lower utilization of the existing bandwidth for successfully acknowledged transmissions, and should be avoided by limiting the sending rate to no more than the available bandwidth. At the same time, it is imperative for TCP senders to attempt to send as much data as possible within the available bandwidth limits, so as to achieve better throughput levels. However, since communication takes place over a network which may be shared with other flows, the bandwidth availability may vary over time. Therefore, dynamic estimation of the available bandwidth is necessary in order to avoid congestion, achieve high bandwidth utilization and sustain high-speed data transfers.

To address this issue, TCP was enhanced with a congestion control mechanism, based on an *Additive Increase, Multiplicative Decrease (AIMD)* mechanism to manage the sender-side *congestion window* (*cwnd*) — the amount of new data the sender may push into the network. The basic idea of the AIMD algorithm can be described as follows. Initially, during a *slow-start* phase, the sender starts with a small congestion window of just one data segment. After this segment is acknowledged, i.e., after one round trip time (RTT), the sender starts increasing the amount of data it sends out (i.e., the size of the congestion window), typically by sending two packets for each acknowledgment (ACK), thereby doubling the amount of new data per RTT. During this phase the sender is trying to more aggressively approach the available bandwidth limit using this additive increase mechanism. This behavior continues up to a certain threshold[1], at which point it enters a *congestion avoidance* phase.

---

[1]Or until the first packet loss is detected.

In this phase the sender more carefully probes how much more bandwidth is available along the data path, by increasing the congestion window size more slowly, e.g., a single data segment at a time. All packets are tagged with a sequence numbers, and ACKs indicate the highest sequence number in a consecutive array of packets (i.e., there might be received packets with higher sequence numbers, but separated by gaps of missing packets). The lack of an acknowledgment within a retransmission timeout interval (RTO) or the repeated receipt of duplicate acknowledgment for the same packet, indicates potential packet loss and, hence, congestion. After congestion is detected, the sender enters a *recovery phase*. The size of the congestion window is reduced in a multiplicative manner (e.g., halved) and the unacknowledged, potentially lost packets are retransmitted.

It is clear that the TCP congestion control mechanism has direct implication of the attainable throughput levels. The size of the congestion window limits the amount of data which can be sent during a RTT interval. If its size is too low, it can underutilize the available bandwidth and, more importantly, fail to achieve the attainable throughput. Towards this end, many improvements have been proposed to the original AIMD solution, and we will next discuss some of the key ones among them.

7.2.3   TCP Variants

7.2.3.1   Loss-based Congestion Control

We will first describe a series of variations of the original AIMD algorithm in TCP, which use the occurrence of a packet loss as a main indication that congestion has occurred.

TCP Tahoe. The simplest form of congestion control determines packet loss when a retransmit timer expires before an acknowledgment (ACK) is received. TCP Tahoe improves this algorithm by considering duplicate acknowledgments as an indication of a packet loss and potential congestion. If n (typically $n = 3$) duplicate ACKs are received, TCP Tahoe enters a *Fast Retransmit* phase, and initiates the retransmission of the lost packet [140]. In this manner the packet loss/congestion is detected more quickly and the "right data", i.e., the lost packet, as opposed to the next packet in the congestion window, is transmitted more promptly. Since packet loss is interpreted as a congestion indication, Tahoo reduces the rate at which data is sent by halving the congestion window, thereby drastically emptying the data path pipeline in the next phase. This may not have been necessary, since the packet loss could have occured as a result of an isolated packet corruption.

TCP Reno. In order to address the "emptied pipeline" issue, TCP Reno extends the Fast Retransmit algorithm with *Fast Recovery*. The main idea in Fast Recovery is that a duplicate ACK is an indication that a packet has been received by the receiver, and therefore has created additional bandwidth in the network pipeline. Therefore, Reno uses the number of duplicate ACKs to inflate the size of the congestion window beyond the value which it had in TCP Tahoe, thereby ensuring that more data will flow into the network after the recovery phase is complete. In both Tahoe and Reno the congestion window is reduced to one if a timeout occurs.

TCP NewReno. In TCP Reno, if a partial ACK, acknowledging some but not all of the data outstanding at the start of Fast Recovery, is received during the recovery process (indicating multiple lost packets), the sender exits the Fast Recovery phase, and ultimately a timeout occurs. TCP NewReno treats these partial acknowledgments as indication that the subsequent packets have been lost, and initiates their retransmission too. This helps avoid the occurrence of a timeout, and the negative performance implications of it to both delay (since the timer must expire before packets are retransmitted) and throughput (since timeouts deflate the congestion window to one). NewReno is the most widely spread version of TCP.

TCP SACK. The mechanisms included in the TCP versions described above help the sender determine that it should send the unacknowledged data sooner than the original version. A *Select Acknowledgment — SACK* extension, can be used in conjunction with the above protocols to help the sender determine which data packets are missing in the event of multiple losses during an RTT interval. SACKs are commonly supported as an optional extension to the TCP header, and the receiver uses them to specify ranges of acknowledged data. The sender than retransmits only the missing packets, thereby not wasting bandwidth resources for retransmission of already received data [140]. Supporting SACKs adds to the complexity of the TCP processing engine, and in low-loss networks it will have negative performance implications. However, this option is particularly well suited for environments where multiple consecutive packet losses are not unlikely.

7.2.3.2   Loss-based Congestion Management in LFNs

The AIMD algorithms used to manage the congestion window size in the TCP variants described so far can be particularly inefficient for high speed, long distance networks, or in general for high speed networks with large *Bandwidth Delay Product — (BDP)*. BDP is an important parameter of the data path — it indicates how much data can be pushed along the data path in an RTT, which depends on the path's capacity (bandwidth) and delay ($\frac{1}{2}$RTT). For networks with large BDPs, first the congestion window will need to grow to large values before it reaches near-congestion level throughputs. Next, when congestion is detected, halving the congestion window, and thereby the data being pushed into the data path pipeline, will drastically reduce the achievable throughput, likely more than what is necessary to just avoid further congestion. Furthermore, AIMD depends on the rate of acknowledgments to "clock" itself to further increase the congestion window. The acknowledgment rate is inversely proportional to the RTT, so for long distance high speed networks, also referred to as *Long Fat Networks*, once reduced, the congestion window will grow more slowly compared to networks with lower RTTs, and it will take longer time to recover from the loss and achieve high throughputs again.

TCP-BIC and TCP-CUBIC. A more recent version of the TCP protocol proposed in [469], TCP Binary Increase Congestion control (TCP-BIC), uses a different function to manage the congestion window size. The main idea in TCP-BIC is to extend AIMD with a range within which a binary search-like logarithmic function is used to explore the set of congestion window sizes between the window size which caused the congestion (as an upper bound) and the window size as determined by the multiplicative decrease algorithm (lower bound). Similar binary exploration takes place when TCP attempts to increase its window size. This type of more aggressive approach to determining the appropriate congestion window size and the rate at which data is pushed into the network is particularly well suited for high speed networks with large BDPs. TCP-BIC is trying to address a key deficiency in the original AIMD function with respect to networks with high BDPs: the additive increase is too slow and the multiplicative decrease is too drastic for these environments. The logarithmic function in TCP-BIC results in larger increments the farther off the sender is from the target bandwidth, and smaller ones when its estimate of the target value gets closer.

In 2007, a team lead by University of Tokyo was awarded the Internet2 Land Speed Record, for longest lasting TCP transmission over longest distance and at highest bandwidth rate [474]. Their testbed implementation was based on TCP-BIC, with some additional optimizations as described in Section 7.3, and it sustained 9.08Gbps constant data rates for 5hrs along a data path spanning 30,000 kilometers. TCP-BIC was also the default TCP implementation in the mainstream Linux kernel up until Linux 2.6.18. It was replaced in Linux 2.6.19 with TCP-CUBIC, a variant developed by the same team as TCP-BIC, which uses a less aggressive (cubic) window management function, more suitable for (more common) environments with shorter RTT.

Other variants. Numerous other variants of TCP exist which specifically target high BDP network paths. Most of them are based on a similar idea regarding the congestion management algorithm, albeit implementing it with different window management functions. These include **High Speed TCP**, **H-TCP**, **Scalable TCP**, and others [183]. Some of these algorithms use the size of the congestion window and the RTT to estimate the BDP of the end-to-end data path, and to dynamically switch their behavior, i.e., their aggressiveness, between one that is more suitable for high vs. low BDP networks. Finally, the use of the TCP SACK option is well suited for data paths in LFNs, due to the increased possibility of multiple packet losses within a single RTT.

7.2.3.3   Delay-Based Congestion Control

There are several performance-related issues with the loss-based TCP flavors described so far. First, they don't operate well in high packet loss scenarios, since repeated packet losses will continue to reduce the congestion size window, and result in lower bandwidth utilization. Second, they are reactive, in the sense that they wait until congestion occurs, i.e., a packet is lost, before trying to recover from it and avoid future losses. This creates an oscillating behavior where the network repeatedly goes in and out of congestion, which results in lower bandwidth utilization.

TCP Vegas. One alternative to this is to dynamically estimate the round-trip time (RTT) and use it as an indication of the congestion level in the network. A large difference between the current and the minimum observed RTT indicates increased queueing delays in the network and increase in the overall network load. TCP Vegas [55] uses the measured RTT to dynamically adjust the values of the congestion window as well as the retransmission timeout. Because of the more proactive approach to tracking the current network load, congestion is more likely avoided in the first place. One downside of delay-based approaches such as TCP Vegas is that they tend to be less aggressive in the event of high network loads with multiple TCP streams. Data flows for which the sender and receiver use a loss-based approach will keep trying to consume more bandwidth, whereas those using delay-based approaches will see that the load levels increase and will reduce how much extra bandwidth they consume by transmitting fewer new packets.

FAST TCP. Another delay-based approach is FAST — FAST AQM (Active Queue Management) Scalable TCP. Its main improvement over TCP Vegas is in the way it increases the congestion window. Similarly to TCP-BIC, FAST TCP is more aggressive if the RTT measurements indicate a large difference between the current and the target rate, and less aggressive otherwise. Because of this, it tends to converge more quickly to stable transmission rates, thereby achieving more quickly the high data rates possible in long fat networks. In fact, a well-tuned implementation of TCP FAST was the transport protocol used by the winner of the Supercomputing Bandwidth Challenge in three consecutive years [451].

Delay/Loss-based Variants. Several protocols rely on the AIMD algorithm, but use the delay information and bandwidth estimates as a way to tune the AIMD parameters for increase and decrease of the *cwnd*. In **TCP Illinois**, as the delay increases, i.e., the network congestion increases, the *cwnd* gets increased more slowly. Upon loss, if the RTT is close to the maximum observed RTT (i.e., indication of a congested network), the *cwnd* is reduced more drastically, e.g., by 1/2, whereas if the RTT is small, and the loss is more likely a result of packet corruption than congestion, *cwnd* is reduced by smaller amount, e.g., by 1/8 [183]. **Compound TCP** [255, 423] is another algorithm which uses delay and loss information. It maintains two windows, a

Congestion management type	TCP Variant	cwnd management function
Loss-based	TCP Tahoe	AIMD with Fast Retransmit or Fast Recovery optimizations
	TCP Reno	
	TCP New Reno	
Loss-based for large BDP	TCP-BIC	AIMD & non-linear probing in regions near target bandwidth
	TCP-CUBIC	
	HighSpeed TCP	
	H-TCP (Hamilton)	
	Scalable TCP	
Delay-based	TCP Vegas	cwnd is function or RTT estimate
	FAST	
Delay/loss-based	Compound TCP	AIMD parameters are function of RTT estimate
	TCP Illinois	cwnd sized based on AIMD; dwnd sized based on RTT

FIGURE 7.1: Congestion management functions in popular TCP variants.

congestion window which behaves in the same manner as AIMD, and a delay window which behaves more similarly to the logarithmic functions used in TCP-BIC and FAST.

It is important to note that the above TCP variants are based on participation of the connection end points only — the sender and receiver. Additional enhancements to the congestion management in TCP can be achieved with participation of the intermediate nodes along the data path, i.e., network routers, to perform tasks such as early congestion detection and active queue management (e.g., eXplicit Congestion control Protocol (XCP) [223]).

Figure 7.1 summarizes the protocols discussed in this section. This set is by no means exhaustive. In fact, over the last three decades hundreds of papers have been published advocating certain types of optimizations in the implementation of the core TCP mechanisms. Many of these variants have become available options in commercial operating systems and have been used in real-world environments, many more others have remained research projects explored through simulation or in smaller-scale laboratory settings. Many experimental studies include performance comparisons among different TCP variants for high-speed networks [255, 361]. However, there is no definite agreement as to which TCP flavor is the winner. The reasons for this are that different variants tend to be better suited for different testbeds, workloads or operating conditions. Furthermore, the performance metric may differ for different application domains. In light of this, the main objective of this section is to illustrate the types of performance challenges related to the congestion control mechanism in TCP and the main approaches which address them.

7.3   TCP Performance Tuning

The primary goal of the TCP variants discussed above is to improve the manner in which bandwidth is utilized — avoid congestion, losses, and better estimate the available bandwidth so as to immediately send out in the data path pipeline as much data as possible. In addition to making adjustments of the TCP congestion control mechanism, other opportunities exist to better tune the behavior of the TCP processing engine, to improve the amount of data the sender is delivering into the network, or to remove CPU processing loads from the connections end-points, thereby increasing their capacity to handle higher processing rates.

7.3.1   Improving Bandwidth Utilization

TCP window and socket buffer sizes. A straightforward way to ensure that all available bandwidth can be utilized whenever possible is to properly configure the maximum size of the TCP window so that it is greater than the bandwidth-delay product on the anticipated data paths. TCP will always send out less if the congestion window (or the advertised window of available buffer space on the receiver side) requires so. Similarly, the socket buffer sizes should be increased in a similar manner, or ideally, dynamically tuned to the value appropriate for the current BDP on the data path. Dynamic Right Sizing (DRS) is an auto-tuning mechanism which provides dynamic right-sizing of TCP buffers [155,452], and is today typically enabled in most current OSs. Several detailed studies of the TCP behavior in high-performance environments illustrate the impact of these, as well as some of the other optimizations discussed in the remainder of this section [200,474].

Packet pacing. The congestion control mechanism attempts to estimate the available bandwidth on the data path and use that as an indication of how much data it should send next. However, the bandwidth estimate is computed on a RTT time scale. On a microscopic level, pushing out the next set of data segments may create bursts at bottleneck points in the network, and result in packet loss. Consider a situation where the host is attached to a high bandwidth link, say 10GigE, which in turn connects to a OC-192 SONET/SDH link at 9.6Gbps. When the sender pushes out data segments into the 10GigE link, the packets will arrive at the 10GigE-SONET interface as a burst at a 10Gbps rate, and may cause buffer overflow and packet losses. Nakamura et al. [302] observe this phenomenon and address it by adjusting the inter-packet gap (IPG) length maintained by the NIC's MAC interface. By increasing the IPG interval, TCP cannot just blast all of its available data at the maximum rate, instead the packets are "paced" at the appropriate rate. With this technique, even at the microscopic level, the data rate values are within the bandwidth estimates, and the occurrence of bursts and packet losses at the bottleneck links is likely prevented. The mechanism was used by the same group to achieve the Internet2 land speed record [474].

Parallel TCP streams. Finally, many applications with high bandwidth requirements, such as for bulk data movements in long-distance large file transfers, rely on multiple TCP streams between the source and the destination. These approaches exploit the fairness properties of TCP, which tries to evenly distribute the network bandwidth across all flows. Dedicating more than one flow to a single data transfer is then likely to attain a larger portion of the available bandwidth, at the cost of diminished TCP-friendliness. GridFTP, a component of the Globus toolkit [163] for Grid-based services, is a popular example of such transfer, which also supports striped data transfers by using multiple TCP streams across a number of sources or destinations.

7.3.2   Reducing Host Loads

Offload. Another well-understood approach to improving TCP performance is through offloading protocol processing elements on the networking hardware. Chapter 3 discusses some of the common offload features present in modern Ethernet NIC devices and Chapter 5 more generally discusses the issues surrounding offload approaches. The main benefits of offload include:

1. reducing host CPU loads so it can sustain higher rates of application-level processing,

2. benefiting from specialized hardware accelerators better suited for a given set of operations, or

3. benefiting from the increased opportunities to overlap different stages of the protocol processing pipeline thereby achieving higher throughputs.

With respect to TCP, checksum computation on both ingress and egress path is one obvious choice for offload, since hardware accelerators can perform these types of computations much more efficiently compared to their software counterparts. On transmission, TCP Segmentation Offload (TSO) is another useful technique — the application-provided data can be passed directly to the network card, where it is segmented into MTU-sized packets. Large Receive Offload (LRO) is a technique used on the receive side to coalesce multiple packets into a larger TCP segment before generating an interrupt and delivering the packets to the protocol stack. Both TSO and LRO reduce the number of copies required on the data path, and thereby the overall memory bandwidth consumed by the TCP processing engine.

At the farthest end of the spectrum, full TCP offload can be performed on *TCP Offload Engines (TOEs)*. The benefits of TOEs have been debated for several reasons [296, 398]. First, the processor architecture in embedded platforms such as TOEs will typically lag behind mainstream CPU capabilities. This may be offset by the fact that TOE engines are more specialized to execute the packet processing operations encountered in the TCP stack, thereby resulting in a more efficient, and more power-friendly implementation [152]. Second, TOEs may lack the memory resources needed for maintaining TCP buffers and window sizes as required to utilize high bandwidth pipelines. This restriction is recently reduced through innovation at the I/O interconnect level, first through interconnection technologies such as PCI-Express and HT, and more recently through tighter integration of the I/O and memory controllers, thereby allowing the TOE engine faster access to the entire host memory. Finally, offloading the TCP stack may require more expensive coordination operations across the host OS-TOE NIC boundary, particularly related to control-plane operations of the protocol processing [226,395]. Depending on the workload properties, the costs associated with these control-plane operations can dominate and result in decrease in overall performance [152]. Finally, in order to fully benefit from an offload engine it may be necessary to make changes to the application-protocol interface, e.g., to replace the regular socket interface with an asynchronous interface such as SDP [152]. This may not be trivial or even possible for certain types of applications, particularly for those based on legacy codes.

MTU size. By coalescing multiple packets into a larger segment, LRO helps eliminate other sources of overhead. First, the host's TCP stack sees the larger segment as a single TCP segment, and the TCP processing costs (which include window management, generating acknowledgments, etc.) are incurred per larger number of bytes, thereby improving the achievable throughput levels. Second, it reduces the number of interrupts being generated, and with that it reduces the interrupt handling overheads. Another way to achieve the same benefit is to use larger sizes for the maximum transmission unit (MTU) supported by the link level protocol, e.g., jumbo Ethernet frames, provided that the interconnect device (i.e., NIC) has support for such options.

DMA transfer size. For similar reasons as above, to amortize over a larger amount of data the overhead of configuring transactions over the NIC-host interconnect (nowadays typically PCI-Express interface for high end devices), it is important to configure the PCIe transactions to use the maximum possible request and payload size. Hurwitz and Feng [200] experimentally demonstrate the bandwidth improvements resulting from adjustments in the MTU and DMA transfer size.

Interrupt coalescing. The total number of interrupts can be reduced implicitly, through use of larger MTU sizes or techniques such as LRO, or explicitly, by configuring the NIC hardware to coalesce all packets up to a preconfigured threshold, before generating an interrupt. When the interrupt is finally generated, its handling overheads can potentially be amortized over larger number of packets, and therefore, at least in theory, higher data rates may be achieved.

This latter conclusion proves not to be always accurate for TCP communication, and large interrupt coalescing parameters can have disastrous effects on protocol performance. Yoshino et al. [474] present fine grain measurements to illustrate these effects. Namely, by coalescing packets, we are potentially delaying ACK packets, and as explained earlier, the ACK packets play a key role in estimating RTTs and bandwidth and "clocking" the TCP congestion control mechanism to push more data into the network. Aggressive interrupt coalescing is in fact shown to result in significant throughput degradation [474].

In addition, if the NIC combines LRO with interrupt coalescing, the congestion control mechanism must be modified not to merely count the number of acknowledgments, but rather to consider the amount of data which is being acknowledged, by performing *Appropriate Byte Counting (ABC)* [9].

7.4 UDP-Based Transport Protocols

An alternative to addressing the performance limitations caused by TCP's congestion control mechanism, is to use a UDP-based transport. The use of a UDP-based transport offers additional flexibility, since it does not enforce strict ordering and reliability, which may not be required for some applications. This is true for large bulk file transfers, where the order can be subsequently established from file metadata, or for streaming applications which have natural ability to deal with periodic data losses or have built-in mechanisms to recover from data errors. TCP and its congestion control mechanism are also an overkill in environments with dedicated links, such as in high speed networks where bandwidth reservations are possible.

Towards this end, there are many UDP-based protocol alternatives used in the above settings. However, basing the communications solely on UDP, and just blasting out data at highest rates without any mechanism to estimate the available bandwidth can lead to high data loss rates. In addition, such TCP-unfriendly behavior can starve other TCP flows who share the same network links and adjust their data rates as a result of the observed congestion. Therefore, the UDP-based transport solutions typically include an additional rate control layer. In some cases the transport solution relies on a separate TCP channel used solely for control functionality, including bandwidth estimation. This information is then used to control the rate at which UDP sends data into the network. Examples of such protocols include Tsunami [282], Hurricane [463], Reliable Blast User Data Protocol [182], and others.

In other cases, the congestion control is modeled after the AIMD mechanism in TCP, but implemented as a separate module on top of UDP. The approach offers greater flexibility, since by separating this functionality, first it becomes more easy to modify it, and second, it can be more easily tuned to the application needs. For instance, IQ-RUDP [184] takes advantage of the message-oriented nature of UDP, and allows the use of application-specific handlers to determine which application data should be sent out in case the available bandwidth level is insufficient to handle the entire outstanding window. This creates opportunities to better tune the protocol behavior to satisfy application-specific performance metrics.

The exact congestion control mechanism used in purely UDP-based protocols may use packet loss information similar to some of the TCP variants, as is the case with UDP-based Data Transfer (UDT) [173], or more complex models for bandwidth estimation, such as the stochastic modeling used in RUNAT [462]. UDT uses a mix of regular ACKs, SACKs, and negative acknowledgments (NAKs), and adjusts the amount of new data pushed into the network by controlling the duration of the packet-sending period. A suite of UDT-based data intensive applications was the winner of the 2008 Bandwidth Challenge, reaching, for one of the three applications, data rates of up to 8Gbps for distance from Austin, TX to Kitakyushu, Japan [173].

7.5 SCTP

The Stream Control Transport Protocol [414] is an IP-based protocol, developed by the Internet Engineering Task Force (IETF), and today it is available in most standard operating systems. It is a TCP-friendly, IP-based transport, fully supported in the public Internet infrastructure. It is a message-oriented transport, and within a single connection there may be multiple "message streams." Reliability and ordering constrains are optional, and can be specified on per stream basis. That is, within a single communication, one stream of messages of one type may be ordered and reliable (like TCP), another may be unordered and reliable, and a third one may be neither (like UDP). This feature eliminates "head-of-line blocking," which occurs in TCP, by allowing progress to be made on a per stream basis, even if there is an "earlier" message or message segment still missing in one of the other streams. Other benefits include the fact that a connection is represented by sets of IP address-port pairs, i.e., it is multi-homed, and that SCTP is designed to be easily extensible — its header format includes control elements (chunks) which may be used to specify new types of behaviors. The core of the congestion management engine in SCTP is based on the AIMD loss-based congestion management in TCP.

The message-based semantics of SCTP map more easily to the types of application APIs used in scientific HPC settings, such as MPI and RDMA-based I/O, as well as to the interfaces of full procol offload engines with OS-bypass and direct data-placement capabilities. In fact, SCTP was originally envisioned as a key layer in iWARP - a wide-area RDMA-based transport discussed in more detail in Chapter 8. One of the main reasons for the slow pace of its adoption is the relative complexity of the SCTP protocol processing engine, compared to TCP or UDP-based protocols.

7.6 Chapter Summary

This chapter focuses on IP-based transport protocols, targeting high performance requirements of applications deployed in distributed environments over the public networking infrastructures. The focus is mostly on the most dominant transport protocol in the Internet today — TCP. We present several variants of the original TCP, addressing mainly the manner in which TCP flows utilize the available bandwidth and achieve maximum performance levels. Some of the variants specifically target high speed, long distance communication paths, often used by distributed scientific collaborations and large-scale, data intensive applications. We also point out several important implementation issues, targeted at removing various bottlenecks and sources of overhead in the TCP processing engine.

Chapter 8

Remote Direct Memory Access and iWARP

Dennis Dalessandro
Ohio Supercomputer Center
Pete Wyckoff
Ohio Supercomputer Center

8.1   Introduction

As modern networks become faster, and the demands for increased bandwidth and low latency become more prevalent, computational resources are unable to keep pace. The problem is the way in which networking is traditionally handled. It is inefficient, not only in terms of CPU utilization, but also in access to memory. Today's CPU and memory subsystem can handle networks on the order of 1 Gbps, but as network speeds reach 10 Gbps and beyond, problems begin to arise. The crux of the matter is that as networks become faster, so must the CPU in order to process the associated protocols. The rate of increase in CPU clock rate has fallen off dramatically in recent years, while the networking rate has continued to increase. Squeezing ever more cores onto a single CPU and adding more processors to motherboards helps to mitigate the problems, but only to a limited extent.

A relatively new approach to network communications known as Remote Direct Memory Access provides a mechanism to transfer data between two hosts with minimal CPU usage and very little overhead. Coupled with protocol offload, RDMA is well equipped to handle today's high speed networks, as well as the networks of tomorrow.

While one form of RDMA, known as InfiniBand, is commonplace in high performance computing, a new approach to RDMA known as iWARP is emerging as a top notch competitor. iWARP is quite simply RDMA, but over the familiar TCP/IP networking stack. iWARP provides all the benefits of RDMA but over a transport that is compatible with existing network infrastructure.

8.2   RDMA

Traditionally, network communication is conducted in a very inefficient manner. When a user application wants to send data, it must first be copied from user space buffers to kernel space buffers. The CPU is responsible for making this copy. In addition to copying data the operating system must also undergo a context switch in order to operate in privileged mode and use the network hardware. Once the data is copied into the kernel buffers, the operating system can process the functions in the network protocol stack, which is usually TCP/IP. This processing of the network protocol stack is prohibitively expensive in terms of CPU usage. The problem is that while the CPU is busy doing computations on behalf of communication it cannot be utilized for other work. This problem is made worse by the ever increasing network speeds. Currently 1 Gbps TCP/IP-based networks are commonplace, and 10 Gbps networks are beginning to appear more often. As we move to 10 Gbps and faster networks the amount of time the CPU spends doing network processing increases greatly, thus starving the real workload.

In order to combat this, a common approach known as offloading can be utilized. Offloading involves using a separate piece of specialized hardware to handle some of the work of the CPU. This is similar to the approach taken with math co-processors in the early days of the i386 architecture, and modern day graphics accelerator cards. It may seem that this is a solved problem with the increasing CPU count in multi-core systems; however, the addition of multiple processing elements only exacerbates the other troublesome and extensively studied architectural limitation: the Memory Wall [464]. The rate at which processors can access memory frequently becomes the compelling factor that justifies the use of offloading.

In general, the drawback of offloading technologies is getting data to and from the accelerating hardware. This is where Direct Memory Access, or DMA comes in. DMA has long been utilized in hardware such as disk drives. As the name implies DMA allows a device to have direct access to memory without needing to utilize the CPU.

Naturally utilizing these two techniques in tandem can make possible specialized hardware that gives the best possible performance. Extending these ideas to remote hosts via an interconnection network gives rise to what is known as Remote Direct Memory Access, or more simply RDMA. RDMA enables efficient access to memory of a remote host, as well as processing offload in order to conduct high throughput and low latency data transfers between two remote hosts. Some common examples of RDMA include iWARP (Section 8.3), and InfiniBand (Chapter 2).

With increased performance come certain drawbacks related to how applications are programmed. This stems from a change in semantic of how data

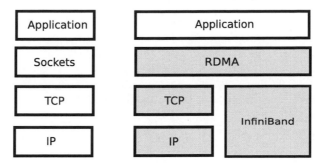

FIGURE 8.1: Comparison of traditional TCP network stack (left) and RDMA (right). Shaded areas indicate functionality typically handled in hardware.

is transferred. Instead of sending and receiving messages, applications now simply read and write each other's memory. But there is a significant amount of work that applications must do to take advantage of these functions.

Though the techniques involved in RDMA are not new in themselves, RDMA is a relatively new form of network communication. Due to the entrenched nature of TCP/IP and its associated API, transition to something new is extremely difficult. It is almost a certainty that the early adapters of RDMA will continue to be in sectors where performance advantages are key, such as high performance technical computing. Eventually RDMA may filter down into the broad consumer market, but it is a long way off from that. Further advances in standardization and library and programming support are needed.

8.2.1   High-Level Overview of RDMA

Even though RDMA can be utilized over TCP/IP, we will use the moniker "TCP" to refer to how most programs use the sockets interface and TCP/IP for their networking needs. Figure 8.1 shows the network stack comparison between RDMA and the traditional way in which TCP is used. If we consider the OSI network stack [211], then RDMA actually spans multiple layers; it is not a single entity that can be classified into any particular layer. This is a common trend in many modern networking technologies, though to aid in understanding, it helps to reason about RDMA in a layered context.

At the highest level, RDMA applications can be any network related application. However, there are significant enough architectural differences between RDMA and traditional TCP that make it difficult to change an application from TCP to RDMA. RDMA requires the use of calls that are somewhat unfamiliar compared to those used for traditional sockets-based networking, widely used by applications to implement TCP/IP and other communications. TCP/IP-based sockets programming is built around the idea of a bidirectional data stream, where writes to the network push more bytes into the stream, and reads from the network pull out whatever data has arrived. It is not necessary to match up the reads and the writes, and some applications will, for instance, read a header, then decide how much more data to read. RDMA is very much message oriented. The sender must gather up all the parts of a message to do a single send, and the receiver must commit to receiving the entire message before peeking at any of its contents.

RDMA has another mode in which the receiver does not even participate. With what is sometimes called *memory semantics*, the sender moves data directly to a memory region of the remote process and no receive call is ever issued. With sockets, both sides always need to actively participate. With RDMA the remote side does not need to actively participate in exchanging data — it is done by the hardware. This is why RDMA operations are often referred to as one-sided operations. This semantic difference between RDMA and traditional network communications is an important consideration. It means that existing applications that utilize sockets-based APIs can not always simply be ported to an RDMA API.

In both message and memory semantics on RDMA, data buffers for incoming and outgoing data must be registered in advance and associated with an identification *tag*. This memory registration process informs the hardware adapter of the buffer's physical location in memory so that the contents may be accessed directly using DMA. Memory registration is a costly operation, due largely in part to the translation between virtual memory addresses and physical memory addresses [107, 294]. Another requirement for the memory buffers is that they be *pinned*, or marked as invulnerable to operating system interference, such as being swapped out or rearranged to defragment memory for TLB or allocation reasons.

Below the application layer lie the RDMA protocols. The RDMA protocol layer consists of not only the high level protocol for exchanging RDMA messages, but also the protocols needed for directly placing data, i.e., moving it to the right user application without requiring an extra copy or help from the OS. The placement protocol ensures that all messages are self describing of their intended destination. This enables the RDMA hardware device to place data directly in memory as it is received.

At the transport layer, RDMA can utilize a specialized transport, as is the case with InfiniBand, or make use of an existing transport as is the case of iWARP using TCP or SCTP. It is important to realize that if TCP is used as a transport, another protocol is needed in the layer above the transport. Since TCP is stream-based, it does not have the idea of message boundaries that RDMA needs. Packets can be fragmented or merged or arrive out of order, making it difficult to determine where to place incoming data. The headers required to directly place incoming data might be incomplete or not even present when the data arrives, and buffering enough packets to handle potential out-of-order conditions is costly. This problem is solved via a fragmentation and marker protocol that sits on top of TCP and adapts it to provide the message boundaries that RDMA needs. In the case of a network protocol like SCTP which already has message boundary support, this adaption layer is not needed. Since some networks like InfiniBand utilize their own transport they too do not require a fragmentation and marker layer.

Similarly to the transport layer, the network and data link layers for RDMA can utilize IP and thus Ethernet or something else entirely. Thus in the case of iWARP which uses TCP/IP and hence generally Ethernet, it is able to take advantage of existing networking hardware infrastructure. iWARP RDMA traffic can be switched and routed, and can use existing Ethernet cabling. Networks such as InfiniBand use their own special networking hardware, but often use the same cables as Ethernet, at least in the CX4, 10 Gbps variety. InfiniBand switches have the advantage, though, that they can make performance modifications that Ethernet switches can not, such as reliable hardware multicast, although both are able to take advantage of cut-through switching. Current InfiniBand hardware is not directly routable over existing commodity networks, although hardware exists to allow wide-area connections on dedicated fiber [313] and features of the specification do describe an approach to routing.

8.2.2 Architectural Motivations

Since TCP/IP are complex protocols, the traditional TCP networking stack is quite computationally intensive to process. This means that as networking rates increase to 10 Gbps and beyond, an increasing amount of CPU power is required to keep up with the network device and its appetite for CPU cycles. Thus while the CPU is busy doing network processing, it is unavailable for other processing activities. This problem is exacerbated by advances in programming languages which do not require software developers to be concerned with such architectural details. It is common in parallel programming to write code that operates on multiple processing entities in parallel, but no thought need be given to how those processes break up into multiple, physically separate compute hosts.

By offloading the network protocol processing, or portions of it, it is possible to alleviate strain on the CPU and vastly improve performance. As mentioned previously, this is the same principle as using a graphics accelerator, or a math co-processor in older PCs. However, there still remains a problem with the way the offload device gets data to operate on.

Even with utilizing multi-core systems or other methods of offloading the expensive protocol processing, access to memory may still be a bottleneck. As processor speeds increase, and increase in numbers per system as well, the memory wall becomes an increasing problem. This refers to the fact that the speed of memory does not increase as fast as the need for that memory. In other words, CPUs are getting faster at a greater rate than memory is. The net result of this is that for the CPU to access memory simply takes longer

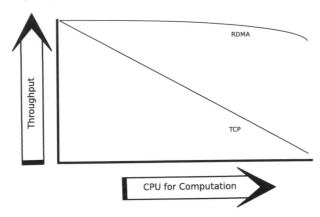

FIGURE 8.2: Effect of overlapping computation and communication.

compared to how fast the CPU can operate. When the number of cores or CPUs is increased this problem simply gets worse.

Another consequence of offloading is the ability to overlap computation and communication. In many compute-intensive applications that also require heavy or frequent communication with peers, having a NIC do some of the work to manage the network frees up the CPU to continue with the calculations at hand. This is enabled by having assistive processing engines on the NIC, but also by the use of APIs that permit and encourage overlap. The common OpenFabrics API offers only asynchronous "post send" and "post receive" calls to initiate transfers, with a separate "poll for completion" call to detect completion. Figure 8.2 illustrates the effect of overlap between communication and computation. Since TCP requires the CPU for all of its functions, as soon as portions of the CPU cycles are used for other work (computation), TCP performance in terms of throughput begins to degrade. The more CPU used for other tasks, the worse the performance of TCP. RDMA on the other hand, which does not need CPU to transfer data is mostly unaffected. The RDMA curve does decrease in throughput a small amount when the CPU is maxed out for computation. This is because CPU is still needed to run application codes, manage transfers, and post descriptors. Posting descriptors and checking for completion are relatively fast, but memory registration is CPU-intensive [107]. This can be alleviated by using pipelining, where a message is broken into smaller pieces, and registration for future chunks is overlapped with sending of current chunks. Using RDMA write operations, the receiver of the data need not be aware that this operation is being used on the sender side. More discussion on memory registration follows in Section 8.2.6.

Aggravating things further is the memory copy problem. In order for a user application to send data via TCP sockets, that data must reside in a kernel buffer. In order to get data into the kernel buffer, the CPU must copy data from user space to kernel space. Not only does this cause slow

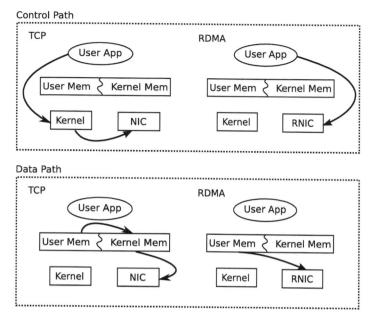

FIGURE 8.3: TCP and RDMA communication architecture.

downs due to the memory copy, but there is also a context switch that must take place. The running process enters privileged mode in the kernel, saving process state and restoring it again when the kernel is finished. Although context switches have become much faster with modern processors, they still appear as significant overheads when considering high-speed communications, partially due to indirect effects such as instruction cache misses and inter-core communication for acquiring spinlocks.

The same memory copy work is necessary on the receive side, where the kernel must buffer incoming data, and then, upon request, copy it to the user buffer. Certainly it is possible to avoid some of these copies for kernel-level services, and specialized operations such as *sendfile()*, but the basic problem of the CPU accessing memory in the critical networking path still exists.

Figure 8.3 shows the basic architecture of traditional TCP compared to RDMA, be it iWARP or InfiniBand. The label *RNIC* in the figure is used to represent a generic RDMA network device and could easily represent an HCA. In non-RDMA communications, data is sent by the initiation of a system call that traverses the user-kernel boundary. The kernel will then copy data out of the user buffer and into its internal buffers for protocol processing. As the TCP/IP protocols are processed, data can be handed to the NIC to be sent. The kernel is responsible for a large amount of processing and bookkeeping. And the kernel memory can not be freed until it has acknowledgment from the receiver that the data was successfully delivered, tying up kernel resources until the transfer is done. With RDMA things are streamlined: the user application will call directly to hardware without the aid of the operating system. The RDMA hardware then directly accesses the physical memory that the user's buffer resides in. There is no need to copy any data into the kernel. After all, the kernel is not even involved in the sending and receiving of data. The RDMA device takes care of all the protocol processing as well.

8.2.3   Fundamental Aspects of RDMA

Often RDMA is described as being made up of two components. Actually it may make more sense to consider RDMA as consisting of three main components. Zero-Copy and OS-bypass are the two commonly referred to mechanisms by which RDMA operates. The third component, protocol offload, is often lumped in with OS-bypass. However, OS-bypass and protocol offload are in fact two very different ideas.

Zero-copy.   Zero-copy refers to the capability to send and receive data without making copies. This means that data is sent directly from user space. As a consequence, care must be taken to ensure that user buffers stay resident in memory and do not get swapped out by the OS. This can be challenging to integrate in a fully-features virtual memory system of modern OSes, but the use of memory "locking" is fairly straightforward. More difficult is to enable zero-copy on the receive side. In order to place data into memory as it arrives, RDMA operations must contain sufficient information to tell the remote side where to put data.

However, send/receive operations do not provide any indication of where to actually place data. Hence there is a requirement to pre-register all receives. The receive registration tells the adapter where to put incoming data, looking at some tags in the incoming message to decide where each packet should go. It is important to note that there are other examples of zero-copy network communications, in fact the commonly used *sendfile()* system call on Linux is a zero-copy transmit operation. While zero-copy is essential for RDMA, it does not by itself constitute RDMA.

OS-bypass.   Another main component of RDMA is operating system bypass. OS-bypass refers to the capability for user applications to have direct access to hardware. In other words to send data, the kernel is not required to intervene on behalf of the user. Normally the kernel would need to be involved for user space applications to interact with hardware, to ensure integrity of the system and privacy of other applications' data. But, for properly designed peripherals, it is possible for the user application to directly access the hardware device. The operating system arranges for certain sections of the memory space of the device to be mapped into the address space of the user application. System libraries provide the necessary functionality for user space applications to use these memory mapped regions. Some areas of device memory act as triggers, or *doorbells*, with which an application can signal the device with very low overhead, for instance to start a new send operation. Certain aspects of NIC functionality require cooperation between the application, device, and operating system. Memory registration is a privileged operation initiated by the application but carried out between the OS and the device. The result of a registration is a "handle" that is provided to the application, which can then use it in future communication with the NIC to reference the registered memory region.

Protocol offload. Finally, RDMA requires protocol offloading. It may seem tempting to label this as just a form of OS-bypass; however, OS-bypass refers to the capability to interact with hardware without the intervention of the operating system. Offload is taking things a step further and moving work, such as protocol processing, to a specialized piece of hardware. This includes RDMA protocol processing as well as transport layer protocol processing. By enabling the hardware to take care of the network processing, the CPU is available for other purposes. The overall system performance is no longer impacted in a major way by the network device. Naturally, offloading network processing requires special hardware. One of the earliest attempts at solving the network processing problem was what is known as TCP Offload Engines, or TOEs. These devices handled all protocol processing, basically supporting the protocol offload concept. However, TOE cards, as they are commonly called, do not address the zero-copy aspect. As such, their performance is somewhat less than RDMA devices. A TOE card still requires data copies from user space into the kernel. Despite these drawbacks, TOE cards can offer a significant performance increase over traditional TCP network devices, using just the protocol offload concept.

8.2.4 RDMA Historical Foundations

One of the earliest precursors to RDMA was a project developed at Cornell University under the name U-Net. U-Net [446] was one of the first attempts at a user level networking interface. Realizing the performance degradation when the operating system handles network processing, U-Net opted for user level access to hardware. This enabled U-Net to be one of the earliest attempts at zero-copy network communications. Since the network processing was done in user space, there is obviously no need to copy data into the kernel. U-Net was one of the first works to introduce the notion of separate send and receive queues that work requests can be submitted to. This concept still exists today in modern RDMA devices.

Another direct milestone in the development of RDMA was the Virtual Interface Architecture, or VIA [448]. VIA was a joint project between industry groups such as Intel, Microsoft, Compaq, and many others. VIA sought to standardize programming of high performance networking by creating a verbs specification. A verbs specification abstractly defines an API such that it outlines the functions and parameters without exactly defining them. Thus multiple APIs can conform to a single verbs specification. VIA served as the basis for the early InfiniBand and iWARP verbs specifications.

Early adopters of what is today recognizable as RDMA were companies like Myrinet and Quadrics, followed soon by InfiniBand architecture (see Chapter 2). Both Myrinet and Quadrics enjoyed success in the RDMA networking market for some time, though in recent years InfiniBand architecture has carved out an increasing presence in the market, including standard support in the Linux kernel.

While these RDMA based architectures were geared for non-TCP types of applications, advances in ordinary NICs were taking place as well. The movement from protocol processing being done by the operating system to processing the protocol stack on the NIC was becoming common. Early attempts at TCP Offload Engines saw good performance improvements, but were ultimately limiting in performance due to memory copies. Despite the shortcomings of TOE cards, the concepts were an important step in the evolution of RDMA. In fact, today it is commonplace to offload certain functionalities to the NIC, such as segmentation. TCP offloading concepts, coupled with the success of RDMA networks, soon gave rise to what is known as iWARP. iWARP is simply RDMA with one important characterization. iWARP uses standard TCP/IP as its reliable transport layer. For a detailed explanation of iWARP see Section 8.3.

8.2.5   Programming Interface

RDMA architectures require a new way of thinking about application programming in order to achieve the best performance. The primitive objects and actions in RDMA are sufficiently different from the familiar TCP ones. Also, not all applications can simply use an RDMA transport without major architectural changes. Sockets based applications always require both sides to participate in the transfer: one sends and the other receives. While under the hood the OS kernel handles physically receiving data, the user must utilize the *read()* system call in order to tell the OS where to put the data, after it has been received. With one-sided RDMA operations, data ends up directly in buffers without any further action required by the user application. These functions seem more like reading and writing memory in an SMP than sending and receiving data over a network.

The actual programming interface is generally adopted from VIA [448], which was just a messaging model and API, but specified no particular wire-level protocol. Two different VIA devices could use the same software, but could not talk to each other. With InfiniBand, a committee specified all details of the wire protocol, down to voltage levels and packaging, but could not agree on an API. Hence the idea of "verbs" to specify the actions of how an application and operating system would interact with the device was introduced. Aspects like the actual names of function calls, organization of data structures and so forth were left to each manufacturer. Each InfiniBand vendor thus produced its own API; and, while different devices could in theory talk to each other, a different program was required to communicate with each device.

As Mellanox [283] came to dominate the market for InfiniBand silicon, its Verbs API, or VAPI, became the *de facto* standard for InfiniBand. iWARP also adopted the verbs approach of InfiniBand — in fact the original iWARP verbs [364] were based loosely on VIA and VAPI, though there are a few important distinctions, such as implied memory access permissions.

In order to solve the issue of unique APIs for every device, the Direct Access Programming Library, or DAPL [115] was developed. DAPL enabled code to be written in its API and compiled on any RDMA architecture that supported DAPL. Another attempted solution was to support a sockets API which mapped to the individual RDMA API. This approach, known as Sockets Direct Protocol [364] (SDP), achieves reasonable performance only in very particular use cases of sockets, largely due to the fundamental semantic mismatches between RDMA and sockets methodologies. There have been multiple implementations of SDP over various transports from academia, and SDP is available in both the OpenFabrics Linux distribution and as part of Windows Sockets Direct.

Another common approach is to simulate an IP transport over an InfiniBand network, called IPoIB. This approach enables any TCP application transparently to take advantage of a faster IB transport. The conceptual problem with IPoIB however, is that an application will use the in-kernel TCP/IP stack to produce individual packets. These packets are then handed to an IB driver, which uses its own reliable transport to move them to a remote host. The remote host then hands the packets to its TCP/IP stack to turn them back into messages for the application. This introduces extra transport processing and multiple memory copies.

As RDMA interconnects gained partial acceptance, it was clear that there needed to be a common software stack to support all RDMA devices. Device vendors (correctly) perceived that the market would not accept yet another custom device with its own API. Thus they banded together, under duress from some of their biggest customers, to form the OpenIB consortium. Their goal was to be the open source initiative to support InfiniBand in the Linux kernel. As other non-IB adapters (such as iWARP) began to appear the OpenIB consortium became OpenFabrics. The goal of OpenFabrics is now to support *all* RDMA devices in the Linux kernel. The common OpenFabrics software stack is layered on top of a vendor-specific driver which is included in the kernel as well. The core stack enables a common API amongst all RDMA devices that support OpenFabrics. Now it is possible to write code for one device and have it work on all RDMA devices be they InfiniBand, iWARP, or other.

Even though there exists operating system support, and a common API, there are still drawbacks to writing applications for RDMA networks. A number of operating system mismatches (explored further in section 8.2.6) imply that simply porting existing network applications is not straightforward.

To date, the use of RDMA technologies by applications has been comparatively limited. Certain regimes, such as message passing in high-performance computing (HPC), have adopted RDMA vigorously. This is made relatively easy by the fact that most HPC applications use the same library to coordinate parallel processes, called MPI. By porting MPI to use RDMA, a large class of HPC applications directly take advantage of RDMA as well.

Enterprise and high performance storage is another area where RDMA has seen great successes. In block storage, iSER and SRP are the commonly used protocols to transmit SCSI commands across RDMA-capable networks. iSER [232] is short for "iSCSI extensions for RDMA." It enables traditional storage deployments utilizing iSCSI over TCP to make use of a higher performance RDMA interconnect such as iWARP or even InfiniBand. In fact the iSER initiator library is present in current day Linux kernels. There are also open source targets available [104]. In addition to iSER, support for SCSI RDMA Protocol (SRP) is available and has enabled InfiniBand to make further inroads in the block storage arena. In addition to the block level, RDMA has enabled performance boosts in transport protocols of particular distributed file systems. Lustre [95], PVFS [461, 476], and even NFS [74] are now supported to run natively over RDMA.

Other specific applications have also embraced RDMA, but in general purpose computing, RDMA is almost nonexistent. It is true that many applications would not see any benefits from using RDMA, especially if their communication requirements are minimal. But there are many that would, especially in the burgeoning web-based service computing field. With iWARP it is now possible for web-based and other server oriented architectures to take advantage of RDMA, even if clients are not so equipped (see software iWARP [102, 103]). Another potential area of impact for RDMA is in peer-to-peer type networks and applications. One such example is in helping to launch massively parallel jobs. Peer-to-peer concepts can be coupled with RDMA and result in distributing executable code to many computational nodes much faster.

Currently, uptake of RDMA in today's applications is certainly hampered by a usage model that is somewhat new and harder to master. However, as network speeds continue to increase faster than other computational resources, the benefits of RDMA will likely become worth the extra effort.

8.2.6   Operating System Interactions

One of the major complicating factors in using RDMA networks is the need to register memory. Before buffers can be used, they must be pinned down in the kernel, and translated to physical addresses to hand to the network card. The cost of registration increases with the size of the buffer and is significant:

for a 1 MB message, InfiniBand 4X SDR achieves a transfer rate of roughly 900 MB/s, but when registration is included in the timing loop, this drops to around 600 MB/s [466].

Memory registration. Memory registration can certainly be done by the application, either statically for buffers that are used for communication multiple times, or dynamically for each new transfer. Caching of frequently used registrations is often done in libraries such as MPI. Techniques to pipeline registration with communication are also possible [107]. But the operating system services to register and deregister memory for RDMA are not at all integrated with the rest of the memory management systems. This has the potential to cause all sorts of interesting problems and unexpected behavior [466].

The standard POSIX system calls, `mmap` and `munmap`, are used generally to manipulate the virtual memory space of a process. System libraries (such as libc) call these functions as well as applications. Other calls can manipulate the virtual memory space too; `fork` can mark existing regions as "copy on write," whereby future access of the memory will cause the OS to copy the memory page to a new free page. Not all of these calls honor the presence of existing memory registrations. Changes in the VM layout of the process will result in different virtual to physical translations, but the RDMA network card will still have the old registrations. In Linux, at least, per-page reference counts ensure that a page will not be reused until all registrations are explicitly undone, causing at worst memory leaks but not security-sensitive writes to arbitrary physical memory. But inside a single application, the result of RDMA transfers may read from the wrong pages, or write into an area that is not visible by the process. This problem is worsened by the presence of memory registration caching in libraries, where application map/unmap behavior is not communicated to the library, causing it to reuse a now-invalid registration cache entry.

Interaction with existing APIs. Another common stumbling block is in the mechanisms used to initiate communication between two machines. The socket calls are used for most TCP or UDP communication today. Originally, RDMA libraries had their own custom and unfamiliar connection routines. Lately, OpenFabrics has a connection management interface that resembles the socket calls, using the idea of listen, connect, accept or reject; however, there are extra steps and different states to consider. Producing a correct application that handles all possible connection management states and errors is complex.

The most frustrating aspect of connection management is that currently the RDMA port space is completely separated from the in-kernel TCP port space. For non-RDMA applications, the OS enforces that each port is used by no more than one application at a time. But there is no way to prevent an RDMA application from listening on the same port as a non-RDMA application. This leads to interesting bugs, and prevents a common usage model for iWARP devices. Portable server applications would like to listen for a TCP connection to a well-known port, and if the connecting client sends the right handshake indicating it can use RDMA, respond with RDMA, otherwise fall back to non-RDMA communications. The work-around adopted by many applications (including NFS-RDMA) is to use a separate port number for the RDMA version of the protocol.

A pervasive point of mismatch is that RDMA devices do not present file descriptors to the system. Most other modalities of communication in Unix revolve around the manipulation of file descriptors: network sockets, file I/O, keyboard and mouse input, inter-process communication, and others. RDMA communications use a Queue Pair and Event Queue abstraction, and do not support the usual commands of "read" and "write," but worse, do not support the use of "select" or "poll."

A standard application design pattern is to use a single blocking `select` call to determine what to do next: read from the network, write to the disk, or respond to another thread, for instance. This lets the application sleep when there is nothing to do, and relies on the operating system to wake it up, using the common file descriptor methods for all its communication needs. With RDMA, non-blocking polling must be used for fastest access to the network, but a blocking interface is also available. It has been designed to present a file descriptor, which is good, but the semantics are not quite the same.

Another common tweak on the select loop application model is to pay attention to file descriptors selectively, for instance, if there is no data ready to send to the network, do not poll waiting until the network is writable. This method directly connects the flow control mechanism of the network back into the main processing loop of applications in a very natural way. File descriptors for RDMA event notification can not be selectively activated like this; they only indicate that "something happened," and require inspection and manipulation of the RDMA device state to determine what happened and how that impacts flow control requirements of the application.

Another handy feature of the file descriptor abstraction is that it represents a single in-kernel data structure by which all aspects of the communication channel can be accessed. RDMA, on the other hand, represents state in many ways: queue pairs, event queues, protection domains, registered memory regions. Each of these items is addressed by a separate handle, just as a file descriptor is a handle. Unlike file descriptors, though, RDMA handles are specific to the entity that created them, such as a single user application or the operating system itself. The upshot of this is that common semantics for sharing file descriptors among processes, or between the kernel and an application, are not possible with RDMA devices. The reason for this is related to how some aspects of RDMA communication are offloaded to the NIC, and an application can directly talk to the NIC without the central mediator that is the operating system. In practice, this aspect is most often encountered when trying to use RDMA state variables in a process that has been `fork`-ed from the one that created the RDMA state. Passing file descriptors through Unix-domain sockets is of course not possible. And the Linux-specific `sendfile` and `splice` system calls have no parallels for RDMA [107]. For instance, it was shown that RDMA actually adds additional overhead when shipping file data to clients [106], such as is done by a web server. The reason is that in order to do so, the web server needs to copy data from the kernel, which is in charge of reading file data from disk, into user space, where data will be sent to the network. However with a sockets based API, there is a *sendfile* mechanism, which allows the operating system to transfer a file directly from the page cache, omitting the requirement for the web server to read in the file to userspace. RDMA can not use this feature, and thus an *extra* copy is required when using RDMA devices for serving files. The important concept to remember is that just because the network communication is more efficient, hidden applications considerations can sometimes result.

Despite the drawbacks due to the mismatch between queue pairs and file descriptors and memory registration requirements, RDMA has had great success. The de facto standard parallel programming interface, MPI, has successfully utilized RDMA as a communication mechanism. MPI over RDMA is in use on some of the biggest and most powerful computational clusters in the world. Not only is the implementation of MPI efficient over RDMA, there are MPI operations which are themselves one sided, and as such see great performance gains with RDMA. RDMA is an absolute necessity in many applications, particularly in the high performance computing arena where every bit of CPU power is needed for computation. So far the front-runner for clustering interconnects has been InfiniBand due to its widespread availability, low cost, and periodic releases of faster hardware. Numerous iWARP devices are available and iWARP technology may likely be commonplace in future high-speed Ethernet adopters, riding the curves of continued Ethernet market dominance and shrinking ASIC feature size.

8.3   iWARP

RDMA is a beneficial technology in that it can greatly increase performance in terms of lowering latency and increasing throughput. One of the drawbacks to, RDMA, however, is that it has in the past been relegated to non-traditional networks. In other words RDMA networks such as InfiniBand have not been built on TCP/IP. This means such RDMA networks are not compatible with the existing networking infrastructure commonly found in most environments. This can, and certainly has, hindered the adoption of RDMA. This is also the reason that RDMA is only just now beginning to make inroads in the enterprise world. TCP-based RDMA adapters have the ability to make the transition to RDMA technology easier, in that it can use large amounts of existing hardware and software. This is where iWARP comes in.

8.3.1   High-Level Overview of iWARP

iWARP is most simply described as RDMA over TCP/IP.[1] Originally the term iWARP did not stand for anything. However, a number of names attempting to fit the acronym have been thrown around, such as Internet WARP, or Internet Wide Area RDMA Protocol, and the like. Despite the confusion about the name, iWARP is beginning to gain ground in the commodity cluster and enterprise server market.

Clearly one of the most important features of iWARP is that it is not a new technology in itself. iWARP is simply taking the performance enhancing characteristics of RDMA-based networks like InfiniBand, and making them available over the common TCP infrastructure. The net result is a network that has performance comparable to InfiniBand in terms of latency and throughput, running over the most popular network transport. iWARP aims to be a central player in the RDMA market in the near and foreseeable future.

As mentioned previously iWARP is simply RDMA over TCP. iWARP includes all three fundamental components of RDMA (see Section 8.2.3). It is important to realize the difference between iWARP and simple TCP Offload Engines, or TOE cards. In both cases the underlying network is TCP based. Also both iWARP and TOE fully offload the TCP/IP protocol stack to the NIC, thus reducing CPU load. However, iWARP addresses the problem of memory copies and direct access to hardware. With iWARP, data is able to be moved to and from the card without the need to make intermediate copies. TOE cards still require data to be passed through the kernel, hence introducing copies. iWARP devices are also able to be directly accessed from user applications. TCP offload engines rely on the kernel to communicate with the network device. And perhaps worst of all, TOE cards still suffer from the *read()* socket semantic on the receive side, which always inserts an implicit data copy.

iWARP is basically taking all the things about RDMA networks such as InfiniBand and doing them over ordinary TCP/IP. iWARP, being TCP based, is able to take advantage of the current network infrastructure. Ordinary Ethernet switches and routers are able to be used with iWARP. 10 Gbps iWARP uses ordinary 10 Gbps capable Ethernet cables, such as CX4. Actually InfiniBand also can use the same CX4 cables, but InfiniBand is not compatible with the Ethernet switching and routing infrastructure, nor with any of the associated software services such as ARP, SNMP, or DNS. Thus InfiniBand can not be compatible with iWARP, although they share many characteristics.

---

[1]Note that the historical parallel computing platform, iWarp [172], is completely unrelated.

iWARP devices have the other nice feature that they also tend to function as TOE devices, or as ordinary Ethernet cards, which means that an iWARP card need not be an additional network device in a machine, but can serve as the single device for all Ethernet communications.

8.3.2   iWARP Device History

The iWARP story begins with the RDMA consortium [364]. The consortium was started by a group of industry leaders and its goal was to craft the early wire and hardware specifications for RDMA over TCP. These early specifications were quickly codified in multiple IETF RFCs. This aspect is explored further in the following sections. Aside from the RDMA protocol specifications, the RDMA consortium came up with an RDMA verbs interface that was meant to serve as a guide for APIs to be developed by the individual device manufacturers. Like their InfiniBand counterparts, iWARP vendors recognized the advantages of agreeing on a single API and abandoned their early RDMA Consortium verbs efforts for the common API of the OpenFabrics project. The OpenFabrics project is the ongoing collaboration to support all RDMA devices in the Linux kernel under a common software stack.

The first commercially available iWARP product was a 1 Gbps offering from Ammasso. The Ammasso 1100 was a standard Ethernet device that managed to drastically reduce latency by avoiding memory copies and offloading the TCP stack to hardware. Being an FPGA-based board though, the Ammasso 1100 was quick to market but somewhat lacked luster in overall performance. Compounding the problem further, the Ammasso 1100 was competing with the likes of InfiniBand which were already at 10 Gbps and cost roughly the same, if not less, per adapter as the Ammasso card. The only real advantage of the Ammasso card was in the switching and wiring. The Ammasso device could use ordinary Ethernet switches, common in any data center. InfiniBand naturally required new, and at the time, very expensive switches and cables, so there was a trade off between price and performance. In the end the price point for the Ammasso 1100 compared to its limited performance gains was too great and Ammasso eventually went out of business.

At the time the Ammasso 1100 was being sold, a 10 Gbps iWARP offering from a company called Siliquent was being tested in a limited capacity. The company was bought out by the much larger networking company Broadcom in 2005, and the iWARP device developed by Siliquent has, as of yet, never made it to market.

It was not until some time later that rumors of the next iWARP devices started to fly. This is where NetEffect and Chelsio came into the fray. These companies both saw the value in a TCP-based RDMA product, but also realized it was absolutely vital to be comparable in performance to InfiniBand. Following the lessons learned by Ammasso, both NetEffect and Chelsio went right into making 10 Gbps iWARP devices. While widespread adoption has

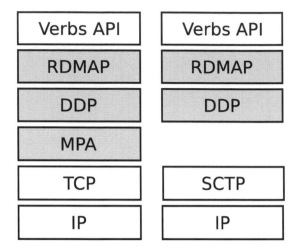

FIGURE 8.4:   iWARP protocols stack.

been somewhat slow, both NetEffect and Chelsio are currently still actively selling products, and leveraging their iWARP business with TOE devices in the case of Chelsio. In fact, both NetEffect and Chelsio are supported in the Linux kernel via the OpenFabrics project. Recently, other vendors have made public their intention to sell iWARP products as well.

8.3.3   iWARP Standardization

There are three main specifications that lay out how iWARP actually works. What started out as a single document of the RDMA Consortium eventually became multiple layers and documents of the Remote Direct Data Placement (RDDP) working group, part of the Internet Engineering Task Force (IETF). Figure 8.4 shows the most important of these standards: RDMAP, DDP, and MPA. After much community involvement, eventually these all became official IETF RFCs.

At the bottom of the protocol stack is the Marker PDU Aligned Framing layer, or MPA [101]. The main goal for MPA is to convert the stream-based TCP into a message-based transport needed by the next higher layer, DDP. Since TCP packets are subject to fragmentation and arriving on the host out of order, steps must be taken to find the upper layer protocol (DDP) record boundaries. This is essential for DDP to be able to determine where to place data. One possible approach is to embed in the stream a known value that can be detected; however, as it is possible for this magic value to be replicated in payload data this is not a good approach. MPA uses what is known as a deterministic framing approach and determines the DDP boundaries by looking at specific areas in the stream. MPA, however, does not specify how two hosts connect or agree to use MPA. The MPA layer takes data from the upper layer, DDP, and places a header, possibly markers, and a more careful checksum field around the payload. This is then passed on to TCP for transmission. Markers are not mandatory, but are required to avoid retransmission if packets are dropped or reordered. The checksum, a CRC32C that is required by MPA due to shortcomings in the 16-bit TCP checksum, is not absolutely mandatory. While markers will only be used if both sides agree, CRC must be used if either side wants, thus all implementations of MPA must at least support CRC. The MPA layer is only necessary when the underlying transport is TCP. iWARP architecture documents also specify how to use SCTP [414] as the reliable transport instead [43]. In this case, SCTP handles the framing and CRC issues and an adaptation like MPA is not needed. Despite the better fit with SCTP, its use is not widespread and support for SCTP is not guaranteed on all networks.

Above MPA is the Direct Data Placement (DDP) layer [393]. The main job of DDP is to place data directly into memory without the need to make copies, using information from the transport layer as well as in the incoming packets. DDP provides two buffer models. One is the named, or tagged buffer. This allows a remote host to send data to a specific buffer on the other side of the connection via a 32-bit steering tag, STag. The other buffer model is the unnamed, or untagged buffer. This requires that the recipient make available an anonymous buffer to receive data prior to that data actually being received. Both of these buffer models are reliable. DDP carefully separates the ideas of data placement from data delivery. Data placement involves figuring out where the data should go, while data delivery is the notification to an upper layer that all the data has arrived. This separation permits reception of data segments in any order and avoids buffering on the network card.

The other main specification is the Remote Direct Memory Access Protocol (RDMAP) [365]. There are four main architectural goals for RDMAP. Two of the most obvious are to provide applications with read and write services, while facilitating a zero-copy and operating system bypass approach. RDMAP provides the capability to read and write up to 4 GB of remote host memory in a single operation. The third architectural goal is to provide the mechanism to send data to a buffer that has not been explicitly advertised. However, in order for the send to be successful, a matching receive operation must have been posted on the other side. The fourth architectural goal of RDMAP is to enable the use of the send operation to inform the remote side that all previous messages have been completed. RDMAP also provides for all operations to be reliable, and in the case of multiple data streams per connection, for each stream to operate independently.

More detailed explanations of these protocols are available in the respective RFCs available from the IETF. There also exists an RFC for dealing with security issues related to iWARP [341], and a general applicability guide [42].

8.3.4   Trade-Offs of Using TCP

Since iWARP operates over traditional TCP/IP based networks, which includes Ethernet, there is a specific advantage over InfiniBand. This advantage is known as *single-sided acceleration*. The easiest way to understand this is by example. Suppose a web server is outfitted with a hardware iWARP device (RNIC). It is possible for this server to take advantage of its local device, reaping the benefits of RDMA even if the clients it is talking to do not have hardware iWARP devices. The clients can use ordinary Ethernet NICs and emulate the iWARP protocols in software [102, 103]. While the clients expend a bit more CPU effort to run the iWARP protocol stack on top of the TCP/IP stack they already were using, their effort enables the web server to save a lot of effort and thus serve more clients [106]. The web server does not have to know that it is talking to software-based clients.

The other benefit of being TCP based, and likewise Ethernet compatible, is that it is a very well understood technology. Management of Ethernet and other TCP networks is standard practice and well supported. A possible future adoption path for iWARP is that it simply becomes standard hardware in all Ethernet devices, even inexpensive built-in devices on motherboards. Applications can choose to take advantage of the hardware feature, or use the device as a bare Ethernet NIC.

There are downsides to being TCP based, however. One such drawback is that TCP is not always the most efficient choice for a network transport. TCP is widely deployed and tuned for an Internet scale environment, with some packet loss and possibly considerable congestion. For other networks these flow and congestion control algorithms are unnecessary and limit performance of the transport. In a machine-room scale cluster, for instance, communication patterns may be used where congestion will never be an issue, and "slow start" is completely unnecessary. Also in very high speed and long distance networks, TCP is overly conservative and can not achieve full throughput. Some modifications to the venerable deployed TCP stacks can alleviate the problems in both these situations, though.

Cost is an undeniably important factor to the success of a technology. Currently 10 Gbps Ethernet is fairly expensive. In the last few years, however, the price of 10 Gbps Ethernet has fallen dramatically, especially when it comes to switch infrastructure. InfiniBand is currently cheaper than 10 Gbps Ethernet in terms of total cost to deploy; however, history has shown us that despite the high initial cost of Ethernet technology it will steadily fall as it becomes adopted. Ethernet is and most probably will always be the dominant form of interconnection network for the local area. This is backed up by the fact that former specialized RDMA device companies have recently moved to begin selling Ethernet based products. Two examples of this are Myrinet and Quadrics, with their relatively low cost per port 10 Gbps Ethernet switches.

8.3.5 Software-Based iWARP

As introduced in the previous section, since iWARP is based upon TCP, it is natural to implement the iWARP protocols in software, leveraging existing TCP/IP stacks. In fact, this has been shown to work in both user [102] and kernel [103] mode. While these examples of software iWARP are based on an unmodified TCP stack, it is also possible to modify and tune the TCP/IP stack in order to provide better performance.

Aside from the single sided acceleration benefit that software iWARP provides, there are other benefits as well. Since most network based applications are currently TCP/IP sockets oriented they require significant code changes to support RDMA networks. Software iWARP enables these codes to be written and tested without the need to purchase additional hardware. A software solution is also useful as a research vehicle for future work and tuning of the protocols. Finally, an open source, software based, solution allows testing of hardware devices to ensure interoperability between devices.

Validation that using software makes sense can be found in historical examples. Qlogic (originally Pathscale) offers an InfiniBand PCI-based NIC that is essentially a "dumb" packet-pushing engine on an IB physical layer. The original device had no DMA support, requiring the host CPU to push all data to and from the NIC. It also supported essentially none of the IB protocols: the host is responsible for all packetization of data and framing. A large amount of kernel-resident software is required to use this device, but it saw great success in spite of its shortcomings. The simpler hardware allowed very low latencies, high message rates, and low cost. The adapter worked exceedingly well for applications that had extra CPU cycles to spare and did not make much use of computation/communication overlap.

Another example is the on-again off-again relationship with TCP offload engines (TOEs). Today one generally thinks of iWARP devices as being appropriate for hardware implementations, but that TCP is naturally implemented in an operating system, not in a device. A few decades ago and persisting up into the 1990s, hardware-based TCP accelerators were common and a required feature for high-throughput communications. With the relentless progress in CPU clock rates, the host processor became able to drive the network just as well, making TOEs largely obsolete. Recently, at the introduction of 10 Gbps Ethernet devices in the early 2000 time frame, standard machines were incapable of achieving much more than 3 Gbps of throughput, opening the window for TOE again. Fast forward another half decade, and current multi-core hardware can sustain 10 Gbps without much trouble. Some of this comes from adding minimal, but significant, chunks of offload capability to traditional "dumb" NICs. However, 40 Gbps and 100 Gbps devices are on near-term industry road-maps and may well require fully offloaded devices to use effectively, at least at first. The debate of specialized hardware vs. software on a general purpose processor will likely continue [296].

8.3.6   Differences between IB and iWARP

Other than architectural differences, InfiniBand and iWARP also differ in a few other areas. Since iWARP and InfiniBand are two of the more popular RDMA implementations, it is interesting to compare their differences.

One of the fundamental differences between iWARP and InfiniBand is in routability. iWARP being based on TCP, and subsequently IP as well, means that iWARP is just ordinary TCP/IP traffic and can thus traverse routers and switches in the same manner as any other TCP stream. InfiniBand however, rather than relying on TCP and IP, utilizes its own transport and network layers. According to the specification, InfiniBand is routable, in the sense that InfiniBand packets are compatible with IPv6 [212]. However, InfiniBand cannot effectively be deployed with existing TCP traffic due to fairness and congestion control issues. There have, however, been successes of deploying InfiniBand in the wide area, but this requires significant hardware investment and a dedicated fiber connection between end points [313].

Another important difference is in implied access to memory regions. Since RDMA offers exposure to host memory by remote machines, memory protection is a central point in RDMA architecture. Before memory can be used to transfer data, it must be registered. Registration provides a means to pin data as well as set the access control that local and remote peers have for the memory area. In both iWARP and InfiniBand memory regions can be marked as readable and writable by remote peers. However, when it comes to specifying local access permissions, iWARP and InfiniBand diverge. With InfiniBand, local read access is always assumed. In other words any, and all, memory regions are available to read from locally. This is in contrast to iWARP, which prohibits local read access unless the memory region is created with local read permissions. Note that access to the underlying buffers through the usual system call interface is not impeded in any way. The permissions of the memory region merely dictate access for operations that utilize an STag or lkey/rkey to refer to memory.

The way in which iWARP and InfiniBand refer to memory regions is also somewhat different. iWARP uses a steering tag (STag) to refer to memory regions. In the current specification [393] this is a 32-bit number that consists of two parts. The most significant 24 bits are the STag index, which is set by the device, and the least significant 8 bits are the STag key which can be provided by the user. Local operations use this STag to refer to memory. The same value is used on the remote side for RDMA operations. InfiniBand, on the other hand, uses two keys to refer to memory regions. There is an lkey and rkey, for local and remote keys. These again are 32 bit numbers. The difference is that the lkey and rkey can be, and generally are, different. The remote host does not know the lkey of the host it wants to do RDMA operations on, it only knows the remote access key, or rkey.

The transport layer is another important difference between iWARP and InfiniBand. With iWARP, there is no architectural requirement that DDP operate over TCP, or SCTP. However, DDP does require that the underlying transport be reliable. This is in contrast to InfiniBand, which provides both reliable and unreliable transport types.

For the most part InfiniBand and iWARP share the same types of basic functions, such as RDMA read/write and send/recv. However, iWARP lacks support for atomic operations. InfiniBand implements both a *fetch-and-add* and a *compare-and-swap* operation. InfiniBand atomics are potentially useful features but are hardly ever used in practice, at least so far. Another feature that is lacking in iWARP is the capability to generate a completion event on the remote host by an RDMA write operation.

Normally an RDMA write message generates a completion event on the *initiating* host, on both networks. With InfiniBand, it is possible to use a *RDMA write with immediate* operation to cause a completion event on the target host when the RDMA write is fully placed. The immediate data, which is 32 bits, is then made available in the completion queue event. This feature is handy to use on the final RDMA write operation to signal that some series of writes has completed. On iWARP a separate "all done" send must follow the writes.

One feature that iWARP has that is a potential benefit over InfiniBand is the support for a *send with invalidate* operation. This allows a remote peer to send a message and pass along a particular steering tag to be invalidated. This is useful for a remote peer to read or write a memory region, then issue a send with invalidate, to have the providing host remove access to that memory region. When a peer receives a request to invalidate a steering tag it must immediately make that steering tag unavailable for use until it is re-advertised (re-registered).

8.4   Chapter Summary

This chapter describes an approach to network communication known as Remote Direct Memory Access (RDMA). RDMA provides a mechanism to transfer data between two hosts with minimal CPU intervention and very low overheads. It relies on device-level support for protocol offload and direct data movement between the network and the host's target memory location, thereby eliminating CPU cycles required for protocol processing and repeated data copies typically needed in most common protocol stacks.

One common type of network with RDMA support is InfiniBand, discussed in greater detail in Chapter 2. In this chapter we also discuss iWARP — an emerging RDMA-based solution based on the prevalent TCP/IP protocol stack and Ethernet networks, and present the standardization efforts surrounding this technology.

Chapter 9

Accelerating Communication Services on Multi-Core Platforms

Ada Gavrilovska

Georgia Institute of Technology

9.1 Introduction

Multi-core platforms are becoming the norm today, and all topics discussed in other parts in this book are presented in that context. It is worthwhile, however, to more explicitly devote some time to the alternative implementation and deployment strategies one can consider for communication stacks on multi- and many-core platforms. This chapter discusses possibilities on how to best leverage the hardware-level parallelism on current and future multi-cores to better support communication services. In particular, we focus on alternatives to the straightforward approach of simply replicating the service execution across all cores, as well as alternatives which consider the use of heterogeneous cores and specialized accelerators for execution of elements of the stack. Regarding the latter, we also include a brief summary of a special class of multi-core communications accelerator — *network processors*. Our discussion applies to all types of communication services, ranging from standard protocol processing engines, e.g., TCP/IP stack, to richer application-level services, such as content-based filtering or routing.

Giving proper consideration to the multi-core nature of current and future hardware is important since these platforms are already becoming ubiquitous, and projections indicate that in the near future platforms will have 10s of cores in a single socket, with multi-socket configurations delivering unprecedented computational capabilities. With such computational power it is worth revising the onload/offload discussion, since it is now possible to move processing tasks onto other, otherwise idle cores, thereby both removing CPU loads from the core(s) running main application processing, and also benefiting from the tight integration and fast(er) synchronization and coordination mechanisms available across cores.

Second, Chapter 4 discusses advances in interconnection technologies which allow tighter coupling between I/O devices and the general purpose host processing complex. This integration makes it possible to attain both the benefit of specialized NIC-resident protocol processing capabilities, as well as of tight coupling which exists in the onload model.

Finally, the same advances in interconnection technology, and standardization efforts fueled under the umbrella of programs such as SIG-PCI, AMD's Torrenza and Intel's QuickAssist (see Chapter 4 for more details on system integration technology), are enabling possibilities to integrate other, arbitrary acceleration units on chip — thereby creating *heterogeneous multi-core* platforms consisting of general purpose and specialized accelerators.

In this chapter we first discuss the possibilities for leveraging the multi-core nature of future general purpose hardware platforms to deliver a rich set of communication services at high performance. We also contrast this with capabilities and models currently used in specialized multi-core accelerators for network processing. Throughout the chapter we highlight several research efforts which demonstrate various mechanisms and tradeoffs regarding these topics. We discuss the benefits of using hardware accelerators and the types of communication services which can be derived through their use. Finally, we summarize with a discussion of emerging trends in heterogeneous multi-core platforms.

9.2 The "Simple" Onload Approach

Several of the previous chapters already raised the important issue of onloading vs. offloading the protocol stack functionality, and pointed out that recent industry developments and trends, such as many-core platforms and tight integration between the main processing complex and I/O devices or other accelerators, blur the boundary between these approaches.

In summary, the benefits of onload include tighter coupling with the OS- and application-level processing, which enables faster synchronization and lower overheads in the interaction with the protocol processing engine. In addition, onloading functionality on a general purpose CPU delivers benefits from the use of familiar hardware technology, availability of development tools and programming environments, etc.

In contrast, the offload approach essentially integrates the communication stack on the network interface device (e.g., Ethernet NIC or InfiniBand HCAs), and either fully in hardware, or in combination with software/firmware processing on the device-level processing cores, executes the communication-related functionality. The key benefits from offload include ability to remove loads from the host CPU for application processing, and ability to benefit from specialized hardware features on the offload engine, optimized for the specific communication tasks.

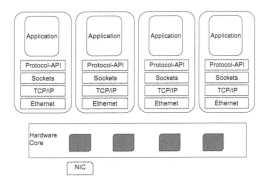

FIGURE 9.1:   Simple protocol onload approach on multi-core platforms.

For multi-core platforms, a simple onload approach, illustrated in Figure 9.1, is one where the entire communication stack is simply executed in software, on the same general purpose processing cores running OS- and application-level processing. This is no different than the manner in which most commodity platforms currently support their communication stacks — the NICs offer fairly limited offload functionality, such as segmentation and checksums, and the remainder of the stack is fully onloaded on the same CPU as the application, and executed in software.

This model has one very important advantage — it requires absolutely no modification to applications, the communication APIs they use, or the underlying communication stack in order to deliver performance gains. It exploits the parallel resources available on modern hardware, and simply replicates the tasks across all cores.

This type of function-level parallelism is very well suited for many classes of applications. For instance server applications are inherently parallel, with multiple threads and/or processes concurrently handling separate requests. By leveraging the hardware-supported parallelism, it is straightforward to deploy the execution of server threads/processes on a per core basis. The aggregate throughput rate delivered by the implementation is significantly improved.

An example of this approach (though combined with additional optimization of the processing performed by each thread) is described in Chapter 13, where the authors describe the implementation of a data decoding and forwarding engine for specific types of messages common in the financial market. The implementation of their engine on commodity multi-core hardware results in impressive performance of 15 million messages/sec. Numerous other cases demonstrate the performance capabilities of commodity multi-core platforms, including for services with significant communication requirements.

9.2.1   Limitations of the "Simple" Onload

In spite of its simplicity, the model described above has several drawbacks, because of which its ability to scale is limited.

First, using the simple onload approach we are able to accelerate throughput, e.g., in terms of number of requests, or messages or packets, being processed. However, individual request processing will only be accelerated due to availability of faster cores, and not due to availability of many cores.

Second, the memory footprint of each of the request/application contexts is essentially the same — each thread/process is performing the same task. There are several downsides to this. First, as each context has to execute everything, if the memory footprint is larger than the available cache sizes, there is no way to "partition" or reduce the contexts' memory requirements and adjust their size based on the available caches. Second, as emerging multi-cores already have complex caching architectures, there will be different levels of sharing of physical caches among the tasks. This implies that, depending on memory sizes, there may be interference and cache trashing across cores and the processing contexts executed on them. If the communication service is implemented as a kernel module, e.g., TCP/IP stack, the per request processing overheads for each context will also include costs of kernel-user space switch.

Next, in spite of the ability to process individual requests independently, depending on the application it is possible that there will be a need to perform synchronization and coordination operations across the parallel contexts. For instance, web server engines offering services such as ticket pricing and airline reservations [316], do process individual user queries in separate contexts; however, they need synchronized access to shared "application-level" cache of recent fares, both to service the current request as well as to update this cache when prefetching data from the back-end database.

Even control-plane protocol operations require access to state which has to be done in a coherent manner, e.g., for lookup and update to routing tables, port management, management of performance data, etc. As a result, proper synchronization constructs must be used across processing contexts and across the cores where these tasks execute. There are many lightweight synchronization constructs, but their applicability is tied to the types of accesses performed on the shared state (read vs. write) and their distribution. Therefore, the simple onload model will be limited by the type and amount of synchronized operation across the replicated communication service instances. As the number of cores increases, this may become a significant challenge.

From the discussion presented so far, the simple onload model can scale with increasing number of hardware supported cores, as long as data can be delivered to each core, and the processing task deployed on each core can proceed independently, with minimal level of coordination and state sharing.

However, even in the absence of shared state and need for coordination across the parallel components, or without any performance degradation due to lack of sufficient memory/cache resources, the simple onload model can still have limited use. One potential limitation to this was present on earlier platforms where interrupts were limited to a single core, partially due to the requirement to deal with synchronization needed to disable/enable interrupts in the OS kernel. As a result this created cores with asymmetric capabilities [387], where the "interrupt core," i.e., the core to which interrupts are delivered, is naturally better equipped to handle interrupts and device-driver/low protocol-level processing. On more recent platforms this limitation is eliminated, however, this model implies a simple NIC, without any additional capabilities to intelligently distribute data packets, and thereby take advantage of any state locality.

More importantly, note that in this model the NIC is responsible for simply pushing network data units, e.g., 1.5kB Ethernet packets, to one of the cores. Therefore, if the application-level unit of data exceeds the MTU size, i.e., the size of the packet delivered from the NIC to each of the cores, additional coordination and synchronization is required to deliver and assemble all data unit components, distributed across multiple packet units, to the appropriate communication service instance residing on a single core. For the application described in Chapter 13, and for many other server applications, such as the aforementioned airline pricing and reservation engine, the 1.5kB limit for regular Ethernet packets, or the 9kB limit for jumbo Ethernet frames is sufficient. However, the requirement that the application-level data unit must fully fit within a network-level packet unit does limit the generality of the approach, and its applicability to all classes of applications.

Therefore, it is unlikely that this approach will be suitable for accelerating all classes of communication-intensive applications, particularly those whose communication services involve data sizes larger than the network-supported packet limits, and which, require more state and more costly synchronization operations, particularly as we scale to larger number of processors.

9.3   Partitioned Communication Stacks

An alternative to the simple onload approach is to adopt a pipelined model, where the application and its communication stack are partitioned across (a subset of) cores, to create an execution pipeline. Figure 9.2 illustrates this: The communication stack, a basic TCP/IP stack in the figure, is onloaded on a single core acting as a *Communication Processing Engine* (*CPE*), whereas all other application components run on the remaining cores. Applications communicate with the CPE using a communication API (e.g., sockets in the event of basic TCP processing), which, via shared memory-based channels, enables movement of data and control (i.e., notification), between the different

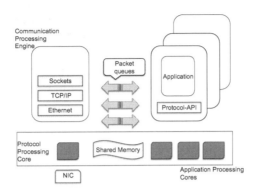

FIGURE 9.2:   Deploying communication stacks on dedicated cores.

processing tasks (communication and application). Similar partitioning can be designed for other, higher level communication APIs. For instance, the CPE can execute RPC or other stacks for communication with file- or other external servers, or it can include additional data processing, such as cryptographic operations, XML-preprocessing on incoming and outgoing data packets, or other tasks common in modern distributed enterprise applications.

In this pipelined, or partitioned model, each core is responsible for subset of the application processing. Particularly, the communication services are "factored out" and deployed on a designated (set of) core(s). Several efforts have demonstrated the possibility of deriving performance gains by designating a core in multi-processor environments for a single task, including specifically for TCP processing [56, 241, 367].

With respect to partitioning basic communication services, such as TCP processing, the use of this approach creates TCP or packet processing engines, onloaded onto a general purpose core. This onload method eliminates some of the challenges present in the TOE approach. The TCP processing engine is now executed on a mainstream CPU with same architectural features and technology advancements as other cores, and same performance costs for performing memory or synchronization operations.

Splitting application in this manner — where the TCP, or the communication service processing in general, is dedicated to a core (or subset of cores), and the remaining platform resources are utilized for other application level tasks — may result in more efficient utilization of resources and higher throughputs, particularly due to overlapping the protocol and application processing tasks. Whether or not the approach will indeed result in better performance depends on multiple factors, including the ratio of application level computation compared to communications-related processing, communication loads (in terms of connections, rates, and sizes, as there are per connection/packets/byte overheads to be considered), and the characteristics of the physical hardware — such as interprocessor communication latencies, memory hierarchy and caching architecture.

Experimental measurements present in [56], conducted on less integrated CPUs in an SMP system, demonstrate that as long as both cores have similar load levels and the system is well-balanced, the onload approach can deliver performance benefits due to improved cache locality and reduced number of cache misses and context switches. This is further enabled due to the fact that each of the components, the CPE and the core application component have smaller memory footprints, as compared to the simple onload model. When the application tasks are much more computationally intensive compared to the TCP stack processing, or when the rate at which the application tasks are invoked is significantly less compared to the TCP packet rate, this type of static core partitioning results in lower performance, as the system's resources are not well utilized, i.e., one of the CPUs is left idling.

However, as the number of cores continues to grow on next generation platforms, it is likely that there will be increased opportunities to partition tasks across cores so as to address load imbalance-related issues, while still benefiting from exploiting core affinity. First, it is likely that there will be idle cores, since the application-level degree of parallelism may be lesser than the hardware-supported one (e.g., the number of application threads is smaller than the number of cores), so it will be easier to justify the dedicated use of cores for certain tasks, including communications.

Furthermore, as the core count increases to several 10s and 100s of cores, this type of partitioning may be necessary so as to reduce the requirement of platform-wide synchronization operations needed to update shared state of the communication engine across all cores. For instance, for TCP stacks it is already demonstrated that control-plane operations requiring synchronization across multiple connections and multiple network interfaces with protocol offload support does present challenges [226,395]. The same type of operations will impact performance on large-scale multi-cores unless the need for synchronization is eliminated, which may require redesign of the communication stack engine, and therefore may not be favored, or, more likely, will require that related processing is localized on a single or a subset of cores, so as to avoid the need for "global" synchronization.

In such partitioned model, when determining the task placement onto physical cores, careful consideration must be made of the platforms' caching hierarchy. As Scogland et al. point out [387], there is a performance asymmetry among the cores, because of the manner in which they share the underlying resources with the protocol processing engine executing the communication services. The deployment strategy should be to exploit better cache locality for both the application code as well state maintained by the communication processing engine. In this manner, the partitioned model can help achieve not just improved throughput levels, but also reduced latencies.

9.3.1   API Considerations

In order to truly benefit from the pipelined deployment of communication stacks, one must consider the features of the communication API used between the application component, executing on application processing cores and the communications engine running on its dedicated core(s). The majority of applications outside of the HPC domain depend mainly on standard POSIX socket APIs, with synchronous and blocking semantics. When considering traditional socket-based services in the partitioned model, their API is best supported by a communication channel with message passing semantics. However, strict message passing, synchronous semantics imply the need for blocking and/or copying data. More importantly, this API does not allow us to exploit benefits due to overlapping communication and computation processing, and in fact may result in lower costs, as we must now pay for the increased core-to-core messaging/communication latency.

One important observation is that the partitioned model simply onloads on one/some of the host's cores the computation otherwise executed on the offload-capable NIC. For the offload based approaches for InfiniBand communication stacks or RNICs for iWARP, Chapters 2 and 8 already point out that the best performance for socket-based applications is if they are ported to an asynchronous version of the socket API - the Socket Direct Protocol (SDP). More generally stated, partitioning applications so as to designate the execution of their communication service on a set of dedicated cores will be most effective if the communication APIs used by the application are asynchronous ones.

For TCP stacks, several efforts have demonstrated this. The basic idea in the "Embedded TCP Architecture" engine [367], is to leverage asynchronous APIs (e.g., AIO), for interfacing application processing with the communication engine onloaded onto other cores. The exact performance benefits depend upon workload characteristics, message sizes and types, processing costs and inter-core messaging costs (i.e., for notification and shared memory operations).

The realization of such asynchronous APIs should include similar optimizations as those present in other asynchronous communication APIs, including in InfiniBand, or others reviewed in Chapter 6. With respect to data movement, its implementation should include mechanisms to enable copy and aggregation for small data — so as to amortize the costs associated with interactions with the communication processing engine. For large data transfers, mechanisms for data remapping should be available. Finally, to hide the cost of mapping a large memory region a combination of copy and remap operations can be considered, as already used in MPI transports. With respect to notification, the alternatives are signaling (based on Inter-Processor Interrupts) supported by the hardware, or polling-based (based on use of dedicated memory locations and doorbell registers). The tradeoffs between the two are often platform and workload specific.

9.4   Specialized Network Multi-Cores

Before we shift our discussion away from platforms with homogeneous cores we will briefly discuss a special class of multi-core engines, *network processors* (*NPs*), which have demonstrated great potential in their ability to execute communications related services.

9.4.1   The (Original) Case for Network Processors

Long before the multi-core renaissance overtook the general purpose commodity market, network processors, as well as other classes of specialized hardware platforms (e.g., graphics processing units), had high degree of hardware-supported parallelism, achieved through a mix of processing engines — cores, and hardware multithreading. In the case of network processors these cores are specialized for network processing tasks.

The motivation for the emergence of NPs was rather straightforward. Internet services, and the classes of applications dependent upon them, were becoming more diverse, more complex, and more high speed. Sample services ranged from network-level services, performing operations on protocol headers, such as network address translation (NAT), network monitoring, intrusion detection and routing, to middleware and application-level services depending on access to application-level headers and content, including content-based routers, or service differentiation based on request types. Furthermore, the pace of the emergence of new service types, and the evolution of their requirements was extremely fast.

In order to deal with this service diversity and the end-users' (primarily from the enterprise domain) demand to customize the service behavior to their needs, internal data formats, or business logic, it was necessary to have a network processing platform that can easily be customized. The general purpose platforms of the day offered the necessary programmability and tool chains, however, lacked the capacity to deal with the high data rates that these applications needed to deal with.

At the other end of the spectrum, it was feasible to develop custom ASICs (application-specific integrated circuits) which could address the exact performance and functionality requirements of any application setting. In fact, in come cases companies did explore this approach. For instance, in order to deal with the need for diverse and high speed data services, companies running mainframe-based servers expanded their operational facilities with server-based clusters. They then relied on custom ASIC boxes to "in-transit" deal with the needed data translation into internal company formats, and the data replication and aggregation necessary to distribute incoming data streams to/from both, the mainframe and server-based components. With this approach it was possible to continue to seamlessly offer the core business services but to also enable support for value added high data-rate emerging web services [316]. This was particularly needed for some classes of mainframe-based classes of applications, such as those running on legacy runtime environments such as the Transactional Processing Facility (TPF), used by large-scale enterprise operational information systems (OIS), which did not even have support for the Internet protocol stack until more recently.

Solutions based on networking ASICs did offer the customized functionality and support for high data rates, however at substantially increased development costs, and, more critically, extended development times. This was a particular challenge, as service evolution was, and still is, fast, with companies deploying new or evolving existing services within shorter time spans that what is needed to economically justify the ASIC development.

Driven by industry desires to offer new capabilities and services as part of basic network infrastructures, device-level research evaluated the tradeoffs in cost/performance vs. utility of networking devices. As a results, *programmable network processors (NPs)* emerged as an attractive vehicle for deploying new functionality "into" the network infrastructure, with shorter development times than custom-designed ASICs and with levels of cost/performance far exceeding that of purely server-based infrastructures. The key reason was that NP hardware was optimized to efficiently move large volumes of packets between their incoming and outgoing ports, and typically, there was an excess of cycles available on the packets' fast path through the NP. Such "headroom" was successfully used to implement both *network-centric* services like software routing, network monitoring, intrusion detection, service differentiation, and others [159, 262, 372, 411], as well as application-level services including content-based load-balancing for cluster servers, the proxy forwarding functionality servers need, and payload caching [16, 473].

One outcome of these developments was that increasingly, networking vendors were deploying NPs into their products and "opening" them to developers to facilitate rapid service development. At the peak of the market, in the late 1990s/early 2000s, over 40 vendors sold various network processing units [394, 455] with a range of capabilities, architectural features, or types of services they could support. Today the diversity in the NP products is significantly reduced. Most of the major hardware companies sold or abandoned their investments in their NP line of products [201, 208, 345]. However, there are still a number of platforms — TCP offload engines, network appliances and customizable network flow engines which internally contain network processors [14, 83, 90, 308, 424, 467], and continue to generate annual profit levels in excess of $300 million dollars [455].

9.4.2   Network Processors Features

In spite of the diversity in products and vendors, there are a number of unifying features among the types of NP engines which were and are available on the market.

Their main architectural feature is that they consist of many on-chip processing elements, or cores. The capabilities of the processing cores, as well as other on-chip and board level features are chosen based on the application/service processing requirements that these platforms are targeting.

Given the types of tasks typically found in a network processing pipeline, such as header manipulation, byte- and bit-wise operations, lookups and hash functions, etc., it is not necessary for the NP cores to be rich, general purpose processors. Rather, the majority are based on RISC cores, augmented with instructions specialized for elements of the network processing pipeline. While this was not the only approach, and there were examples of NPs with complex or even dynamically reconfigurable instruction sets, the majority of products were and are in fact based on derivatives of ARM, MIPS, or Power-based cores. The type and number of cores does vary across products, from solutions with many (188!) simple cores in Cisco's Silicon Packet Processor (a.k.a. Metro) [130], to fewer more complex ones, like Intel's IXP series of NPs [208], to solutions with just few dedicated cores, each customized for different types of functionality (e.g., a classification vs. a forwarding core) [424].

In addition to the the main processing cores, most NPs typically include additional hardware support for network-related functionality, such as fabric management, look-up table management and queue management, different types and sizes of on-chip memory and controllers for additional memory, and hardware support for a number of atomic operations including enqueue and dequeue for hardware supported queues. Some higher-end NP products offer additional hardware support such as for packet traffic management or crypto functionality [308, 355].

Furthermore, NPs have interfaces for multiple types of memories, each of which is best leveraged for specific tasks in the communication pipeline. While there are some exceptions, a large number of NP products do not have on-chip cache. The rationale for this decision is due to the fact that there is little locality in packet processing operations, and that caches require area and power resources which are not being justified with better performance/area/watt metric. Instead most NPs have some fast on-chip memory which is used for temporary storage of state and is managed explicitly in software. In general, the memory hierarchy available on these platforms is often customized for the target applications and deployment context. Most of the target applications require header-based accesses and manipulations — for this reason NPs typically have access to some fairly fast memory sufficient to store queues of packet descriptor which include the header portion (e.g., SRAM). The rest of the packet payload is typically stored in larger, slower memory (e.g., DRAM).

In addition, to support the pipelined deployment of service components across cores, NPs rely on fast internal buses, and in some cases dedicated core-to-core communication paths, specifically designed for fast communication across pipeline stages. This is the case with the Intel IXP 2xxx network processors, which have rings of next-neighbor registers between consequtive cores. Finally, depending on the context where they are used, it may be useful for NPs to perform very efficient lookup operations. For this reason some NPs at least have the option to be configured with a TCAM memory.

The above decisions, in terms of types of core, memory hierarchy and presence of specialized hardware accelerators, keep NPs extremely power efficient and allow high performance execution of network operations.

9.4.2.1   Programming Models

The type and number of cores ultimately determine the types of communication services which can be supported on NPs. "Network-centric" services, performing network-header manipulations, can be easily implemented in 100s of instructions. On the other hand, richer, content-based application-level services may require processing of higher level protocol stacks and interpretation (i.e., parsing) of XML encoded data before any of the service logic can be executed, resulting in very large (several thousands) instruction counts.

Network processors with features such as those described above deliver opportunities for high-speed data services due to the programmability of the NP cores and the ability to deploy communication stacks on them. The typical deployments are similar to those in the general-purpose multi-core environments — functionally replicate the same functionality across all cores, and rely on hardware support for fast dispatching of packets to any of the cores on the platform, or create computational pipelines in which each core is responsible for a set of processing actions. For instance, a core performs the IP stack processing to form IP packets, a second core performs look-ups in routing tables to make a forward decision, along with any updates, if necessary, and a third core is responsible for the lower-level execution of the IP stack as the packet is being sent out. Or, for a richer service, this pipeline can be extended with modules performing accounting, auditing, security (e.g., firewalling or encryption/decryption operations), or XML parsing of payload data to interpret and access application-level content and support services such as content-based load balancing. For "rich" services the pipelined model may be the only possiblity, as NP cores may have a limit on their instruction store, and therefore, a limit on the complexity of the processing that can be depoyed on them. In addition, some NPs have hardware support for replicated pipelines [275]. Incoming packets are dispatched to one of several pipelines, where each pipeline consists of multiple processing cores, each executing a fixed module of the target service.

Finally, it is worth commenting on the NPs' openness and programmability.

NPs consist of programmable processing engines, oftentimes requiring use of specialized toolchains, development environments and languages [6, 208]. For this reason, the development of new services, and the validation of high-performance implementation of these services proved to be a non-trivial task. Therefore, more frequently the business model is to package NPs as *content processors* or *appliances* which deliver a specialized service — from classification, protocol acceleration, and multicast, to VoIP, content-based load balancing, security-related services and XML-preprocessing, or in general for regular-expression processing needed for these and other types of protocol parsing functionality.

9.4.2.2   IXP family of NPs

One example of network processors widely available in the research community is the Intel IXP family of network processors. The Intel IXP NPs form an ecosystem of programmable network processors that can be used across the entire range of moderate data rate embedded systems to high-end data streaming at 10Gbps link speeds [208, 308]. At the low end, the IXP 425 targets embedded systems applications like residential gateways and in fact, is already used in Linksys' and other similar products. The former generation Intel IXP1200 was used as a mid-level product, and its commercial use has included Video over IP [331] applications. From the recent generation, the IXP2350 and IXP2400 are examples of mid-level products with commercial applications in a range of network appliances and broad academic research use. At the high end, the IXP2800 and IXP2850 operate at 10 Gigabit speeds, and include features such as on-chip content addressable memory and a powerful crypto engine.

The IXPs consist of a number of *microengines* — specialized hyperthreaded RISC cores optimized for communications, up to 16 8-way microengines in the case of the IXP2850. They include fast on-chip ScratchPad memory, and interfaces for off-chip SRAM, DRAM and/or TCAM memories. Additional features, particularly for the high-end IXPs include accelerators for packet processing functions (e.g., Ethernet and IP header manipulations), CRC and hash computation, on-chip CAM and crypto unit, fast internal communication network including next-neighbor ring for efficient support of processing pipelines. Finally, the IXPs include a general purpose XScale core for executing management and initialization, and control path functionality. In addition, Intel made a substantial investment in programming and development tools for these platforms, including higher-level language constructs (microC and a rich set of microcode libraries), and cycle accurate simulators and software architecture design tools.

While Intel has discontinued its commitment to the IXP network processors, the technology remains relevant. On one end of the spectrum, the microengine technology is used to develop a 40-core programmable appliance engine [308], which can support a wide array of network- and application-level services. On

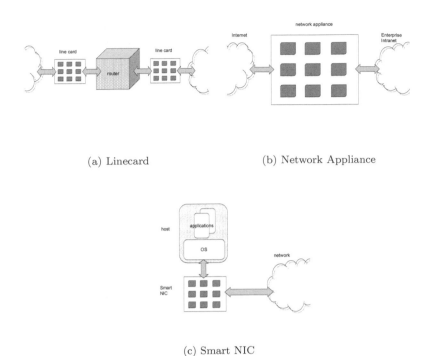

(a) Linecard   (b) Network Appliance

(c) Smart NIC

FIGURE 9.3:   Deployment alternatives for network processors.

the other end, this technology is being integrated with general purpose CPUs
to form heterogeneous multi-core platforms [432], as discussed in Section 9.5.

9.4.3   Application Diversity

Over the years, NPs have been demonstrated to have capability to support
a rich set of services, and to be used in many different environments.

First, NPs can be easily used as programmable routing engines, either as a
standalone router, or as a line card, which modifies the router inputs/outputs at
line rate (see Figure 9.3). This usage mode has been exploited in many settings.
Most recently, one of the building blocks in the GENI program — aimed at
creating a testbed for experimentation with new architectural features and
protocols for a next generation Internet — includes a reprogrammable routing
engine based on a NP (an Intel IXP2850) [458]. The platform implements the
core data path elements — receive (Rx) and transmit (Tx) blocks — on a
subset of the available hardware resources, and makes the remaining NP cores
(microengines) available for user customized or provided programs — most
typically different types of routing algorithms.

NPs are particularly efficient for in-transit inspection of packet headers so as to provide value added services, such as firewalls, differentiated QoS based on e.g., VLAN tags [262], fastpath monitoring and anonymization (e.g., for accounting or management purposes) [396], traffic shaping and scheduling [175], etc. Typically, these services are realized as processing pipelines, with dedicated cores performing Rx/Tx processing, and the remainder of the service modules executed on other NP cores.

Furthermore, NPs can efficiently support services which depend on true packet content, i.e., *deep packet inspection* services. Examples include intrusion detection systems searching for Snort rules across streams of packets [93], packet engines processing XML-formatted data or merging data streams based on true packet content [374], content-based routers [480], etc.

In addition to being used as standalone appliances or as extensions to other networking platforms (e.g., linecards in routers), NPs can be used in combination with general purpose processing platforms. Our own work discussed the use on NPs as "smart, programmable, NICs," onto which arbitrary layers of protocol- or application-level processing can be offloaded [158], including multicast, data filtering, and content distribution and load balancing [480], or payload caching, as in web server front-ends [473]. The last examples can be generalized into a case which argues for the use of NPs as a component in appliance platforms consisting of general purpose hosts and NP accelerators [427].

Finally, network processor platforms have been shown to be well suited not just for networking applications, but in general for accelerating application kernels with high degree of parallelism. Our early work demonstrated the feasibility of performing ultra high-rates image manipulation operations on older generation IXPs (we are not advocating the use of NPs over GPUs, merely pointing out their suitability for this type of parallelizable codes), and others have demonstrated impressive performance capabilities for other applications, including protein folding [465].

9.5   Toward Heterogeneous Multi-Cores

So far we have discussed the issues related to implementing communication services on general purpose, homogeneous multi-cores, on one end, and on special purpose network-processing multi-cores on the other end. However, industry trends indicate that next generation platforms will be increasingly more *heterogeneous*, and in the remainder of this chapter we briefly outline the possibilities that such heterogeneity will present.

One type of heterogeneity exists on platforms with single-ISA, functionally asymmetric or heterogeneous cores. As the number of cores increases to several tens and hundreds, it is more likely that not all cores will be equal with respect

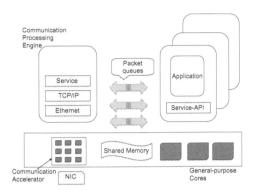

FIGURE 9.4:   Heterogeneous multi-core platform.

to their distance from caches or memory units, their performance, reliability, or power capabilities, presence of certain functional units, such as floating point, etc. In these environments, the operating system and the runtime layer must include mechanisms to map the threads and application processes onto those cores best suited for their execution. One example is to "sequester" cores and dedicate them for execution of specific types of operations — ranging from protocol stacks [366], to video encoding/decoding operations [376], or support for other OS- and application-level services [242].

Other types of heterogeneous multi-cores are those which consist of a mix of general purpose cores (e.g., IA cores) and hardware accelerators specialized for certain tasks, such as graphics, or more relevant in this context, communications. As mentioned in the previous section, in our own research effort we consider platforms consisting of general purpose multi-cores and PCIe attached network processors (Intel IXPs, but also other accelerators — GPUs, FPGAs, STI Cells) as prototypes of future closely coupled heterogeneous multi-cores. Chapter 4 discusses the ongoing efforts and technology advancements which make such integration possible, and tighter. These efforts include system interconnection standards, such as AMD's HyperTransport (HTX) and Intel's QuickPath Interconnect (QPI), which allow third party accelerator cores to be seamlessly integrated with homogeneous, general purpose components, and software- and API-level efforts aimed at improving the applications' ability to benefit from the hardware accelerator.

Hardware accelerators have been long used in the HPC community; however, the recent technology trends are giving raise to their broadened popularity. The demand for high performance and low power is driving the use of specialized low-power accelerators over software implementations on general purpose cores, and technologies such as QPI and HTX enable tight integration of specialized accelerators with general purpose multi-cores. In [384], Schlansker et al. use an experimental platform, designed with an FPGA-based communications processing accelerator, to demonstrate some of the possibilities in this type of environment. The use of the FPGA-based accelerator results in latency and throughput improvements, not just for the low-level packet transmission/receipt operations, but also for higher level protocol components, including TCP and MPI. Other research efforts have also demonstrated the feasibility and the utility of tightly coupled heterogeneous multi-core platforms [87,427].

More interestingly, Intel is leveraging the same microengine technology used in their family of IXP network processors to develop communication accelerators and to integrate them on the same chip with general purpose IA cores. The first outcome of this effort is their "Tolapai" platform [432, 479], targeting the embedded market, however, similar developments are likely for high-end server platforms with high data rate requirements. The benefits of leveraging the specialized engines for accelerating communications-related tasks results in significant benefits — Intel reports an almost order of magnitude performance gains for an IPSec service, with lower energy requirement, when comparing the execution of this service on the integrated heterogeneous platform vs. one which has the same type of general purpose and off-chip accelerator. In the latter case, the general purpose CPU is fully utilized, indicating that performance would have been significantly lowered in the absence of the accelerator [479].

It is important to note that we already raised the issue of programmability of specialized accelerators, such as network processing engines. The question then remains how will these heterogeneous platforms be programmed, and how will applications leverage the underlying platform resources? In the HPC community, where accelerators have been long used, the standard approach has been to hide them behind domain-specific libraries. For communications services, where NP accelerators are already being presented as "content processors," the use of library APIs allows application developers to easily benefit from the presence of the hardware accelerator by linking the "right" library. For platforms without accelerators, the content processing function is executed in software, either synchronously, on the same core, or perhaps asynchronously, on another CPU — depending on library implementation and underlying platform resources.

For heterogeneous platforms, the content processing, such as TCP acceleration, crypto- or security-related processing, or XML-processing functionality, is executed on the accelerator, and ideally invoked with asynchronous APIs, so as to attain benefits from the opportunity to overlap the execution of the (accelerated) communication service with other application-processing. Intel is developing its QuickAssist Technology [356] to standardize the APIs exported by different types of accelerators so as to streamline their integration in next generation heterogeneous platforms. Currently, over 20 companies developing a range of accelerators — for high speed communications, security, or other compute-intensive tasks — are developing support for this API [349]. In other contexts, and using other types of accelerators as driving examples (e.g., CUDA and OpenCL, derived originally from the use of GPU-based computational accelerators, IBM's AAL based on the API for the Cell accelerator, and others) a number of other APIs are being positioned as general solutions to ease the development of applications for heterogeneous multi-cores and to aid with the development of systems tools and technologies which will manage such platforms.

It is yet to be seen which of these alternatives will prevail; however, it is clear, that at least for the foreseeable future, in order to meet the data rate, power, and performance requirements of many classes of applications, a purely homogeneous platform will not be as efficient.

It is important to note that NP-based accelerators are not the only relevant class of accelerators for communication services. Numerous research efforts over the years have demonstrated the flexibility and performance which can be delivered with the use of FPGA cores for accelerating communication services ranging from protocol conversion, to packet scheduling and QoS for multimedia services, to pattern matching for instruction detection systems at 100Gbps! More recently, other accelerators have also been used to demonstrate their suitability for communication-related processing, e.g., for streaming applications using the Cell processor [339] or for remote storage using GPUs.

9.5.1 Impact on Systems Software

Supporting a model where application components are partitioned across a number of potentially heterogeneous cores requires additional operating system and runtime support for several reasons.

First, in order to enable scalable execution of core system services, and to benefit from CPU and cache locality and avoid costly platform-wide coordination, it may be necessary to redesign key OS components. Several research efforts advocate different OS design approaches. For instance, Boyd-Wickizer et al. [53] develop a prototype OS, Corey, and demonstrate the importance of per-core OS data structures, so as to minimize the need for sharing and synchronization. Similar ideas are present in the Barrelfish OS [33], where multiple kernels are concurrently deployed across the platforms' (potentially heterogeneous) cores. Each of the Barrelfish kernels, consisting of a core-specific "CPU driver" and an application-specific "monitor" component, utilizes local state whenever possible, and communicates with other kernels using asynchronous, message-passing mechanisms. With respect to communication services, these approaches allow the existence and concurrent execution of multiple network stacks, either copies of the same communication stack (e.g., TCP/IP) or distinct ones, potentially customized based on the CPU features or the application needs. Each stack is deployed on a separate (set of) core(s), thereby increasing the achievable level of concurrency and improving performance.

An earlier approach developed by our group at Georgia Tech proposes the use of *sidecores* [241] — cores designated for the execution of some subset of OS functionality. In this manner, the systems software is "decomposed," with select services, such as interrupt processing, device driver operations, protocol stack execution, etc., bound to these "sequestered" cores [376]. This is particularly useful since some cores may be better suited for execution of certain tasks.

One issue arising in both of these cases, is that different cores may execute different types of applications or system software components, requiring different resource management policies, such as for CPU scheduling or memory management [33, 142, 239]. Furthermore, the differences can be such that they require memory to be treated differently for different processing tasks. One example of this is the use of non-coherent memory, either between accelerators and general purpose cores [384, 479], or even among same-ISA general purpose cores in next generation multi-cores [278]. Another example is the use of adaptive caching mechanisms [81], which can be helpful in allowing the use of both regular, cachable memory, and non-cache-coherent scratch-pad memory for services such as protocol processing tasks. This is already demonstrated as necessary for heterogeneous platforms with communications accelerators [384, 479].

In general, as we scale to larger number of cores, it is likely that we are going to have multiple, distributed resource managers, even when the behavior of these resource managers is the same. It is simply unlikely that platform management operations, such as scheduling, will be able to continue to operate globally and centrally. Rather, a scalable resource management solution will require active coordination across multiple independent schedulers, each responsible for a subset of processing resources. In such environments, in order to better meet the service-level requirements of applications partitioned across multiple cores — such as applications relying on communication stacks executed on other cores, as described in Section 9.3 — it is necessary to enable mechanisms for active coordination across the separate resource managers. Therefore, the model is one in which distributed schedulers each manage certain resources but also coordinate and "activate" certain actions across their different scheduling domains. For instance, on a homogeneous platform, where one type of scheduler is responsible for scheduling the communication-intensive processing tasks on the communication engine, and another type of scheduler, perhaps with support for different scheduling classes and notions of priorities, is responsible for the remaining cores executing application services or VMs, coordination actions can be used to indicate a burst in certain type of network packets from one end, or an change in the application QoS requirements or its behavior, from the other.

Similar coordination is required among schedulers responsible for accelerator-level resources and general-purpose schedulers. In fact, distributing the resource management tasks in this manner will likely be necessary for future heterogeneous multi-cores, since the proliferation of different types of accelerators should not imply the need for general purpose OS- and hypervisor-level resource managers to understand the details of all possible types of accelerator management operations.

Even in the presence of per-accelerator resource managers, the systems software overall will have to be made aware of the platform heterogeneity. On one end of the spectrum, such "awareness" can be introduced via *architecture desriptions* (e.g., specified using an architectuer definition language [350,470]), which may contain information regarding the core's features and capabilities (e.g., ISA, existence of floating point unit, frequency, distance from other cores, etc.), or can consist of performance profiles for select application components. The descriptions can be available a priori, potentially derived during a profiling phase. On the other end, the systems software can be equipped with mechanisms to derive this information through observations [260], by continuously gathering platform events regarding faults or performance. By being aware of the heterogeneity, the OS can place communication intensive components on cores specialized for communications, or on cores "closer" to I/O devices and memory, depending on the platform's memory architecture and on-chip interconnection network [278].

Finally, heterogeneous multi-cores, particularly those with complex memory hierarchies, raise interesting challenges with respect to direct cache access and cache injection. These present a range of tradeoffs in terms of data placement and computation (co-)location, which need further investigation with realistic prototypes and workloads [258,259].

These are just few of the challenges related to high-performance communication services in heterogeneous multi-core platforms. As these types of systems are becoming available, numerous other issues remain open, ranging from ensuring quality of service and performance isolation guarantees, to sharing models for accelerators, to mechanisms to "account" or "bill" for resource utilization across the aggregate platform resources, to programmability and cross-core migration, and many others.

9.6   Chapter Summary

In this chapter we discuss alternatives and challenges regarding implementation of communication stacks on platforms with hardware-level parallelism. We consider homogeneous multi-cores, multiprocessor network accelerators, and hybrid platforms consisting of general purpose cores and specialized accelerators. We discuss the tradeoffs presented by the various alternatives, and conclude with a brief discussion of some of the challenges in emerging heterogeneous platforms.

Chapter 10

Virtualized I/O

Ada Gavrilovska, Adit Ranadive,
Dulloor Rao, Karsten Schwan
Georgia Institute of Technology

10.1 Introduction

Virtualization technologies like VMWare [445] and Xen [124] have now become a de facto standard in the IT industry. A main focus of these technologies is to enable multiple Virtual Machines (VMs), each running their own OS and applications, to share the same set of physical resources. VMs can be consolidated to concurrently run on the same platform, in which case the virtualization layer, a Virtual Machine Monitor (VMM), is responsible for their management, in order to ensure the appropriate allocation of platform resources to each VM, as well as to provide for isolation across multiple VMs. While virtualization has gained significant momentum in the enterprise domain, for consolidation, operating cost and power reduction, ease of management, improved fault-tolerance and RAS properties, etc., its adoption in settings with high performance requirements, particularly with respect to I/O performance, is still lagging.

The reasons for this are multifold. First, in the traditional HPC domain the use of virtualization in general is lagging, primarily due to a belief that it will introduce prohibitive overheads, particularly for capability-class high-end computing. Therefore, the class of HPC applications with high performance I/O needs did not "drive" the need for high speed virtualized I/O solutions. On the other end, in the enterprise domain, where virtualization plays a significant role, the lack of "drive" came from the fact that high-speed networking solutions were not commoditized, and therefore not ubiquitously present, and that not all enterprise workloads executing on virtualized platforms had high-performance I/O requirements.

However, we are witnessing a shift in the above trends. First, advances in hardware-support for virtualization significantly reduce virtualization-related overheads. This makes virtualization a viable technology for certain HPC settings which can benefit from its support for more general management, fault-tolerance and migration tools, or for isolation across capacity platforms or shared service nodes [157]. Next, high performance I/O technologies such as 10GigE and Infiniband are becoming significant components of the commodity market, and are increasingly used in enterprise environments. In addition, the multi-core nature of current hardware platforms as well as the need for reduced energy costs are further pushing the reliance on virtualization. This is due to the dynamic management mechanisms offered in virtualized platforms, which help ensure better workload consolidation and resource utilization, improved power efficiency, and reduced operating costs. As the number of cores per node continues to grow, the aggregate I/O needs of the consolidated workloads continue to grow. Furthermore, as virtualization-related overheads continue to be addressed by hardware- and software-level techniques, the possibility for reducing operating costs through virtualization makes its use attractive even for classes of applications typically executed in native environments on dedicated hardware resources. These, as well as the new emerging classes of applications with high data rate/low latency requirements, depend on the availability of high performance I/O services.

All of these factors have pushed the development, and continued advancement of I/O virtualization techniques with minimal performance costs. In this chapter, we first illustrate the main approaches to dealing with I/O in modern virtualized platforms, particularly focusing on network I/O. For each, we present a series of software- and hardware-supported performance optimization opportunities which help address applications' I/O needs, and discuss their limitations and tradeoffs.

10.1.1   Virtualization Overview

Virtualization is a technique which enables multiple operating systems to share the underlying physical resources. The physical platform is under the control of a single software layer — either a regular host operating system, also known as *hosted or host-based virtualization*, or a thin resource management layer termed *Virtual Machine Monitor (VMM)* or *hypervisor*, known as *hypervisor-based virtualization*. Since the scope of this chapter is to discuss issues related to improving the performance of virtualized I/O services, we will focus our discussion on the latter, hypervisor-based virtualization, as this type of virtualization does result in higher overall performance. Many of the techniques described here, however, can also be applied to improve the performance of virtualized I/O in hosted virtualization solutions.

The virtualization layer, either host OS or hypervisor, has the ability to multiplex the underlying hardware resources under its control across multiple containers, termed *Virtual Machines (VMs)* or *domains*, each of which may consist of a *guest* operating system and more than one application process. The guest OS must continue to support all applications it natively supports when running on a non-virtualized platform. It also must be presented with a view that it is executing on top of the real hardware and is still in control of all hardware resources — such as CPUs, memory, page table registers, devices, etc. The responsibility of the virtualization layer then is to (1) maintain shadow copies of the physical resources whose state matches the guest OS's expectations, (2) to intercept the guest OS's attempts to modify the physical resources and both redirect those operations towards the shadow states as well as initiate, when appropriate, the actual operation on the physical hardware, and (3) coordinate across all guest OSs so as to ensure isolation and fairness in their utilization of the underlying resources.

This can be achieved in two ways.

10.1.1.1 Paravirtualization

In the *paravirtualization* approach the guest operating system is modified so as to make it better suited for execution in the virtual environment. One reason for paravirtualization is feasibility-driven. Namely, one way in which the VMM intercepts the guest OS's attempts to modify the physical hardware is by leveraging the protection rings supported on modern platforms. The guest OS is no longer executing in privileged mode, and therefore when issuing privileged instructions, these trap and pass the control to the VMM or host OS, which are running in the most privileged mode. However, on older generations of the popular x86 platforms, several key instructions (e.g., the POPF instruction which pops flags from the CPU stack and is needed to enable/disable interrupts) when invoked without the right privileges, fail silently, without causing a trap and switching to the VMM. Therefore, paravirtualization is used to modify the guest OS so as to avoid the use of these instructions, and instead make explicit calls to the hypervisor (*hypercalls*).

A second reason for paravirtualization is performance-driven. Even in the event when all OS services can be supported on top of a virtualized platform without additional modifications, it is very likely that by modifying the software and explicitly leveraging the information that the guest is executing on top of a VMM, one can achieve higher performance benefits. This is particularly true for I/O service, since to support unmodified OSs VMMs rely on device emulation techniques, which tend to be significantly slower compared to the use of paravirtualized device drivers.

10.1.1.2 Full virtualization

Full virtualization implies that there are no modifications to the guest OS. For earlier generations of popular x86 platforms, this approach was possible only when used with binary rewriting techniques — the virtualization layer was responsible for dynamically detecting the "dangerous" instructions in the guest OS execution stream, and rewriting them, i.e., replacing them with safe, virtualization-friendly binary sequences.

Recent hardware advancements from major vendors such as Intel and AMD have eliminated these challenges, by modifying the hardware to better support virtualization. Intel's VT Vanderpool technology and AMD's SVM (Secure VM) Pacifica technology include hardware level enhancements such as:

- an additional privilege mode for the hypervisor software, so that the guest OS can continue to run at the same protection level (however, in a non-root protection mode),

- a more efficient implementation of additional instructions which allows more efficient execution of key VMM-level tasks, such as switching between VM and hypervisor mode, or

- hardware support for "shadow" copies of the platform resources managed by the guest OS (e.g., page-tables), so as to reduce the need for switching into hypervisor mode.

With respect to I/O virtualization, Intel, AMD and the SIG-PCI standardization body which sets standards for the host-device interface technologies, are heavily invested in identifying host-level and device-level techniques to reduce the need for or the cost of software-based solutions. These enhancements include support for remapping of DMA operations and interrupts and device-level silicon to make devices more virtualization-friendly.

10.1.2   Challenges with I/O Virtualization

There are several issues which present key challenges with respect to I/O virtualization.

First, there is the issue of device-type dependency or the issue of the device interface virtualization. The physical platform onto which a VM is deployed need not necessarily have the exact same type of I/O devices as those originally expected by the guest OS. This implies that the virtualization software (VMM or host OS) must provide the capabilities to translate the original guest VM's I/O operation into the appropriate type of request as expected by the physical device, and provide adequate translation of the device-initiated I/O requests for a given VM. The straightforward approach here is to emulate the missing physical device in software, likely at substantial performance costs.

In practice, this is really feasible for only the more common types of devices. In the more general case, virtualization relies on the same type of generic device driver interfaces available even in proprietary OSs, and virtualizes the I/O stack at this level. The generic I/O operations are then translated by components in the virtualized I/O stack to the appropriate physical device interface. Enabling device-type independence and interoperability between the physical and virtual device interface in this manner may require significant resources and impact performance. The same mechanism, however, opens additional opportunities in terms of allowing *device morphing* or enabling virtual device interfaces with features not otherwise supported by existing physical devices.

The next issue is the manner in which the device is shared across guest VMs. Is it dedicated to a single VM or do multiple VMs share the device resources (e.g., bandwidth, I/O buffers, etc.)? How is this sharing achieved — are the VMs' device accesses interleaved, so that at any given point of time the device executes on behalf of at most a single VM (i.e., time-sharing), or are concurrent accesses allowed by partitioning the aggregate device resources in some manner (i.e., space-sharing)?

In either case, the virtualization layer is responsible to (1) first authorize the actual accessibility of the device to the VM. Once the device is made available to the VM, the VMM must provide mechanisms to (2) authorize device accesses, enforce sharing policies, coordinate concurrent VM-device interactions, and ensure isolation and fairness among them. The first, initialization-related task results in one time costs which are not part of the I/O fast path and are not critical for achieving high performance levels. The later set of tasks must be continuously performed and therefore the manner in which these operations are supported by the virtualization layer, by the VMM or host OS (i.e., the virtualization software) or via hardware-level virtualization support, is of particular importance for attaining high performance virtualized I/O services. In the remainder of this chapter we consider several techniques aimed at reducing the runtime overheads associated with these tasks.

Finally, there is the issue of signaling or notification between the physical device and the guest VMs. Some device interfaces are *polling* based — the device driver periodically polls a predefined memory location to determine that there are device requests it needs to service (e.g., packets to be received). In most other cases, the device-host interactions rely on the ability of the host OS to receive *interrupts* generated by the device. On the physical platform these interrupts are received by the software virtualization layer (e.g., VMM), and must be first routed to the appropriate VM and then handled by its guest OS. The ability to perform efficient interpretation of these interrupts so as to determine which VM are they targeted for, and then to route them to the appropriate VM — which may be executing on a different CPU, or may not be currently scheduled on any core — presents significant performance challenges. We later discuss both technology advances which help eliminate some of the costs associated with *interrupt virtualization* and also discuss opportunities to leverage this information in the VMM-level scheduling of guest VMs.

10.1.3 I/O Virtualization Approaches

The above challenges can be addressed differently depending on the approach to I/O virtualization supported by a given VMM and a target device. The two main approaches to virtualize I/O are referred to as the *split device driver model* and the *direct device access model*. The first method relies on the virtualization

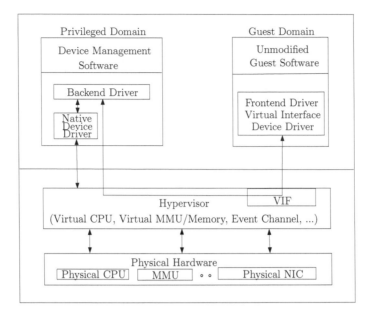

FIGURE 10.1:   Split device driver model.

layer to redirect all I/O operations to a centralized software entity on the host platform, which in turn performs the physical device access or routes the device requests/responses to the appropriate VM. This method is described in the following section. The second method, described in Section 10.3, is based on the VMM's ability to provide VMs with direct access of the physical I/O device(s).

A third approach based on emulation is used for fully virtualized guest OSs, including for Hardware Virtual Machines (HVMs) or for unmodified device drivers, but its overheads are currently so significant that, at present, it is not relevant to the discussion of high performance virtualized communication services.

10.2   Split Device Driver Model

10.2.1   Overview

In both virtualization approaches described in the previous section, paravirtualization and full virtualization, the most common way to provide access to devices is the so called *split device driver model* [346], illustrated in Figure 10.1.

Applications and guest OSs within the VMs initiate device accesses through a VM-resident driver, often termed *front-end*. The front-end driver is not the real device driver, and has no direct control of the physical device. Instead, it is paravirtualized, i.e., modified, so as to redirect the VM's request to a software entity part of the platform's I/O virtualization software stack, termed *back-end*. The back-end's responsibility is to translate and redirect the requests between the native device driver and the guest VM's front-end.

For network I/O, the guest VM's networking stack passes packets to its front-end (called netfront in Xen), which moves those packets to the back-end (netback), along with a notification event that a network request needs to be serviced. Upon receipt of the notification event, the back-end retrieves these packets, and initiates the physical device driver to perform the actual transmission. Similar steps are followed in the opposite direction for receive operations. In addition, a software bridging/routing entity is also involved on the request path. Based on address information in the packet's network header, this entity ensures that the request is redirected to their target VM[1]. For instance, in the event of an Ethernet network adapter, the routing is performed by a software bridge which dispatches the Ethernet packets to their target VM based on the MAC address in the Ethernet header. In this case, each VM has its own (virtual) MAC address, assigned to it by the hypervisor during boot time, and the physical device is set in promiscuous mode so as to capture Ethernet packets with any MAC address, and not just the real MAC address of the physical Ethernet NIC.

The communication between the front-end and back-end depends on I/O channels consisting of an event notification mechanism and bidirectional shared memory rings containing *request/response descriptors*. These descriptors identify the type of request to be performed on behalf of a VM, as well as its data attributes (e.g., packets to be sent, data block to be written, etc.). Similar information is carried in the descriptors from the back-end to the front-end driver in the VM. Since the front-end and back-end are part of two different address spaces, the actual data movement across these domains includes copy or page remapping operations, and associated with them are the usual costs and tradeoffs of copy vs. memory mapping, which are also discussed in other chapters in this book (e.g., see Chapters 2, 5, and 11).

A common approach is to deploy the back-end and the physical device driver in a special privileged domain, as illustrated in Figure 10.1. This domain has additional privileges compared to other guest domains. Specifically, it is allowed to interact with the physical device directly. Xen uses this approach, and places all native device drivers and back-end modules in a single privileged

---

[1]The need for software routing is eliminated for emerging multi-queue types of devices further discussed later in this chapter.

domain — *dom0*. Dom0 runs a standard operating system, typically Linux, though other OS options do exist, and, for the most part, it can support the original device drivers, without requiring special device drivers to be available for the VMM software environment. Given its importance, dom0 is given higher priority with respect to its access to platform resources. For instance, whenever possible, it is given dedicated access to a CPU (at least a logical CPU on hyperthreaded platforms), so as to eliminate the need for context switches and to improve the latency of execution of the control functionality that it is responsible for. This is particularly useful with respect to I/O virtualization, as the invocation of this domain is on the fast path of each I/O operation.

Alternatively, the I/O virtualization stack may be part of the proper VMM kernel, as is the case with the I/O virtualization stack in VMware's ESX server. This approach relies on the availability of device drivers for the target VMM environment, by specifying and exposing similar type of device-driver APIs as available in operating systems, so that third party devices can be easily supported. The integration of the device domain in the hypervisor layer can result in performance improvements as it removes a layer of indirection by not requiring interactions with the privileged domain. However, it may expose the VMM to additional sources of faults as the the device drivers are no longer isolated in a separate domain.

Finally, a third alternative is to deploy each back-end and its associated physical device driver in a *dedicated driver domain* — a privileged domain responsible only for a specific device. The utility of this approach is that it addresses the issue that device drivers contribute to the largest percentage of systems software bugs. Therefore, isolating faulty drivers in a separate VM (driver domain) will only affect those VMs which actually use that device, and can more easily be dealt with by hotswapping the device and/or restarting the faulting driver domain[2]. In addition to running the native device driver and the back-end drivers, a driver domain is responsible for coordinating across I/O operations from multiple VMs, such as enforcing priority, fairness, and QoS policies.

Note that for brevity, in the remainder of the text we will use the term "driver domain" to refer to any of the above alternative deployment scenarios, unless it is necessary to point out the distinctions among them.

10.2.2   Performance Optimization Opportunities

In all of the aforementioned deployment alternatives for the device driver functionality, there are performance costs associated with the VM's access to the I/O virtualization stack. These include changes in privilege level

---

[2]Note that in order to truly guarantee that the faults are contained within the driver domain, additional memory protection mechanisms (IOMMUs) are needed to ensure the validity of the addresses used as I/O parameters.

(VM exit and entry operations), scheduling of other domains, and costs associated due to cache pollution, data movement and signaling. These costs can lead to a multifold reduction in the VM's I/O performance levels (throughput degradation and increase in latency) [286, 287], which may lead to the conclusion that virtualization is not a suitable candidate for environments with high performance communication requirements. However, numerous techniques have been proposed, and integrated in commercial virtualization products, which do provide performance optimization by eliminating some of these sources of overhead. We next focus on describing some of these features.

10.2.2.1   Data movement

One of the main sources of overhead in the split device-driver I/O model is the need to move data across multiple locations — from the application to the guest OS's protocol stack buffers (e.g., socket buffers), from the front-end to the back-end address space, and finally to the device itself. Similar steps are needed for data received from the device. One approach, as originally used in the early Xen versions, is to dynamically remap pages containing packets across the guest and driver domain [124]. However, when used for Ethernet sized packets, limited to 1.5kB in the event of regular Ethernet packets, the cost of these map/unmap operations may exceed the copy overheads. In order to improve performance, Xen reverted to include support for copy-based data movement.

Relying on a purely copy-based approach can lead to multiple inefficiencies, particularly with respect to the inability to take advantage of advanced NIC features, such as offload of TCP segmentation, Large Receive Offload (LRO), checksum computation, scatter-gather DMA support, or support for larger MTU sizes (e.g., 9kB Ethernet jumbo frames). NICs with these features allow exchanges of larger data units between the device and the native driver. In a copy-based approach, the copying of such large data items may result in high CPU loads and far greater fast-path overheads than the remapping costs, which will degrade both latency and throughput. Also, the implementation of the approach may require that data is segmented before being moved across the VM-driver domain boundary, because of limits in the size of the shared buffer elements used for inter-domain communication. This may offset or prevent any potential benefits from hardware-supported features for dealing with large data fragments.

Therefore, the I/O virtualization stack should have capability to support copy- as well as memory remapping-based data movement across domains. This is facilitated by including information in the request descriptors to specify which alternative is being used, and whether additional memory regions are being used for carrying *fragments* of the data. One challenge with enabling support for large data transfers is that these fragments may need to be represented as contiguous memory regions, in both the guest VM's memory view as well as in real physical memory. This implies that the VMM memory management subsystem must provide guests with the ability to make explicit requests regarding the guest memory mappings and their layout in the platform's physical memory.

The ability to support large data movements between the guest and driver domain creates additional performance improvement opportunities. First, the per request overheads are now amortized over larger data sizes, which improves sustainable throughput levels. As a result, even in the absence of NIC-level support for some of these large-data-block based protocol optimizations, it is still beneficial to perform data movement at larger granularity, and enhance the driver domain with software implementations of the offload features. This solution also results in greater portability, as now the front-end can be written to assume the existence of hardware offload, which can be transparently provided through software emulation in the driver domain in the event the physical NIC does not support such features.

Second, for NICs with scatter-gather DMA capabilities which operate with physical addresses, the need to remap data fragments into the driver domain can be mostly eliminated. We would still need to map, or copy, a portion of the packet containing protocol header information, but to initiate the NIC operation we would only need to establish the physical addresses of the fragments' locations. These addresses are accessible through the page table structures maintaining the mappings from the guest's view of the physical memory into physical memory pages. Performance gains can be derived by maintaining persistent, i.e., cached mappings across the driver and guest domains, provided that the communication protocol semantics and the device driver operations permit the reuse of the designated memory locations.

In addition to the above optimizations targeting larger-sized data movements, Santos et al. [380] recently demonstrated several optimization opportunities for copy-based data movement operations. First, they point out the importance of proper alignment, as hardware such as the 64bit x86 platforms used in their evaluation have different copy costs for copying data across addresses aligned at cache-line boundary vs. at 64 bit boundary vs. without any careful alignment. Provided that one can take advantage of cache-line aligned data copies, the back-end could copy data directly into the guests' network stack buffers (e.g., socket buffers). Similarly to the previous argument, these buffer pools, once mapped in the driver domain, could be reused, thereby amortizing those costs as well.

A second important optimization pointed out by the same authors is due to the caching hierarchy present in modern multi-core platforms. When the driver domain performs a copy into a shared page provided by the guest domain, the copy operation pollutes the cache used by the driver domain. It is very unlikely that the driver domain itself needs the same data — it is merely being passed to a guest, where it will be (hopefully) used by some application. Depending on the underlying physical platform, this cache may not be shared with the CPU where the guest domain and its application execute, therefore, the subsequent move of the data into the application space will not benefit from the driver domain's cache contents. The results in [380] demonstrate very significant performance gains which can be attained by modifying the page sharing protocol (i.e., grant mechanism in Xen) so that the guest front-end is responsible for executing the copy operation — the cached contents from that copy will likely be delivered to the application very soon, while they are still present in the guest VM's local cache. As memory distances continue to increase, the importance of such cache-conserving optimizations will continue to grow.

10.2.2.2 Scheduling

The last paragraph raises an important point, and that is that scheduling of domains on the physical platform resources also plays a role in the attainable communication performance in virtualized environments. First, collocating VMs on cores which share the cache with the control/driver domain will result in cache trashing and will impact the virtualized I/O performance levels for all remaining VMs on the platform. In virtualized environments, in general, the VM scheduling must be done with careful consideration of the underlying memory hierarchy in order to limit the performance interference across domains.

Second, for high-end devices the ability of the driver domain to respond and execute as soon as an I/O operation is issued is critical as the context/domain switching overheads can be significant relative to the per packet latency. Delaying the servicing of an I/O request can also cause buffer overflow and dropped packets, particularly when one considers multi-Gbps data rates. Ideally, the driver domain should be pinned to a dedicated CPU and not subject to the VMM scheduling policy. For multi-device platforms with dedicated driver domains (as opposed to a single privileged control domain such as dom0), this may be a challenge given current 4-8 core systems. However, as the number of cores continues to increase, it will be more feasible to support specialized driver domains for high-end devices.

10.2.2.3 Interrupt processing

Several studies have investigated the impact of interrupt processing in virtualized environments. Major reason for the overheads associated with this is that the cost of interrupt handling is increased three-fold in virtualized platforms following the driver domain model: the interrupt is initially intercepted by the VMM layer, next it is handled by the driver domain, which in turn generates a virtual interrupt to be handled by the guest domain. This leads to an important observation that interrupt coalescing techniques, which are nowadays supported on most devices, can have three times larger impact in virtualized settings [380]. The costs associated with interrupt processing and routing in virtualized platforms is one of the key reasons for advancement in hardware- and board interconnect-level technologies and standards. Intel, AMD and the SIG-PCI community are investing heavily in these efforts.

10.2.2.4   Specialized Driver Domain

Finally, the last set of optimization opportunities with respect to the "split device driver" model that we will mention is related to the driver domain's guest OS. A driver domain can execute a standard operating system which supports drivers for a given device. However, because of the reduced type of tasks which need to be executed in the driver domain (e.g., run the device driver and back-ends, and perform the required multiplexing and demultiplexing actions), it is possible to strip the guest OS of most of the unnecessary functionality and turn it into a specialized driver domain. By executing a minimal set of OS services needed to support only the network virtualization stack components, including the physical device driver, we minimize the driver domain's complexity and memory footprint, and eliminate potential internal noise sources. Multiple efforts have demonstrated the utility of such *mini OSes* or *specialized execution environments* for situations where only a well defined set of services or applications need to be supported [15, 177, 247, 379].

10.3   Direct Device Access Model

While there are multiple techniques to reduce the overheads incurred in the I/O virtualization approach described in the previous section, all of them still require a level of indirection on the fast-path of the VM-device interactions. When it comes to attaining high performance, VMs require direct access to the hardware device in order to experience near-native performance levels. Allowing VMs direct device access can be achieved in several ways.

A most straightforward way is for the VMM to give a VM exclusive access to a device, map the device solely in that particular VM and prevent all other VMs from using it. This approach clearly has significant limitations as it breaks one of the basic goals virtualization aims to provide by enabling consolidation and shared access to the same physical hardware by multiple VMs. The solution of giving each VM direct access to its own device has limited impact as it reduces the platform's scalability in terms of number of VMs which can be concurrently supported. This can be particularly limiting for blade-based systems, as these typically support a smaller number of devices per node.

Therefore, a practical solution is sought in a manner which allows partitioning of the device resources, so that each VM gets direct access to some subset of them. The model is illustrated in Figure 10.2. At a coarse-grain level, the

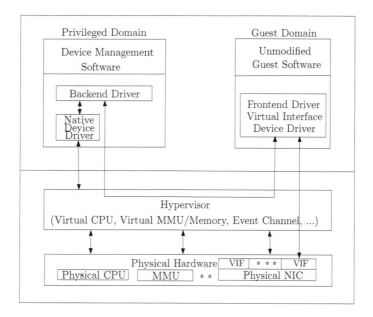

FIGURE 10.2: Direct device access model.

device access is still under the control of the VMM. As the figure illustrates, the control path which initializes the guest's device driver (i.e., the front-end) follows the split driver model and causes control to be transfered to the main driver domain. Only the native device driver, present in the driver domain, is allowed to perform any control plane operations which determine the set of device resources (i.e, DMA-memory, device interrupts, virtual identifier or lane, etc.) initialized and allocated for the guest's use. However, after the initialization phase is complete, all data plane operations (receive and send), are applied from the front-end directly to the device, thereby bypassing the VMM. For this reason this approach is referred to as *VMM-bypass*, or *pass-through*, since the VM-device interactions are allowed to bypass the VMM or to pass through the virtualization layer without being intercepted.

Enabling this type of I/O virtualization requires several device-level capabilities, discussed in the remainder of this section. Devices which have these capabilities are also referred to as *self-virtualizing devices* [357].

10.3.1 Multi-Queue Devices

Traditional devices export a single interface, in the case of communication devices consisting of set of receive and transmit queues, containing either the actual network packets, or containing packet descriptors which point to a memory location accessible to both host and device, where the packet is stored.

Requests from/for multiple VMs are multiplexed and are sharing this pair of queues. It is therefore necessary to virtualize them using the split device driver model — the back-end presents an interface to the front-end similar to this single queue pair interface, and the send/receive operations on the real device queue pair are multiplexed/demultiplexed in the driver domain.

However, many types of high-end devices, currently mostly with high performance capabilities, support multi-queue interfaces consisting of sets of queue pairs. Each of these queue pairs can then present an interface to a virtual device equivalent to the physical one, but with reduced levels of available resources. During initialization, the VMM and the driver domain map a queue pair instance in the guest VM and configure the (virtual) device interface so it can be uniquely used as a dedicated device for the given guest. Future data movement operations are performed directly between the VM's front-end and the virtual device instance represented by the mapped queue pair.

In spite of the fact that the virtual device is one with more limited resources, e.g., less memory available for packet queues, the communications capabilities provided by it, the latency and bandwidth which can be delivered to the VMs, can match those of the physical device. In fact, multiple research efforts which explore the performance capabilities of VMM-bypass type of I/O virtualization demonstrate the near-native performance levels, including for 10Gbps network devices [264, 360, 456]. Naturally, this is true only when the device is not concurrently used by other VMs, in which case it must multiplex I/O operations from all VMs onto the same communication channel, thereby dividing the maximum bandwidth among all VMs. Figure 10.3 illustrates this by showing that under increased network loads the aggregate bandwidth is divided evenly across all VMs.

The measurements shown in this figure are gathered for an InfiniBand network adapter, which is an example of an I/O device with multi-queue capabilities (for more details on InfiniBand see Chapter 2). Liu et al. [264] describe the implementation details of virtualizing these devices in the Xen VMM environment, and report negligible performance overheads compared to native execution. Several other groups [358, 457], as well as our own group [357], have come to similar conclusions using programmable hardware, FPGAs and programmable Intel IXP network processors, to prototype Ethernet NICs with similar capabilities.

More recently, with the commoditization of high-performance technologies such as 10GigE, multi-queue capabilities are enabled on high-end commodity Ethernet devices [209]. At both the OS and the VMM layer, solutions have emerged which take advantage of the hardware supported queues, in order to better utilize and share communication resources. Companies like Intel have gone further so as to more explicitly define *Virtual Multi Device Queues* (*VMDq*) as one of the cornerstone components in their Intel Virtualization VT technology [444]. In all of these cases, however, the existence of multiple hardware supported queue pair interfaces is only one of the device-level features

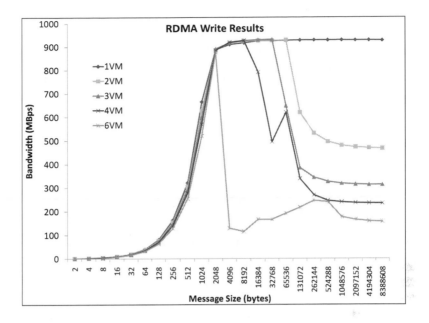

FIGURE 10.3: RDMA write bandwidth divided among VMs.

needed to support this type of VMM-bypass virtualized I/O operations. We next discuss other necessary device capabilities.

10.3.2 Device-Level Packet Classification

Clearly, one of the key requirements for multi-queue devices is that the device itself must be capable of immediately classifying network packets into the hardware-supported queues. Stated differently, the device must be capable of interpreting sufficient portions of the network header so as to classify and route incoming packets to the correct virtual queue.

For multi-queue Ethernet devices this means that the software bridging executed in the NIC's driver domain now must be offloaded on the device itself, either as a software module executing on the device-resident processing cores, or supported directly by the NIC hardware. Recently, 10GigE devices became available which offload on the device the packet classification functionality performed by the NetQueue component of VMware's ESX's network virtualization stack, and demultiplex packets into one of many queues based on the MAC address or the VLAN tag. Offloading the packet classification capability and enabling multi-queue support, boosts virtualized I/O performance — with industry-wide measurements indicating boosts from 4Gbps to near maximum rates for 10GigE devices.

Similar capabilities have been supported on InfiniBand devices all along. Note that in the case of InfiniBand HCAs, in addition to routing incoming requests and data into the appropriate queues, the device implements full transport termination.

10.3.3 Signaling

The existence of multiple queues for data and request descriptors can eliminate many of the costs associated with data movement in the split-driver model. However, whether the VM-device interaction can fully bypass the VMM and eliminate any fast-path overheads of VMexit/entry and scheduling operations will depend on the signaling interface supported by the device. Asynchronous, polling-based interfaces can easily benefit from the VMM-bypass channels. In this case, a memory mapped (set of) *doorbell* registers or event queues (e.g., the completion queues for InfiniBand adapters) can be used between the VM and the device to signal new requests or the completion of issued ones.

For synchronous, interrupt-driven interfaces the situation is different. In the absence of hardware support to directly route the interrupt to the appropriate VM, the interrupt still has to be handled first by the driver domain, and then rerouted to the target VM, in a manner similar to the split-driver model. However, the data movement costs associate with this model can be avoided, because the request queue and data buffers can be directly mapped in the target VM.

The use of message-based interrupts, like MSI and MSI-X, and the ongoing innovations for better virtualization support at the PCI/HyperTransport interconnect, and better interrupt renaming and routing capabilities in upcoming Intel and AMD platforms, will likely soon eliminate the need for driver domain intervention in the interrupt handling phase. Additional benefits can be derived by leveraging the multi-core nature of current and future platforms to enable the concurrent processing of multiple interrupts by co-scheduling the target VMs on all available cores. Our group's prior work [357], as well as others [457] have demonstrated the utility of using a bit vector to encode all pending interrupts from the virtual device interfaces, and then exploiting that information to guide VMM-level scheduling decisions.

10.3.4 IOMMU

While establishing the memory mappings among the virtual device queue and the VM allows for direct VM-device interactions, it does not on its own guarantee that DMA operations invoked during data movement operations are legal. This is the task of the I/O memory management unit, responsible for translation and validation of DMA address targets. In the absence of hardware-supported IOMMU the address translations must be performed by a software IOMMU entity part of the driver domain, and will result in significant fast-path overheads associated with each request.

Hardware IOMMUs may alleviate this issue and more efficiently enforce protection guarantees for the virtualized I/O operations, but their exact impact on performance is still not fully assessed. Certain measurements report as much as 45% degradation in communication performance in the presence of a hardware IOMMU on a prototype platform [38]. Intel VT-d IOMMU and AMD's GART memory management unit can improve these overheads since the basic costs associated with them are shown to be significantly lower compared to the prototype platform used in [38], but the overheads will remain non-negligible. At the cost of reduced protection one can eliminate the IOMMU operations from the fast-path in favor of improved performance, or at least partially eliminate them by reducing the granularity of protection. This may be a valid option when aiming to achieve line-rate throughput with virtualized high end devices.

Finally, the protection may rely on efficient on-device memory management components, as is the case with InfiniBand adapters. The memory management and protection features supported on these devices are described in greater detail in Section 2.3.3.

10.4 Opportunities and Trade-Offs

From the previous sections it is clear that enabling direct device access in virtualized environments is going to be a winning approach when it comes to attaining raw I/O performance. However, there are other factors which may further impact the overall application performance. We next discuss some of these factors.

10.4.1 Scalability

As with all offload approaches, offloading (parts of) the driver domain functionality onto the device, and making the device virtualization-aware or self-virtualizing [264, 357], will result in lower CPU utilization, and will reduce data copies and the memory load in the system, which in turn translates into improved host capacity, higher communication rates, and ability to scale to a larger number of concurrent VMs.

It is likely, however, that at least certain types of multi-queue devices will have hardware limits on the number of virtual interfaces they can present, which in turn can limit the maximum number of VMs per node. For InfiniBand devices this number is large, and in fact in our own work we have only been limited by the memory available on the host system to a total of 32 VMs.

Current 10GigE VMDq NICs have a smaller number of virtual queues supported by the device hardware, so some level of multiplexing of VMs per device queue is needed, and the use of host-based split device driver model is unavoidable.

10.4.2   Migration

By allowing direct device access, the VM is truly exposed to and dependent on the physical device. This first violates one of the objectives virtualization aims to achieve — and that is to decouple VMs and their guest OSs and applications from the underlying hardware. In addition, it presents challenges with respect to supporting migration, as the target node must support the same type of physical device. The split driver model eliminates these challenges. Furthermore, the split driver model may be used with multi-queue devices — so a VM can transparently benefit from reduced data movement costs in the event a multi-queue device is indeed present on the physical platform, but not be affected if one is not.

This does not imply that migration cannot be supported on direct bypass devices. The work in [195, 196, 382] demonstrates that even though there are significant challenges in migrating VMs across bypass devices they can benefit'from the high performance like Infiniband. Direct bypass devices rely on direct access to the state and memory addresses stored in the I/O queues of each VM for transfer of data. For migration to work without the knowledge of the VM/application, these buffer queues need to be remapped on the target machine to the same guest virtual addresses. Also, they need to be registered with the device, so it can continue to have direct access to host memory (i.e., the hardware IOMMU engine or the device-resident memory management unit must be configured to allow these accesses). The migration process and the copying of the VM's memory state can begin only after this registration completes. The migration is enabled by a VMM-level process, and Huang et al. [195] explain its implementation for the Xen VMM and Infiniband interconnects. In terms of performance, their results indicate that VM downtime during migration is substantially lower when the device is virtualized using the VMM-bypass direct device access model (when it can take full advantage of RDMA operations), compared to when it is used as an IP device virtualized using the split device driver model.

10.4.3   Higher-Level Interfaces

While a single VM will always perform better in the presence of VMM-bypass capable devices, for platforms with many consolidated VMs, the lack of a shared centralized resource may eliminate certain optimization opportunities. Higher level I/O APIs, such as file system services or other application-level services, such as search and web services or other RPC services, benefit from shared file caches or caches of application specific data and computation.

If multiple collocated VMs are relying on the same protocol stack, including application- or middleware-level processing associated with that I/O, it is possible that the driver domain can be customized to support this type of higher level interface (e.g., socket, file, or application library interface). In this case concrete performance benefits can be attained by implementing such caching services in the driver domain, as well as a result of using a dedicated domain for the target service and exploiting CPU and hardware cache locality benefits during its execution [292, 429].

In the event of pass through virtualization these types of optimization opportunities are lost. However, the overall impact on the performance of the communication services depends on the types of collocated workloads and their access patterns, i.e., the benefits which can be derived through shared use of service- or application-level cache depend on the degree of sharing and the distribution of accesses to shared data and state.

10.4.4 Monitoring and Management

Many workloads exhibit dynamic behavior, with periods of increased I/O activity or bursts in their communication patterns. In order to address this we rely on platform-level management mechanisms to continuously monitor resource utilization levels and make dynamic adjustments in the resource allocations. Examples of management actions include increasing CPU credits, physical memory used for the application/VM or its I/O queues. These are important not just to improve the aggregate utilization of platform resources, but also to enforce quality of service (QoS) policies and support SLA guarantees. In virtualized environments, hypervisor-level schedulers and memory managers carry out these tasks. The main reason for these QoS/SLA guarantees is the performance isolation of one VM from another. Also, in a data center environment, different customers need different levels of response for the application running inside the VM. It is the job of the management utility to interact with the hypervisor scheduler, memory managers, and I/O device queues to meet the SLAs of the VMs running on the physical server. There are various ways to set the QoS depending on the type of device, as discussed below.

In the split device driver model the driver domain has a centralized view of the VMs' utilization of their I/O resources, and can enforce sharing policies and QoS guarantees. In addition, as a privileged domain it can interact with the VMM-level CPU scheduler to request the corresponding CPU adjustment. This is important since it is insufficient to address the increased I/O needs of a VM by solely increasing its usage of device-level resources and share of I/O buffers. Instead, it is necessary to enable coordinated resource management policies which trigger adequate adjustments in other types of resources, such as VM CPU scheduling or memory allocation. Our own work [225], as well as other efforts [91, 319] demonstrate the importance of such coordination in improving the VMs' communications-related performance metrics, such as throughput or request rate, as well as their overall execution time. These results also show that the lack of coordination leads to performance degradation and significant levels of variability in the VMs' I/O rates.

For VMM-bypass devices, the situation is different. Once the VM is given access to its instance of the virtualized device, the VMM does not further control how that device is used. Since neither the VMM nor some dedicated driver domain intercept the VM-device interactions, we lose the opportunity to monitor these interactions, and, if necessary, to trigger any management tasks needed to make appropriate adjustments in the VM's usage of the aggregate platform resources (i.e., CPU, Memory and I/O).

It may still be possible to obtain monitoring information and have some ability to manage the I/O resources by relying on external entities, such as QoS-aware switches and subnet managers, though at reduced frequency or granularity. For instance, the network device can have the ability to enforce QoS policies across the hardware-supported queues. As Chapter 2 explains, InfiniBand HCAs can map the queue pairs (i.e., the virtualized interfaces) to one of several fabric-supported *Virtual Lanes* (VLs). By controlling the mapping of queue pairs to VLs in the HCA, as well as the fabric-level mapping of *Service Levels* (i.e., priorities) to VLs, it is possible to enforce certain types of QoS policies at a Virtual Lane granularity. However, this may not be sufficient to differentiate between all VM-VM communications. The *Channel I/O Virtualization* [284] support present in newer Mellanox HCAs allows VMs direct access to the device while eliminating any need for setting up the control path through the VMM. The enforcement of QoS is still acheived via the hardware-supported queues but now each VM is associated with a VL. This provides more monitoring data regarding VMs at the fabric level. By specifying these SL-VL mappings at the fabric level, a certain degree of co-ordination is needed between the management utility running on each physical server and the switch to enforce these on that physical server.

Another approach used in our group leverages the capabilities of a smart fabric switch [467], built using specialized network processors and FPGA components, which leverages IB's credit-based link-level flow control to provide QoS guarantees at a queue pair/virtual interface granularity. However, the use of an external, fabric-level entity to enforce QoS policies, and to also provide monitoring information, does not allow for tight coupling between the VMs' behavior and the management processes, and as such may not be adequate in all situations.

Our group recently introduced an approach to address these needs, and demonstrated its utility via a prototype implementation targeting InfiniBand devices virtualized with the Xen VMM [359]. Our tool, termed IBMon, uses memory introspection techniques [156, 332] to asynchronously monitor the updates to the IB queue pairs mapped in a VM's address space, and to observe the VM's I/O utilization, request patterns, and even data being sent or received, if necessary. This work makes several key observations. First, that in order to enable the coordinated resource management mechanisms needed to deal with the inherent dynamism in virtualized platforms (various types of workloads, consolidation and migration, etc.) the VMM-level management tools must be able to obtain device utilization information at the virtual interface granularity. It is common for devices to maintain and export aggregate resource utilization data, but not necessarily at finer, per virtual interface, granularity. Therefore, for VMM-bypass devices, the hypervisor or the management tools must rely on gray-box monitoring techniques. In the case of IBmon we must know the in-memory layout of the IB queue pairs in order to determine the memory region on which the memory introspection techniques are being applied, and we must understand sufficient details about the IB protocols in order to interpret the utilization information incorporated in the work requests placed in those queues. Finally, our results highlight the utility of extending the request descriptors with some timing related information, such as timestamps, as it can greatly help improve the accuracy of the monitored information and dramatically reduce the overheads of the monitoring process.

10.5 Chapter Summary

This chapter discusses the main approaches to achieving virtualized I/O services in modern virtualized environments. We discuss in greater detail the mechanisms involved in supporting each of the two main approaches: the split device-driver model, and the direct device access model. For both approaches we discuss possibilities to improve the runtime overheads incurred in the virtualized I/O stack through a variety of software and hardware-supported techniques, at the virtualization layer (i.e., in the hypervisor or in the accompanying device driver domain), or at the device itself. In addition we present a brief discussion of additional performance improvement opportunities which can be enabled in conjunction with each approach.

Chapter 11

The Message Passing Interface (MPI)

Jeff Squyres
Cisco Systems

11.1 Introduction

Since the mid-1990s, the Message Passing Interface has become the de facto standard for parallel computing and High Performance Computing (HPC). Originally written as a codification and standardization of decades of parallel programming experience, the MPI standards documents have grown and evolved over time. The original MPI-1.0 specification was passed in May of 1994 [288]. Work on extensions to MPI-1 began shortly thereafter; the MPI-2.0 specification was passed in July of 1997 [160]. Annotated versions of the two standards documents have also been published ([408], and [170], respectively). After a long hiatus, MPI-2.1 was published in June of 2008 [149].[1]

11.1.1 Chapter Scope

This chapter provides some historical perspective on MPI and outlines some of the challenges in creating a software and/or hardware implementation of the MPI standard. These challenges are directly related to MPI applications; understanding the issues involved can help programmers write better MPI-based codes.

This chapter assumes a basic familiarity with MPI; it does not serve as an MPI tutorial.

---

[1]As of this writing, work continues on the MPI-2.2 and MPI-3.0 documents. MPI-2.2 is expected to be finalized in September of 2009, while the first parts of MPI-3.0 are not expected to be ready until at least 2010.

11.1.2 MPI Implementations

After MPI-2.0 was published, the focus of the MPI community generally shifted away from evolving the standard; most effort concentrated on creating, "hardening," and featurizing MPI implementations. A number of different MPI implementations appeared throughout the late 1990s and early 2000s, each with its own set of goals and objectives. The following list of goals is by no means comprehensive, but it provides a summary of some of the more popular rationale for creating a new MPI implementation.

- Platform for research: Academics and researchers used MPI as a vehicle for research; these MPI implementations investigated experimental aspects of providing MPI services such as run-time issues, network transport issues, collective algorithms, etc. The focus of these implementations was to extend the state of the art, not necessarily provide robust implementations intended for general use.

- Portability: Some MPI implementation projects aimed to provide a portable MPI implementation that worked across a variety of platforms, networks, and/or operating systems, usually in the open source community.

- Support a specific network: As MPI became more popular in HPC applications, customers began requiring MPI implementations adapted and tuned for specific network types from their network vendors. This model was particularly common in the commodity cluster market, where niche network types such as Myrinet, InfiniBand, Quadrics, and others were popular.

- Support a specific platform: Similar to network vendors in the commodity cluster market, HPC system integrators such as SGI, Cray, IBM, and others also needed to provide an MPI implementation specifically tuned for their platforms.

- Make money: Some vendor-provided MPI implementations are not free, either in their initial purchase price and/or support contract costs.

In 2008, many MPI implementations still exist, most embodying one or more of the goals listed above. These implementations remain distinct from each other for technical, political, and monetary reasons.

11.1.2.1 Technical Reasons

At its core, an MPI implementation is about message passing: the seemingly simple act of moving bytes from one place to another. However, there are a large number of other services and data structures that accompany this core functionality. The MPI specification contains over 300 API functions and dozens of pre-defined constants. Each of these API functions has specific, defined behavior (frequently related to other API functions) to which an implementation must conform. Enforcing this complex defined behavior requires an intricate web of supporting data structures. The creation and run-time maintenance of these structures is a difficult task, requiring careful coding and painstaking debugging.

Even though the MPI documents are a standard, there are many places — both deliberate and unfortunately unintentional — where the text is ambiguous, and an MPI developer has to make a choice in the implementation. Should MPI_SEND block or return immediately? Should a given message be sent eagerly or use a rendezvous protocol? Should progress occur on asynchronous or polling basis? Are simultaneous user threads supported? Are errors handled? And if so, how? And so on — the list is lengthy.

Each implementer may answer these questions differently, largely depending on the goals of their specific implementation. Additionally, research-quality implementations are likely to take shortcuts in areas that are unrelated to their particular research topic(s). Other implementations are hardened production quality products that must be able to run large parallel jobs for weeks at a time without leaking resources or crashing.

As such, MPI implementation code bases tend to be both large and complex. Merging them into a single implementation (or perhaps just a small number of implementations) is simply not feasible; it would be far easier to start a new implementation from scratch with the explicit goal of supporting a large number of platforms and environments.[2]

11.1.2.2 Political Reasons

The HPC and MPI communities are fairly small; many members of these communities have known and worked with each other for long periods of time. But even in these small communities, there is a wide variety of different opinions and strongly-held core beliefs. For example, in the highly-competitive research community, it is common practice (and necessary) for pioneering new approaches to definitely prove that prior approaches were not good enough.

As a direct result, since the research community is important to both the HPC and MPI communities, different funding-worthy ideas can necessitate that distinct research efforts each spawn their own MPI implementations.

11.1.2.3 Monetary Reasons

Some vendor-created MPI implementations serve as a revenue stream. These MPI implementations are typically closed source and provide proprietary

---

[2]One such MPI project, Open MPI [154], was initiated with exactly this concept. Its base design was explicitly designed to be able to support a variety of environments via plugins, thereby making it significantly easier to resolve inter-platform differences in a single code base.

"secret sauce" for high performance, such as optimized MPI collective algorithms, tuned resource utilization techniques, and accelerators for specific applications. Support contracts may also be offered; enterprise-class customers in particular tend to prefer contractual customer support options rather than relying on the goodwill of the open source community for addressing business-critical problems.

It is not in the interests of these vendors to combine their technologies with others, nor to provide proprietary details on how they achieve performance that beats their competitors. The situation is similar to that of the research community: competition is fierce.

11.1.3   MPI Standard Evolution

A number of errata items for the MPI specifications were raised throughout the late 1990s and early 2000s. Some were voted to "official" status via email consensus by those who remained on the official Forum e-mail lists, but a new document containing the errata was not formally published.

The Forum reconvened in early 2008 for three reasons:

- to produce an MPI-2.1 document that incorporated the email-voted errata, finalized numerous new errata, and combined the previously-separate MPI-1 and MPI-2 documents. After nine months of work, the MPI-2.1 document was published in June, 2008 [149].

- to produce an MPI-2.2 document containing fixes and errata that were deemed "too large" for the relatively quick release of the MPI-2.1 document.

- to start work on MPI-3, extending the MPI-2.x series with entirely new functionality.

11.1.3.1   Portability

An explicit reason for the creation of the MPI standard was to provide application source code portability between a variety of parallel platforms. Prior to MPI, each HPC platform had its own programming API with its own unique requirements and abstractions. Writing portable parallel applications was difficult at best. The MPI-1 API solved this problem by liberally importing ideas from many prior parallel programming systems, and by effectively codifying the best practices of the previous two decades of parallel computing.

MPI-1 therefore achieved a giant leap forward in the state of the art by providing a common API and source code portability for library and application program developers. However, an unfortunate practical reality arose: *performance portability* was not guaranteed between different MPI implementations. As previously mentioned, MPI implementations are large, complex code bases, each of which makes different decisions in both design and implementation.

These differences can lead to significantly different run-time behavior in some applications.

Indeed, some users would find that their application ran fine with MPI implementation X, but failed in some way with MPI implementation Y. Sometimes these problems were due to a faulty or a non-conformant use of the MPI API. For example, an incorrect application may just *happen* to work fine with MPI implementation X, but fail on implementation Y. The fact that the application is incorrect is usually lost on the user; in such large code bases, it can be difficult to pinpoint whether it is the application or the MPI implementation that is faulty. In other cases, the application itself is correct, but implementations X and Y chose to effect some portion of the MPI standard differently, and therefore exhibit different run-time characteristics for the same application. This situation can be extremely frustrating for users.

11.1.4 Chapter Overview

The rest of this chapter is broken down as follows: Section 11.2 discusses MPI's place in the OSI network model, and MPI's relationship with underlying networks. Section 11.3 outlines challenges and rewards associated with multi-threading MPI applications. Section 11.4 details issues surrounding MPI's point-to-point operations, while Section 11.5 briefly describes some issues with MPI collective operations. Finally, Section 11.6 discusses some strategies for creating an MPI implementation.

11.2 MPI's Layer in the Network Stack

The MPI API deliberately hides underlying platform, architecture, and hardware abstractions from the application. While the entire 300+ MPI API functions are quite complex, a small set of basic functions can be used for message passing that are significantly simpler than creating and using TCP sockets. The following functions are typically referred to as "6 function MPI," and can be used to implement a large class of MPI applications:

- MPI_INIT: Initialize the MPI implementation.

- MPI_COMM_SIZE: Query how many processes are in a communicator.

- MPI_COMM_RANK: Query this process' unique identity within a communicator.

- MPI_SEND: Blocking send.

- MPI_RECV: Blocking receive.

- MPI_FINALIZE: Shut down the MPI layer.

The simplicity of these functions lies in the fact that they free the application from issues such as connection establishment and management, network resource management, platform heterogeneity, and unit message transmission and receipt. As an inter-process communication (IPC) mechanism, simplistic use of MPI is targeted at scientists and engineers who neither know nor care how the underlying network functions. They want to concentrate the majority of their work on the application's problem domain and use MPI as a straightforward IPC channel between processes.

More advanced MPI usage can be used to enhance overall application performance. Non-blocking point-to-point communications and collective communications are among the most common enhancements used by applications beyond "6 function MPI."

Note that the application is responsible for all message buffer management. Although an MPI implementation manages all access to networking hardware (which may include some level of buffer management), the source and target message buffers are solely managed by the application. This presents additional complexity in some network models, such as those requiring "registered" memory (see Section 11.6.2 for more details on registered memory).

11.2.1   OSI Network Stack

An MPI implementation can span multiple layers of the OSI network layer stack. Depending on the specific network model exhibited by the underlying hardware, an MPI implementation may range from a simplistic bit-pusher to a complex, network-routing entity. Indeed, some MPI implementations even have components inside the operating system or the network hardware. As such, it is difficult to characterize MPI's position in the OSI model other than to define that its top layer provides users application-level access to some underlying network.

11.2.2   Networks That Provide MPI-Like Interfaces

Several network models were designed specifically to support MPI. These include Portals, Quadrics' Tports interface, Myrinet's MX interface, and Infinipath's PSM interface. All of these network models export an interface that is quite similar to the user-level point-to-point MPI API. In most cases, the network hardware itself (perhaps through firmware) supports abstractions that are quite similar to MPI's core concepts such as process identification, unit message transmission and receipt, and tag-based message matching.

A direct consequence is that an MPI API implementation over these interfaces can be quite "thin" in the sense that very little data structure and control translation needs to occur between the MPI application and the underlying network layer. For example, an MPI implementation over the Portals interface discussed in Chapter 6 can simply allocate a large "slab" of memory for incoming network messages. The underlying network layer efficiently uses the memory based on the size of each incoming message, freeing the MPI implementation from having to deal with many memory allocation issues.

However, even for networks that export MPI-like interface, it is not always possible to optimize same-node communication — typically implemented in shared memory — with hardware-based acceleration. For example, the overheads incurred by fusing shared memory and network hardware MPI wildcard message matching typically negate any speed advantage that hardware matching alone could provide. As such, MPI implementations based on these interfaces must typically perform all MPI message matching in software.

11.2.3   Networks That Provide Non-MPI-Like Interfaces

Other network interfaces and transports such as TCP sockets, Ethernet frames, OpenFabrics verbs, and shared memory all present abstraction models that do not exactly match MPI's requirements. An MPI implementation therefore must translate MPI communicators, tags, and message matching and progress concepts to the underlying network. This translation process is typically affected both by what abstractions the network can support and run-time resource management.

An MPI implementation based on TCP sockets, for example, must translate successive partial transmissions and receipts into discrete MPI messages. OpenFabrics-based MPI implementations must manage their own incoming message buffers (vs. the "slab" Portals approach), and must take great pains to ensure that incoming messages efficiently utilize precious network-registered memory.

11.2.4   Resource Management

An MPI implementation (or the network layer below it) typically has to effect a careful balance between resource utilization and delivered performance. Two examples of common resource allocation issues are unexpected message buffer management and coping with disparate on-host and off-host processing speeds.

11.2.4.1   Unexpected Messages

"Unexpected" messages are where an MPI process receives a message for which the application has not posted a receive buffer with a matching MPI message signature. MPI implementations typically buffer unexpected messages in a separate queue such that when/if the receiver application posts a matching receive request, it can be satisfied immediately by copying the message from the unexpected queue to the target buffer.

However, a fast sender can quickly overwhelm a slow receiver if the slow receiver unconditionally buffers unexpected messages: the receiver can simply run out of memory. However, effecting limits on unexpected messages entails at least two complex issues:

- A "pause" handshake must be implemented where the receiver can tell the sender to temporarily stop sending. However, this has an unfortunate side-effect that effectively forces an MPI implementation to acknowledge every message (which may be redundant with network hardware acknowledgements, leading to further network overhead). Sliding window protocols are therefore typically used to offset network latency — eagerly sending as much data as possible and aggregating receiver ACKs to notify the sender when corresponding sends have completed. This necessarily means that the sender may need to wait for ACKs from the receiver before marking an MPI send request as complete. However, this is problematic for MPI implementations without asynchronous progress (e.g., single-threaded MPI implementations that only make message passing progress when the application invokes MPI functions): what if the receiver does not call an MPI function for a long, long time? The sender will be left in limbo, waiting for an ACK despite the fact that the message may be safely sitting in the receiver's network hardware, simply waiting for the MPI implementation to process it.

- A degenerate case that happens in at least some real-world MPI applications is where N MPI processes overwhelm a slow receiver with unexpected messages. Sliding windows are effective when dealing with a small number of peers, but consider the resources required to support N sliding windows with a depth of M messages, each of size S. Even if M is relatively small, the overall buffering required ($M \times N \times S$) quickly becomes dominated by N (assuming that S is kept small, perhaps by the MPI implementation fragmenting large messages and using rendezvous protocols in an attempt to not overwhelm receivers before the MPI application posts a matching receive). The overall buffering required may simply be too large for a receiver to handle.

Solving these issues quickly becomes incredibly complex, both from a theoretical and practical perspective. Consider the following seemingly simplistic case: a single set of receive buffers is used for all senders. Flow control is implemented by the receiver sending a "pause" message to all senders when its pool of receive buffers is nearly empty.

Broadcasting the "pause" message from process A to all of its peers is an obviously scalability bottleneck. Not only does the broadcast take a linear amount of time to send to N peers, it can cause congestion on the network, consume resources on both the sender and the receiver, and likely needlessly involve peer processes who are not currently communicating with process A. For example, what happens if one of A's peers cannot receive the "pause" message because it, too, is out of receive buffers?

As this example shows, flow control, particularly with large-scale MPI jobs, can be a complex issue.

11.2.4.2 Processing Speed Disparity

The main processor in a typical HPC computation host may be significantly faster than the processor in its network interface. As such, even with today's low latency, high bandwidth networks, an MPI application that frequently sends small messages can overwhelm the network's ability to actually transmit them. At some point, the MPI implementation must actually block any new send requests until at least some of the pending requests can be transmitted.

An obvious solution for this problem is for the MPI implementation to allocate more queueing buffers at the network layer (e.g., increase the network interface's send queue depth), perhaps in conjunction with a higher-level queue (e.g., have a "pending send" queue in the MPI layer). Larger queues means that the MPI implementation can queue up more send requests which, assuming the network hardware can process them asynchronously from the MPI software layer, translates into efficient network use. But increasing queue depth consumes memory resources (and potentially other kinds of resources) that then become unavailable to both the sending process and potentially to other processes on the same host.[3]

Exactly what limits an MPI implementation should impose is therefore a complex issue. Increasing the queue depth can improve one process' network throughput but at the cost of consuming resources that could be used by other entities (including the application itself), which may therefore cause a corresponding drop in performance as measured by a different metric.

This section has only briefly touched on several resource management issues that an MPI implementation must face, and has not even provided any definitive solutions to the problems presented. It is beyond the scope of this chapter to fully address these issues, and is left as an exercise to the reader to explore workable solutions.

---

[3]Core counts in hosts are increasing; an MPI implementation cannot assume that it is the only process running on a host. It therefore must assume that it has to share resources with others (potentially other MPI processes in the same parallel job, or potentially even MPI processes in different parallel jobs).

11.3   Threading and MPI

The MPI-1 specification included minimal discussion of threading issues; MPI-2 added explicit descriptions and API function regarding threading issues with MPI applications. MPI supports four levels of threading, usually indicated by passing an option to the MPI_INIT_THREAD function call:

1. MPI_THREAD_SINGLE: Only one thread will execute in the MPI process.

2. MPI_THREAD_FUNNELED: The process may be multi-threaded, but only the thread which invoked MPI_INIT or MPI_INIT_THREAD will call MPI functions (all MPI calls are "funneled" to the main thread).

3. MPI_THREAD_SERIALIZED: The process may be multi-threaded and multiple threads may make MPI calls, but only one at a time. MPI calls are not made concurrently from two distinct threads (i.e., all MPI calls are "serialized").

4. MPI_THREAD_MULTIPLE: Multiple threads in a single process may simultaneously call MPI functions.

11.3.1   Implementation Complexity

All full MPI-1 implementations inherently support MPI_THREAD_SINGLE. Most full implementations found that the additional logic and infrastructure necessary for MPI_THREAD_FUNNELED and MPI_THREAD_SERIALIZED was relatively straightforward to implement. But only some MPI implementations fully support MPI_THREAD_MULTIPLE. At least one reason for this lack of thread support is because full support of MPI_THREAD_MULTIPLE is extremely difficult to implement. Consider that multi-threaded applications are already difficult to write and debug. Adding a new dimension of asynchronicity — additional concurrent MPI processes, each with multiple threads — significantly increases the complexity of the application's run-time behavior. Race conditions and other complex interactions between threads and MPI processes (both from the standpoint of the MPI application and the MPI implementation) can become prohibitively difficult to debug and maintain.

Many MPI applications are either not multi-threaded or take measures to ensure that their use of the MPI implementation is *effectively* single threaded, meaning that they restrict all MPI calls to a single thread, or use a global lock to ensure that only one thread is calling MPI at any given time. Such schemes allow multi-threaded applications to utilize MPI implementations that only support MPI_THREAD_FUNNELED and MPI_THREAD_SERIALIZED.

The continued lack of MPI_THREAD_MULTIPLE support in MPI implementations is fueled by a paradox: there are few highly threaded MPI applications because there are few MPI implementations that support MPI_-THREAD_MULTIPLE. But there are few MPI implementations that support MPI_THREAD_MULTIPLE because not many applications require it.

11.3.2 Application Simplicity

MPI_THREAD_MULTIPLE support in an MPI implementation is desirable for many of the same reasons that multi-threaded support is useful in non-MPI applications. Not only can it effect good communication and computation overlap (e.g., let a thread block in an MPI function that utilizes I/O hardware while other threads are free to perform other processor-bound application work), some algorithms are naturally suited to blocking patterns rather than non-blocking, re-entrant state machines.

A canonical example showing this effect is a task farm MPI application. Consider a single "manager" MPI process that provides work for and receives results from a group of "worker" MPI processes. Two ways that the manager could be implemented are:

1. Single-threaded manager: the manager loops over sending initial work requests to each child. The manager then loops over posting a blocking MPI receive from MPI_ANY_SOURCE, waiting for results from any worker. When the receive completes, the manager looks up which worker sent the messages, pairs the results with the corresponding outstanding work request, and then looks up a suitable pending work request to send to the now-idle worker.

2. Multi-threaded manager: the manager spawns off N threads, each of which is responsible for a single worker. Each thread simply loops over a blocking send of the work request to the worker and a blocking receive from the worker to get the results. No additional lookups or state tables are necessary because each thread is only interacting with a single worker.

This example is artificial and fairly simplistic; it only shows marginal benefits of a threaded implementation. But consider extending the example to include stateful request and reply protocols between the manager and workers. In order to achieve maximum message passing concurrency to keep the workers as busy as possible, a single-threaded manager would need a re-entrant state machine to know where in the protocol it was with each worker. After each receive/send, the manager would need to go back to the central protocol engine because the next incoming message may be from a different worker. In this case, a threaded model would require no additional state machine; each thread is only dealing with a single worker, and therefore never has to swap that worker's current state with another.

11.3.3 Performance Implications

Threaded MPI implementations have performance implications compared to non-threaded MPI implementations. For example, threaded MPI implementations:

- May have to utilize locks to serialize access to network and I/O devices. This may negatively impact microbenchmark performance, particularly the latency performance of short messages.[4]

- May require threaded IPC mechanisms to allow fair progress of multiple application threads using MPI.

- May be able to implement progress threads and/or be able to provide better communication/computation overlap, particularly for long messages.

For example, the task farm example described in Section 11.3.2 can quickly run into scalability issues if the number of threads managing workers grows large.

11.4 Point-to-Point Communications

While it is possible to write a wide variety of MPI applications using only six function MPI, some applications benefit from more advanced features.

11.4.1 Communication/Computation Overlap

Since communication across a network can take orders of magnitude longer than local computations — even with high bandwidth, low latency networks — a common optimization is to overlap remote communication with local computation. The idea is to structure the MPI application in the following general form:

1. Initiate network communication

2. Perform local computation

3. Poll (or block waiting) for completion of the network communication

---

[4]For this reason, some MPI implementations that support MPI_THREAD_MULTIPLE offer two versions of their MPI library: one that supports MPI_THREAD_MULTIPLE and one that does not. The version that does not support MPI_THREAD_MULTIPLE may exhibit lower latency for short messages because of a lack of threaded IPC and associated coordination (e.g., locks).

Even if the MPI implementation is single threaded, some degree of overlap is possible by allowing the operating system and/or networking hardware to progress message passing "in the background."

MPI offers several different forms of non-blocking communication; the simplest of which are *immediate* sends and receives. The MPI_ISEND and MPI_IRECV functions return "immediately," [5] regardless of whether the communication has completed or not. These functions return an MPI *request* which then must be tested later for completion.

Requests can be tested for completion in a blocking or non-blocking manner via the MPI_WAIT and MPI_TEST functions, respectively. Several variants of these functions are also available, including:

- MPI_WAITALL: Block waiting for all specified requests to complete.

- MPI_WAITANY: Block waiting for any one of several specified requests to complete.

- MPI_WAITSOME: Block waiting for at least one of the specified requests to complete (although many MPI implementations treat this function as a synonym for MPI_WAITANY).

Analogous non-blocking versions are also available in the MPI_TEST × functions:

- MPI_TESTALL: Returns true if all of the specified requests are complete; false otherwise.

- MPI_TESTANY: Returns true if any one of the specified requests is complete; false otherwise.

- MPI_TESTSOME: Returns the indices of any requests that are complete.

11.4.2   Pre-Posting Receive Buffers

Another advanced technique is the receive-side optimization of posting receives for messages that are expected to arrive sometime in the future. It can be advantageous to post such receives when a receiver is certain that the corresponding messages have not yet been sent, thereby guaranteeing that the receive buffers will be posted before the intended messages arrive. Pre-posting receive buffers in this manner offers the following advantages:

---

[5]Most MPI implementations interpret "immediately" to mean "within a short, finite time." Specifically, they will try to make progress on any outstanding message passing requests before returning. Interpretations vary between MPI implementations; the only characteristics they share is that they will not block indefinitely.

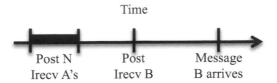

Post N   Post   Message
Irecv A's   Irecv B   B arrives

FIGURE 11.1:   The application first posts N non-blocking receives of signature A, creating a lengthy message envelope list inside the MPI implementation. When the application later posts a single receive of signature B and a message matching B arrives, the entire envelope list may need to be checked before finding the match.

- Incoming messages will be "expected," meaning that they already have resources allocated to hold the received data. Temporary resource allocation, a common side-effect of unexpected message management (described in Section 11.2.4.1), is therefore avoided.

- Incoming messages may be received directly into their target buffers, as opposed to receiving into a temporary buffer and copying to the final destination buffer later.

- Rendezvous protocols for large messages typically wait for a matching receive to be posted by the receiver before transferring the bulk of the message. Pre-posted receives for large messages therefore allow the rendezvous protocol to proceed immediately because the data is expected and the target buffer is available.

- The MPI implementation is given the opportunity to effect message passing progress "in the background," potentially while the application is off performing local computation. This is especially relevant for MPI implementations that provide asynchronous message passing progress, but can also be effected by the application periodically calling MPI_TEST. Consider: even if the message has not been fully transferred when MPI_TEST is invoked, the MPI implementation's progression engine can advance message passing state each time MPI_TEST is invoked.

However, pre-posting receives does not scale well, consuming both application memory (and possibly network resources) and internal MPI resources. Depending on how the MPI implementation handles incoming messages, pre-posting a large number of receives may increase the time necessary to search for a match.

Consider a simplistic MPI implementation that handles posted receives as a linked list of MPI envelope data (communicator ID, peer process ID, message tag). Posting a large number of receives creates a lengthy list of envelopes that may need to be searched in its entirety for each incoming message; see Figure 11.1.

However, since pre-posting receives is a common MPI application technique, MPI implementations have had to devise schemes to accommodate the practice without sacrificing performance. Some techniques include:

- Use pre-allocated memory (e.g., freelists) for pending match envelopes to avoid the overhead of allocating memory from the heap.

- Have each element on the pending envelope match list be a container that is capable of holding multiple envelopes. In this way, not only is the frequency of new match envelope allocations reduced, it also increases cache locality of searching through the list.

- Use a unique envelope match list per communicator. In this way, MPI applications are only penalized if they post large numbers of receives on individual communicators.

- Use a unique envelope match list per peer. This scheme creates complications for wildcard receives however; posting a receive from MPI_ANY_- SOURCE requires adding an entry on each peer list in a communicator. Similarly, matching that envelope requires removing it from all the lists where it was added, which may require additional synchronization in multi-threaded environments.

- Use a hash table for tags; each bucket is an envelope match list for all the tags in the bucket. This method potentially reduces the search space to only envelopes that share the same tag bucket.

These optimizations are not as straightforward as they seem; extreme care must be taken to provide a trivially simple (and fast) model for popular MPI benchmarks. Specifically, blocking ping-pong benchmarks are frequently used to evaluate the efficiency of an MPI implementation. Additionally, care must be taken to ensure that the memory required to post match envelopes does not grow too large, thereby denying the application the resources that it may need.

Using one or more of the above schemes is useful for achieving high scalability (both in terms of number of MPI processes in a job and in the number of simultaneous requests a single MPI process can handle), but sometimes at the cost of increasing overhead — and therefore also increasing message passing latency — for simple MPI applications that only use blocking sends and receives.

11.4.3 Persistent Requests

MPI also provides *persistent requests* that separate the setup of a message passing action from the actual action itself: a communication action can be set up once and then executed many times. The rationale for this separation assumes that iterative applications can reduce the overhead when the same communications are repeatedly executed in a loop.

For example, the MPI_SEND_INIT function can be used to set up a send. This function accepts all the same parameters as MPI_SEND but instead of actually initiating the send, it sets up any relevant internal data structures, potentially pre-allocates any necessary sending resources, and then returns an MPI request representing the already-setup send. This request can then be used repeatedly to start and complete the send, having only incurred the setup overhead once. Specifically, the request can be passed to MPI_START to actually initiate the send (or an array of persistent requests can be started with MPI_STARTALL). Completion is treated just like a non-blocking send: the request must be checked with one of the MPI_TEST or MPI_WAIT variants. Once the request has been completed, the cycle can be started again. The following pseudocode shows an example:

```
/* One-time setup cost */
MPI_Send_init(buf, count, type, dest, tag, comm, &req);

/* Main loop */
for (...) {
    MPI_Start(&req);
    /* ...compute... */
    MPI_Wait(&req, &status);
}

/* Cancel and discard the persistent request */
MPI_Cancel(&req);
MPI_Wait(&req);
```

Similarly, MPI_RECV_INIT can be used to effect persistent receives.

The time required for any given communication can be modeled as $(T_s + T_c)$, where T_s is the time required for local setup and T_c is the time required for the actual communication. Note that T_s and T_c may vary, depending on the specific operation, peer, size of message, and other factors. An iterative application that executes the same N communication actions for M iterations would therefore incur the following cost:

$$M \times \sum_{i=1}^{N} (T_{si} + T_{ci}) \tag{11.1}$$

Using persistent requests, the same communication pattern would only need to setup the communication actions once, and therefore incur the following cost:

$$\sum_{i=1}^{N} T_{si} + M \times \sum_{i=1}^{N} T_{ci} \qquad (11.2)$$

11.4.4 Common Mistakes

Programmers who are new to MPI tend to make the same kinds of mistakes; there are common pitfalls that cause applications to fail, or, even worse, cause degraded performance (sometimes in extremely subtle ways).

11.4.4.1 Orphaning MPI Requests

Non-blocking MPI communication requests must be reaped when their associated communication has completed. If an application creates large numbers of requests and never reaps them — perhaps instead relying on other events to know when a communication has completed — the amount of memory used to hold these requests can reduce the amount available to the application. Applications that only reap requests infrequently may not incur out-of-memory scenarios, but may experience memory fragmentation and cache misses (or, more specifically, fewer than expected cache hits) due to the fact that MPI requests are usually allocated from the heap.

11.4.4.2 Overusing MPI_ANY_SOURCE

The receive source rank wildcard MPI_ANY_SOURCE has long been hated by MPI implementers; its use prevents certain types of optimizations on some platforms. Some MPI implementations utilize these optimizations until a receive with MPI_ANY_SOURCE is posted. The optimizations are then disabled until all pending communications using MPI_ANY_SOURCE have completed. Applications that lazily use MPI_ANY_SOURCE instead of specifically identifying senders (i.e., when the sending rank *could* be known ahead of time) can actually be harmful to performance.

MPI_ANY_SOURCE is not entirely bad, however. It can be necessary for applications to scale to arbitrarily large numbers of processes: posting N receives (e.g., one per sender) can be prohibitive at large scale, as discussed in Section 11.4.2.

11.4.4.3 Misusing MPI_PROBE

The use of MPI_PROBE (and its non-blocking counterpart: MPI_IPROBE) should be avoided when possible. MPI_PROBE can force the MPI implementation to receive an incoming message into a temporary buffer. This causes extra overhead, both when allocating and freeing the temporary buffer, and when copying the message to the final destination buffer when the matching receive is posted.

Additionally, MPI_PROBE likely invokes the MPI implementations progression "engine" to attempt to progress any outgoing messages and check for incoming messages. MPI_RECV may do the same thing. Hence, probing for a message and then receiving it may cost more than just allocating, copying, and freeing a temporary buffer; it can incur multiple passes through the progression engine when only one pass may have been sufficient.

MPI_PROBE is useful in some scenarios, however. MPI does not currently provide a way to post for a receive of an unknown size. A common technique is to not post a receive, but rather probe for incoming messages. The result of the probe can be used to determine the incoming message's size. The application can then allocate a buffer of the appropriate size and post the matching receive. Depending on the application, supporting unknown message sizes may be more important than the potential performance degradation incurred by using probes.

11.4.4.4   Premature Buffer Reuse

It does not make sense to write to a buffer that is being used by an ongoing non-blocking send or receive. However, it can be easy to mistakenly use (or re-use) buffers in this way before their pending communications have completed. This can lead to undefined behavior; in all cases, it is unknown exactly what will end up in the receiver's buffer.

11.4.4.5   Serialization

Communication patterns must be constructed with care to ensure that they do not accidentally serialize the overall computation. Consider the following code:[6]

```
MPI_Comm_rank(comm, &rank);
MPI_Comm_size(comm, &size);
left = (rank == 0) ? MPI_PROC_NULL : rank - 1;
right = (rank == size - 1) ? MPI_PROC_NULL : rank + 1;

for (...) {
  MPI_Recv(rbuf, count, type, left, tag, comm, &stat);
  /* ...compute... */
  MPI_Send(sbuf, count, type, right, tag, comm);
}
```

In the first iteration, the above code creates a synchronization effect where each MPI process must wait for the process to its left to send before it can

---

[6]Recall that sending to or receiving from MPI_PROC_NULL is a no-op. In the example shown, since rank 0 has no left neighbor, setting left = MPI_PROC_NULL makes its MPI_RECV from the left be a no-op. Similarly, setting right = MPI_PROC_NULL in rank $(size - 1)$ makes its send to the right be a no-op.

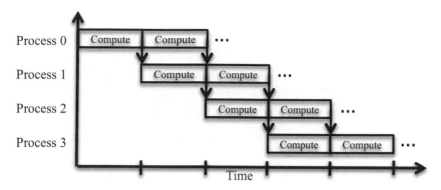

FIGURE 11.2: Two iterations of a sample serialized application. Each process cannot start iteration M until the prior process has sent a message indicating that it has finished iteration $(M - 1)$. This serialization pattern effectively wastes the time required for $(num\_processes - 1)$ iterations.

complete. Specifically, process N must wait for each of the $(N - 1)$ processes before it to complete their computations and send to the right. See Figure 11.2. Successive iterations occur immediately, however, because process $(N - 1)$ will have completed the prior iteration's send, allowing process N to receive immediately. Hence, if the time required to complete one iteration is T, the overall time of the parallel job of N processes to complete M iterations can be expressed as:

$$T \times (M + N - 1) \tag{11.3}$$

The serialized communication pattern effectively causes $T \times (N-1)$ overhead.

Depending on the application and the data being processed, it may be possible to split the computation into what must wait for the receive to complete and that which can be computed before the receive completes:

```
for (...) {
  MPI_Irecv(rbuf, count, type, left, tag, comm, &rreq);
  /* ...compute that which is not dependent on rbuf... */
  MPI_Wait(&rreq, &stat);
  /* ...compute all that is left over... */
  MPI_Send(sbuf, count, type, right, tag, comm);
}
```

Additionally, it may be possible to move the send earlier. Here is a possible decomposition (assuming that **sbuf** is at least partially dependent upon **rbuf**):

```
for (...) {
  MPI_Irecv(rbuf, count, type, left, tag, comm, &rreq);
```

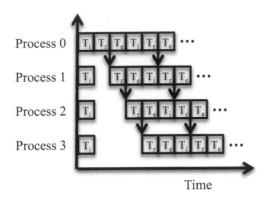

FIGURE 11.3:   Similar to Figure 11.2, two iterations of a sample serialized communication. However, the communication and computation have been overlapped such that the overhead of the serialization is lower. In particular, each process only needs to wait for T_r from the prior process to complete (vs. $(T_i + T_r + T_e)$).

```
    /* ...compute that which is not dependent on rbuf... */
    MPI_Wait(&rreq, &stat);
    /* ...compute that which is necessary for sbuf and
           dependent on rbuf... */
    MPI_Isend(sbuf, count, type, right, tag, comm, &sreq);
    /* ...compute all that is left over... */
    MPI_Wait(&sreq, &stat);
}
```

The end effect is that each process computes as much as it can before it starts computing that which is dependent on `rbuf`, thereby allowing for a communication and computation overlap. It then computes the minimum necessary before it sends out `sbuf`, and then finally computes whatever is left over. While there is still serialization in this pattern, its effect is mitigated as much as possible. The time required for an iteration is now split into three components: T_i is the time required to compute that which is not dependent upon `rbuf`, T_r is the time required to compute that which is dependent on `rbuf`, and T_e is the time required to compute everything else. See Figure 11.3. Hence, the overall time of the parallel job of N processes to complete M iterations can be expressed as:

$$T_r \times (N - 1) + M \times (T_i + T_r + T_e) \qquad (11.4)$$

This pattern is faster than the first pseudocode solution; the difference can be expressed by subtracting equation (11.4) from equation (11.3).

$$T \times (M + N - 1) - (T_r \times (N - 1) + M \times (T_i + T_r + T_e))$$
$$(T_i + T_r + T_e) \times (M + N - 1) - (T_r \times (N - 1) + M \times (T_i + T_r + T_e))$$
$$(T_i + T_r + T_e) \times (N - 1) - T_r \times (N - 1)$$
$$(T_i + T_e) \times (N - 1) \qquad (11.5)$$

The two main optimization opportunities are:

1. Use non-blocking communication to decouple communication and computation

2. Reorder the computation to compute with local data first, then with remote data after it arrives

Depending on the application and data dependencies, further communication and computation overlap may be possible.

11.4.4.6 Assuming Blocking Communication Buffering

As alluded to in Section 11.1.3.1, a common complaint on MPI support forums is "My application works with MPI implementation X, but hangs with MPI implementation Y." More often than not, this is due to a communication pattern similar to the following:

```
MPI_Comm_rank(comm, &rank);
MPI_Comm_size(comm, &size);
left = (rank == 0) ? MPI_PROC_NULL : rank - 1;
right = (rank == size - 1) ? MPI_PROC_NULL : rank + 1;

MPI_Send(sbuf, count, type, right, tag, comm);
MPI_Recv(rbuf, count, type, left, tag, comm, &stat);
```

Specifically, the application is assuming that the call to MPI_SEND will not block and that the MPI implementation will provide enough buffering to cache the entire **sbuf** message and return, possibly even before the message has been sent. Recall that the return from MPI_SEND does *not* guarantee the message has been sent; it only guarantees that the **sbuf** buffer is available for re-use.

Perhaps in MPI implementation X, the entire message is sent eagerly. In this case, MPI_SEND returns and then MPI_RECV is invoked, and the MPI process continues. But in implementation Y, the sending message is long enough to force a rendezvous protocol. In this case, since a matching receive has not yet been posted, all MPI processes will block in the MPI_SEND waiting for a receive that will never occur.

It is worth noting that the MPI standard specifically identifies this communication pattern as unsafe and warns that it should not be used. There are multiple ways to avoid this pattern, including using non-blocking communications, or using pairwise exchanges.

11.5   Collective Operations

MPI offers many different collective patterns of communications, some involving data movement, others involving both computation and communication. Early MPI implementations provided simplistic algorithms for these collective operations that did not perform well.[7] As a direct result, many users ignored the MPI-provided collectives and implemented their own, typically achieving better performance (sometimes dramatically so).

Modern MPI implementations now include the fruits of over a decade of research into MPI collective communications; MPI implementation-provided collective algorithms tend to be much more optimized and well-tuned than what a typical user could implement themselves. Collective communications still remains an active area of research, however. New algorithmic insights, new network architectures and topologies, and new demand from MPI applications (ever since MPI developers have stopped hand-coding their own collective algorithms in their applications) have kept pushing the state of the art in MPI collective communications.

A typical MPI implementation may include several different implementations of a given collective operation. Which one is used at run time can be a combination of many different factors:

- Size of the message being transferred

- Number of processes involved in the collective

- Type, speed, and capacity of the underlying network used to transfer messages

- Locations of the MPI processes in the network

- Whether MPI_THREAD_MULTIPLE is being used or not

11.5.1   Synchronization

A notable point about MPI's collective operations is that only one of them provides any level of guarantee about synchronization: MPI_BARRIER. MPI's barrier operation is defined such that no process will leave the barrier until all processes in the communicator have entered the barrier. This is a fairly weak synchronization guarantee; the definition does *not* provide a guarantee about exactly when processes will leave the barrier. It only indicates a lower bound

---

[7]In the early days of MPI implementations, more emphasis was placed on point-to-point performance than collective operations performance.

for when processes *may* leave the barrier: after the last process has joined the barrier.

It is questionable as to whether any MPI application that relies on a barrier for correctness is, itself, a correct MPI program. Many MPI applications use barriers simply because they were ported from other communication systems that required the use of barrier. Other than the weak synchronization guarantee described above, MPI's barrier provides no additional guarantees about other MPI communications.

None of the other MPI collective operations provide synchronization guarantees. MPI_GATHER is a simple example: it is possible for a non-root process to send its buffer eagerly to the root and then return before the root has called MPI_GATHER.

11.6   Implementation Strategies

As discussed in the prior sections, implementing an MPI progression engine is a complex task. There are many rules dictated by the MPI standard that must be obeyed. The implementer must strike a delicate balance between "good" performance and ancillary issues such as resource management and CPU utilization (where "good" is usually defined differently by each user).

The following sections outline some common implementation strategies used by real-world MPI implementations.

11.6.1   Lazy Connection Setup

On some platforms, connections between MPI processes consume resources at a linear (or near-linear) rate. Before it was common to run hundreds and thousands of MPI processes, MPI implementations would pre-create network connections between all processes during MPI_INIT. Since both the number of processes was relatively small and connections were established in parallel, the time required to pre-connect all the processes was fairly short.

However, MPI execution environments have continued to grow in size. MPI jobs consisting of hundreds and thousands of processes are now commonplace. Hence, MPI implementations on platforms with linear connection costs now typically use a "lazy connection" model. In this model, connections are only made upon the first MPI communication between two processes. For example, TCP sockets, OpenFabrics queue pairs, and shared memory between MPI process peers do not need to be created until necessary (i.e., a message is sent between them). Since many MPI applications exhibit only nearest-neighbor communication patterns, only a small number of connections per MPI process may be required. Therefore, not opening network connections until necessary can conserve memory and network resources that can then be used by the application.

Lazy connection schemes do have some drawbacks:

- The *first* communication between a pair of peers will likely be slower than all following communications. For process peers that exchange many messages, this initial overhead cost becomes negligible over time. But for MPI applications that communicate only a few messages with many different process peers, the overhead can be enormous, especially if the connections are not made in parallel.

- Lazy connection schemes may require complicated state machines, particularly for single-threaded MPI implementations. Pending sends and receives must be queued until connections have been established. Network timeouts can also be difficult to handle, such as where process X calls MPI_SEND, but process Y does not enter the MPI library to complete the connection until after a network timeout. Process X then needs to cancel the first connection attempt and start another. This state machine can get fairly complex; consider unreliable networks where the first connection request may have actually reached process Y, but process Y's reply to process X got lost. All of these complexities need to be handled in a re-entrant state machine rather than a simpler, blocking model (e.g., making all connections during MPI_INIT).

- If an MPI process is eventually going to communicate with every other MPI process, a lazy connection model can be artificially slow. It would be faster to have the MPI make all connections during MPI_INIT using a parallel algorithm. Additionally, memory allocated for network resources will be less fragmented, potentially leading to more cache-friendly connection data accesses and traversal.

Other platforms do not suffer from such restrictions (e.g., those that utilize connectionless network models). The MPI implementations on these platforms can avoid complex lazy connection schemes and perform all (or nearly all) setup during MPI_INIT.

11.6.2　Registered Memory

Some networks require the use of "registered" memory, where the network interface (and possibly the operating system) needs to be notified of exactly which memory will be used for sending and receiving before that memory is used for communication actions. Such networks typically also require pre-posting buffers for *all* receives; it is an error if a message is received and there is no registered buffer available in which to receive it.

Additionally, the cost of the registration (and deregistration) process may be fairly expensive, perhaps involving communications across an I/O channel such

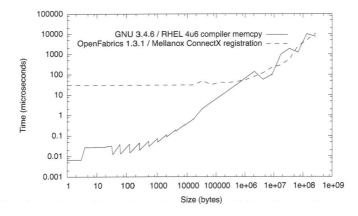

FIGURE 11.4: Graph showing memory copy times vs. OpenFabrics memory registration times for a range of message sizes. The figure data was generated from a Dell 1950 server with a 3.3GHz "Wolfdale" Intel 5260 processor running Redhat Linux 4 update 6, the Gnu C compiler v3.4.6, the OpenFabrics Enterprise Distribution v1.3.1, and Mellanox ConnectX MT_04A0110002 InfiniBand HCAs.

as a PCIe bus, and scale linearly with how much memory is being (de)registered. Finally, the amount of registered memory available on a node is likely to be less than the total memory on the node. Hence, registered memory is a "precious" resource that needs to be conserved.

With these constraints, it is easy to see that the management of registered memory is a collection of complex and subtle issues. For example, the cost of sending a short message can be prohibitively high if the MPI implementation needs to register a buffer, send the message, and then (optionally) unregister the buffer. Instead, most MPI implementations preregister a pool of buffers to send and receive short messages (perhaps during MPI_INIT, or perhaps when a network connection is made). When sending a short message, the message is copied from the user's buffer into a preregistered buffer and sent from there. Similarly, short messages are received into preregistered buffers and copied out to the user's destination buffer.

The size of a "short" message is usually a function of the cost of copying memory vs. the cost of registering memory. Given that memory copying one byte is far faster than registering one byte,[8] one metric for computing the maximum size of a short message is to find the number of bytes where the cost of a memory copy is equal to that of registering memory. All messages smaller

---

[8]Although network interfaces may register in units as small as one byte, operating system typically pin and unpin in units of one page.

than this size are "short" and should use preregistered buffers.

Using this metric, Figure 11.4 suggests that "short" messages should be approximately 1MB. However, if "short" messages were this size, the MPI implementation would not be able to allocate many preregistered buffers. Additionally, received messages that are only a few bytes long would waste large amounts of registered memory because each 1MB buffer would only be fractionally filled. Hence, other metrics, such as allocation fairness and memory registration utilization efficiency must also be taken into consideration when deciding the length of short messages.

Since incoming messages must arrive in preregistered buffers, flow control must be used between process pairs to ensure that messages are only sent when receive buffers are available. After messages are received, receivers must send acknowledgements when new receive buffers have been posted.

Determining the number of buffers to preregister for short messages is a complex resource management issue. If registered buffers are specifically tied to a given network connection, the maximum memory consumed by preregistered buffers can be expressed as $(P \times N \times S)$, where P is the number of peers, N is the number of preregistered buffers for each peer, and S is the size of each buffer. P is usually controlled by the user (i.e., how many MPI processes are launched in the application), but N and S are usually determined by the MPI implementation. Good values strike a balance between not consuming too much memory and being able to sustain transmitting short messages at full speed, even when flow control is being used.

In contrast to short messages, MPI implementations on remote direct memory access (RDMA)-capable networks typically communicate long messages directly to and from the user's buffer. This avoids excessive use of preregistered buffers and extra memory copies. The registration cost in this case may be large, however, so MPI implementations may register each buffer the first time it is used in an MPI communication call, and then *leave it registered*. The next time that buffer is used in an MPI communication call, the MPI can look up its address in a registration cache and see that it is already registered. It can therefore avoid the costly registration and use the buffer immediately. This scheme is obviously most useful in applications that reuse the same communication buffers many times.

Registration caches can be problematic, however, because the MPI application may free a registered buffer. In current POSIX-like operating systems, the MPI implementation does not have direct visibility for when memory is released back to the operating system, and therefore does not know to update its registration cache. Consider the following:

1. Application allocates buffer B from the heap.

2. Application calls MPI_SEND with buffer B.

3. MPI implementation registers buffer B, adds it to its registration cache, and then sends it.

4. Application frees buffer B.

5. Application allocates buffer C from the heap. Buffer C has the same virtual address as the now-freed buffer B, but has a different physical address.

6. Application calls MPI_SEND with buffer C.

7. MPI implementation finds the address in its registration cache, thinks that buffer C is already registered, and therefore sends the buffer immediately.

In this scenario, the network may send from the wrong physical address (i.e., where buffer B was physically located). Typically, the application will either generate an operating system error because the physical page where B resides is no longer part of the process, or worse, the application will silently send the wrong data.

An MPI implementation typically has two choices to fix this problem:

1. Tell the operating system to *never* reclaim memory from the process. Hence, even if the application frees memory, it will still stay part of the process; the scenario outlined above cannot happen because buffers with different physical addresses will always have different virtual addresses.

2. Discover when memory is being returned to the operating system and update its registration cache.[9]

Finally, it is possible to exhaust registered memory. Specifically, the MPI implementation may fail when attempting to register a new buffer and receive an error indicating that no more registered memory is available. In this case, most MPI implementations that support leaving user memory registered (paired with a registration cache) also include some kind of forcible eviction policy from the registration cache. For example, the cache can be searched for the memory that both was used longest ago and is no longer being used by any active MPI communications. This memory can be unregistered and removed from the cache. The MPI implementation can then attempt to re-register the original memory. This process can repeat until the memory registration succeeds and the communication proceeds, or there is no more eligible memory to evict and the MPI implementation must raise an error.

---

[9]The exact method used to intercept "return memory to the operating system" events are dependent upon the operating system. A variety of unofficial "back door" methods exist to intercept these events, many of which cause problems in real-world applications. A real solution will require the operation system to export memory controller hooks to user-level application.

11.6.3   Message Passing Progress

An MPI progress engine is full of many implementation choices:

- Should the progress engine aggressively poll for progress (i.e., never block), or should it use a blocking method? Aggressively polling usually yields the best latency, but consumes CPU cycles and is not generally compatible with the concept of having MPI progress occur in a separate thread. Blocking is more friendly to the concept of a progress thread, but is likely to be slower.

- Should the progress engine poll on any possible incoming message source, or only on sources that have corresponding MPI receives posted? Polling all possible incoming message sources can provide a higher degree of fairness for senders and allow progress on unexpected messages, but may be slower than only checking sources where messages are expected.

- Is the MPI envelope matching done in hardware or software? Hardware matching is likely to be faster than software matching, but less general and may not be possible on all platforms (e.g., where both shared memory and an off-node network is used for MPI communications).

- Does the MPI progress engine support asynchronous progress? Specifically: does the MPI application need to keep invoking MPI functions to make progress on pending communications, or will progress occur in the background, independent of what the MPI application does? True asynchronous progress is usually desired by MPI application developers, but there will likely be at least some cost associated with it (e.g., higher latency, lower message injection rates, lower bandwidth, etc.).

Message passing is usually aided by hardware offload support; some of the best performing MPI environments are where the majority of the message passing logic and progressing are implemented in hardware. In these environments, MPI_SEND is a thin wrapper around passing the buffer to the hardware and letting the hardware do all the work from there.

While many platforms have some kind of network offload support, not all of them match MPI's message passing or matching models. On these platforms, MPI implementations may be limited in terms of what offload capabilities can be usefully utilized.

11.6.4   Trade-Offs

Implementing a point-to-point messaging engine that can perform all of the rules specified by MPI and still provide "good" performance is a difficult task, particularly when "good" depends on which metric(s) are measured. Frequently, microbenchmarks (such as latency- and bandwidth-reporting peer-wise ping-pong) are used to judge the efficiency of an MPI implementation. However, not only are there many other metrics that tend to be more representative of the behaviors of real-world applications, but also the optimization of one metric may sometimes degrade another. For example, optimization of some or all of the following may incur penalties in terms of latency and bandwidth (some of which were described earlier in this chapter):

- Communication and computation overlap

- Support of MPI_THREAD_MULTIPLE

- Management and flow control for unexpected messages

- Message network injection rates

- Scalability to support MPI jobs consisting of hundreds, thousands, or tens of thousands of MPI processes

- Allocation and management of network buffers (and other resources) within a single process

- CPU context switching and utilization within a single process

- Allocation and management of operating system resource utilization within a single process

- Fairness between processes on a single node (memory, network resources, CPU utilization, bus contention, etc.)

- MPI-level error checking and reporting

A good MPI implementation strives to provide a point-to-point message passing engine that both is highly optimized for microbenchmarks and can also handle scalability and corner cases that are not uncommon in real-world applications.

11.7 Chapter Summary

This chapter has briefly discussed many of the issues surrounding MPI implementations. There are many, many other topics surrounding MPI that could not be covered in this chapter due to space limitations. MPI and OpenMP threading models, for example, are becoming popular, but still need much research before they become mature. Network congestion is another example that is fast becoming a major issue for MPI applications — even with modern high speed, high capacity networks. Why? And how can an MPI implementation help avoid network congestion issues? Indeed, several of the individual topics discussed here could be expanded into entire chapters (or books!) themselves.

This chapter has provided the reader with a taste of the complexity of message passing, and at least some insight into how MPI implementations work. Use the information in this chapter to glean insight into how to write better MPI applications. The reader also, hopefully, has gained some appreciation for the types of trade-offs in design, implementation, and run-time choices that have to be made in an MPI implementation. Readers are encouraged to use this chapter as a starting point into exploring the complexity that MPI application developers and end-users take for granted.

Readers should remember that the axiom "there is no such thing as a free lunch" holds true with MPI implementations. Tweaking the highly complex implementation to optimize one metric may well degrade another. It is the job of an MPI implementation to strike a delicate balance between high performance and resource constraints.

MPI implementations will inevitably evolve with changing HPC architectures. As of this publication, the "manycore" computing era is just starting. In some ways, manycore is a return to the days of large SMPs, but with some important differences. For example, with so many processors in one server, when binding them together in a commodity-style cluster, will it be necessary to have multiple (perhaps many) network interfaces in each server? And if so, will MPI implementations need to incorporate router-like characteristics in order to intelligently send messages to peers along "the most optimal path" (where "optimal" may be measured by several different metrics)? In addition to manycore, variations on classic architectures are evolving to support new HPC demands: extremely large amounts of RAM in a single node, multiple levels of NUMA in commodity servers, faster and higher capacity networks, including (potentially) multiple network interfaces in different topology locations in a single server and on the network, etc.

Chapter 12

High Performance Event Communication

Greg Eisenhauer, Matthew Wolf,
Hasan Abbasi, Karsten Schwan
Georgia Institute of Technology

12.1 Introduction

Middleware plays an important role in bridging the high performance communications goals of the application level and the capabilities and capacities of the underlying physical layer. In Chapter 11, we have already seen the effectiveness of MPI in this context. MPI couples both a relatively rich communications middleware infrastructure and a process invocation model that works extraordinarily well for the set of high performance applications that are its focus. However, the need for high performance communications extends beyond that set of applications, including ones that need a less rigid process model, more flexible type handling, or support for higher-level discovery services. This chapter looks at one solution for these sorts of high performance needs, namely the event-based, publish-subscribe middleware approach.

The publish-subscribe paradigm is well-suited to the reactive nature of many novel applications such as collaborative environments, compositions of high performance codes, or enterprise computing. Such applications often evolve from tightly-coupled components running in a single environment to collaborating components shared amongst diverse underlying computational centers. Traditional high performance scientific application components now may also include those responsible for data analysis, temporary and long term storage, data visualization, data preprocessing or staging for input or output. In the business domain, business intelligence flows may include scheduling (airline planning), logistics (package delivery), ROI estimation (just-in-time control), or rule engines (dynamic retail floor space planning). Such components may run on different operating systems and hardware platforms, and they may be written by different organizations in different languages. Complete "applications," then, are constructed by assembling these components in a plug-and-play fashion.

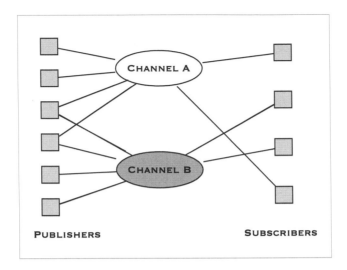

FIGURE 12.1:   A conceptual schematic of a channel-based event delivery system. Logically, data flagged as belonging to channel A or B flows from the publishers through a channel holder and to the subscriber. Particular system implementations may or may not require a centralized channel message center.

The *decoupled* approach to communication offered by event-based systems has been used to aid system adaptability, scalability, and fault-tolerance [138] as it enables the rapid and dynamic integration of legacy software into distributed systems, supports software reuse, facilitates software evolution, and has proved to be a good fit for component-based approaches. This chapter also restricts this space to those example applications and software stacks where *performance* is also a part of the requirements.

In particular, high performance event systems can be categorized along two axes. One axis categorizes the event subscription model — ranging from category-based subscriptions (e.g., all "Passenger:boarding" events) at one end to complex event processing at the other (e.g., event D only delivered if event A is in range $[a1 : a2]$, event B is in $[b1 : b2]$, and there has been no event C since time T_i). In other words, this axis of the design space defines whether delivery is based on metadata or data selection criteria.

This first axis is generally useful for characterizing any event system. The restriction for high performance is addressed in the second axis, with a trade-off between scalability (e.g., millions of publishers or subscribers) and throughput (e.g., handling 10k messages per second). Note that this is not to say that either of these axes are strict dichotomies — particular systems may in fact be able to handle both scalability and throughput, or deal with both metadata and complex data selection rules. However, the design choices made for most systems tend to be focused on one or the other, with very few exceptions. In

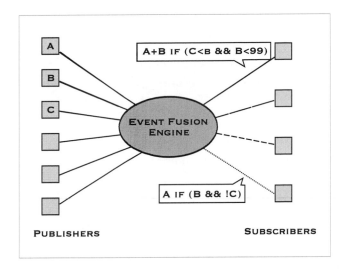

FIGURE 12.2: A conceptual schematic of a complex event processing delivery system. Logically, all data flows into a centralized processing engine that runs each subscriber's personalized subscription request.

the following section, we explore the design trade-offs along these axes in more detail by examining some exemplar event systems.

12.2 Design Points

Historically, event-based publish/subscribe communication owes much to early work on group communication protocols in the distributed systems domain such as [46]. These early systems were largely concerned with abstract properties of group communication, such as fault tolerance, message ordering, and distributed consensus. As group communication grew in popularity a wide variety of systems were developed, each providing a different set of features and performance compromises. Some systems provide features that are incompatible with other potential performance goals. For example, fully-ordered communication requires a very heavy-weight protocol that is unlikely to achieve high performance and may not scale to many subscribers. Because of these incompatibilities, each implementation tended to represent a single point in the possible design space which was intended to be optimal for the types of applications it was attempting to address.

TIBCO's Information Bus [430] was an early commercial offering in the enterprise space used for the delivery of financial data to banks and similar institutions. TIBCO is an example of a throughput-oriented, *subject-* or *topic-based* subscription system, where events are associated with a particular textual "topic" and publishing or subscribing to events associated with a topic constitutes joining a group communication domain. For TIBCO and many such systems, topics are hierarchical, so one subscribes to all NASDAQ trades by subscribing to "/equities/nasdaq/" or narrows the subscription to a specific company by subscribing to "/equities/nasdaq/IBM." Because event routing in these systems depends only upon the relatively static topic information, topic-based systems usually don't require transporting all events to a central location and are well suited for scaling to many subscribers. However, there is no opportunity for subscribers to request some subset of events other than the subdivisions predetermined by the event source by its choice of publishing topic.

Vitria's Velocity server [405] and the CORBA Event Service [318] represent a design point that is similar in flexibility to the TIBCO solution, except that rather than being labeled with textual topics, events are published into "event channels." These channels control the distribution of events so that subscribers associated themselves with a specific channel and receive events that are submitted it to his channel. In CORBA, these channels are objects themselves, implying a certain centralization in event routing, but conceptually the *channel-based* model of event subscription shares much with the *topic-based* model in that one can be used to implement the other with great success. Thus, topic-based and channel-based systems share both the upside and downside of their relatively static event routing models: the potential for scalability, but also the limits of subscriber flexibility.

In *content-based* subscription systems, such as Sienna [80] and Gryphon [419], subscribers can express interest in events based upon certain run-time properties. Subscriptions are expressed as a boolean function over these properties and effectively deliver a filtered version of the overall event stream to the subscriber. Systems supporting content-based subscriptions vary widely in the nature of the properties and the means through which the filtering function is expressed, usually as a function of how event data itself is represented. In traditional message-style systems where the event data is simply a block of data to be moved, properties might be represented as a separate set of name/value pairs associated with the event. This approach is taken by the CORBA Notification Service [317], a variation of the CORBA event service that adds event filtering to its subscription system. Alternatively, if the message is self-describing or supports some kind of introspection on its content, the run-time properties available for filtering can be the data elements of the event itself. For example, an event representing trading data might contain elements such as the company name "IBM" and the trade price "87.54."

At another point in the design space, grid computing allows the interconnection of heterogeneous resources to provide large scale computing resources for high performance applications. Because of its distributed focus, scalability of solutions is at a premium. The Open Grid Services Architecture (OGSA) [169] provides a web service interface to Grid systems. Two separate notification specifications are defined for event updates within the OGSA, WS-Notification [150] and WS-Eventing [52]. WS-Notification defines three interlinked standards, WS-BaseNotifications, WS-BrokeredNotification and WS-Topics. A subscription is made to a specific notification producer, either directly for WS-BaseNotification or through a broker for WS-BrokeredNotification. In addition, WS-Topics allows subscriptions to specific "topics" through the definition of a topic space. By using XML to describe the subscription and the event, the WS-Notification standard provides a high level of interoperability with other open standards. The downside to using XML for such processing is the performance implication of text parsing. Implementations of WS-Notification, such as WS-Messenger [198] trade-off event and query complexity in order to provide for a scalable high throughput event processing system. Similarly Meteor [216] is a content based middleware that provides scalability and high throughput for event processing. Meteor achieves scalability by refining the query into an aggregation tree using a mapping of a space filling curve. By aggressively routing and aggregating queries, Meteor can maintain low computational and messaging overheads.

In the quadrant addressing both scalability and complex event processing, Cayuga [57] is a general purpose stateful event processing system. Using an operator driven query language Cayuga can match event patterns and safety conditions that traditional data stream languages cannot easily address. Cayuga defines a set of operations as part of its algebra allowing the implementation and optimization of complex queries for single events and event streams [117, 118]. Cayuga defines a formal event language through the use of the Cayuga event algebra.

The ECho publish-subscribe system [132] provides a channel-based subscription model, but it also allows subscribers to customize, or *derive*, the publisher's data before delivery. ECho was designed to scale to the high data rates of HPC applications by providing efficient binary transmission of event data with additional features that support runtime data-type discovery and enterprise-scale application evolution. ECho provides this advanced functionality while still delivering full network bandwidth to application-level communications, a critical capability for high-performance codes.

12.2.1 Lessons from Previous Designs

In observing the space of high performance event systems, and in particular some of the authors' experience in developing the ECho system, several lessons were learned. We present them here in order to expose design motivations common in high performance eventing and to lead towards a proposal for a next-generation event infrastructure which is covered in later sections.

First, there was the observation that rigid topic subscriptions, although allowing good scalability, were not flexible enough for many applications. So, to support the performance needs of complex HPC applications, ECho's API allowed receivers to specify event filtering and transformation in "derived event channels" as a means to customize their delivered event stream. However, those transformations could only happen in the context of type-rigid event channels (i.e., channels that could only carry events of a particular data type). Many applications need to send multiple types of data to the same clients (raw data, control, metadata), which because of ECho's type-rigid channel semantics forced them to create multiple event channels with endpoints in the same processes. While the presence of these extra channels created little additional overhead after they were established, channel creation and maintenance costs were substantial.

A second issue that impacted ECho's scalability was that its bookkeeping operations were implemented using a peer-to-peer message passing protocol that required all communication endpoints associated with a channel (sinks and sources) to maintain a fully-connected communication topology. This was a reasonable design decision at the time because ECho's target domain of high-performance computing typically had only a few endpoints. However, as high performance messaging requirements evolved, this design choice made it impractical for ECho to scale to large numbers of endpoints.

Third, ECho's control mechanism was an obstacle to extension to another promising domain, optimizing event delivery by exploiting more recent innovations in peer-to-peer and overlay networks [73]. Specifically, because the ECho protocol was largely peer-to-peer rather than client-server, there was no unique source for knowledge about subscribers. Instead, such knowledge is maintained by the control protocol itself and associated (i.e., "cached locally") with each process involved in a channel. This resulted in a complex protocol for creating and updating subscriber information. Extending it to accommodate the creation and maintenance of internal overlay networks was a daunting task, particularly given the domain mandates to experiment with multiple overlay mechanisms.

12.2.2   Next Generation Event Delivery

In order to more fully explore the space of high performance event systems, the authors began work on an infrastructure called EVPath, for "EVent Path." Having presented the lessons learned from prior designs above, we now explore the means by which those lessons impacted the next generation's system design and its goals.

Instead of offering publish/subscribe services directly, EVPath's goal is to provide a basis for implementing diverse higher level event-based communication models, using the following approach:

- *Separating event-handling from control.* Event systems are largely judged on how efficiently they transport and process events, while the control system may be less performance critical. EVPath's API permits the separation of these two concerns, thereby making it easy to experiment with different control mechanisms and semantics while still preserving high performance event transport.

- *Using overlay networks.* EVPath allows the construction and on-the-fly reconfiguration of overlay networks with embedded processing. Such "computational overlays" are the basis for any attempt to address scalability in event messaging.

- *Runtime protocol evolution.* For flexibility in event handling, protocols can evolve at runtime, as demonstrated with previous work on dynamic "message morphing" [5]. This improves component interoperability through the runtime evolution of the types of events being transported and manipulated.

- *Efficient event filtering/transformation.* Basic event transport abstractions support the implementation of efficient event filtering/transformation, enabling some degree of trade-off between moving processing to events vs. events to processing.

- *Open access to network resources.* EVPath provides a clean mechanism for exposing information about and access to the network at the application level, making it easier for applications to leverage network capabilities, such as by adjusting their event processing or filtering actions [73].

- *Supporting multiple styles of network transports.* Event transport can utilize point-to-point, multicast, or other communication methods, without "breaking" the basic event semantics.

The outcome is a flexible event transport and processing infrastructure that has served as an efficient basis for implementing a wide range of higher level communication models [73, 132, 243, 459].

12.3   The EVPath Architecture

In this section, we explore EVPath as a more in-depth case study of an event system architecture. As a design point, EVPath is an event processing architecture that supports high performance data streaming in overlay networks with internal data processing. EVPath is designed as an event transport

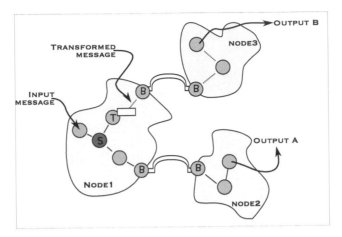

FIGURE 12.3:   A conceptual schematic of an event delivery system built using the EVPath library.

middleware layer that allows for the easy implementation of overlay networks, with active data processing, routing, and management at all points in the overlay. EVPath specifically does not encompass global overlay creation, management or destruction functions. In essence, EVPath is designed as a new basis for building event messaging systems; it does not provide a complete event system itself.

At its core, EVPath implements the parts of the system that are directly involved in message transport, processing and delivery, leaving "management" functions to higher layers. Different implementations of those management layers might support different communications paradigms, management techniques or application domains, all exploiting the base mechanisms provided in EVPath. Because of this, EVPath is envisioned as an infrastructure that can encompass the full range of design points in the event system space.

EVPath represents an entire streaming computation, including data inflow, filtering, transformation, aggregation, and delivery as entities in a dataflow graph. Those entities can then be mapped by an externally provided management layer onto nodes and processes as appropriate, given data locality, processing, and network resource demands, or throughput and availability SLAs. As demands and resource availability change, the computational elements and data flows can be changed by remapping elements or splitting flows to avoid network or computation bottlenecks.

EVPath is built around the concept of *"stones"* (like stepping stones) that are linked to build end-to-end *"paths"*. While the *"path"* is not an explicitly supported construct in EVPath (since the control of transport is key to any particular design solution), the goal is to provide all of the local-level support necessary to accomplish path setup, monitoring, management, modification, and tear-down. Here paths are more like forking, meandering garden paths than like the linear graph-theoretic concept.

Stones in EVPath are lightweight entities. Stones of different types perform data filtering, data transformation, mux and demux of data, as well as the transmission of data between processes over network links. In order to support in-network data filtering and transformation, EVPath also is fully type-aware, marshalling and unmarshalling application messages as required for its operations. Further, as much as possible, EVPath tries to separate the specification of path connectivity (the graph structure) from the specification of operations to be performed on the data.

12.3.1 Taxonomy of Stone Types

Figure 12.4 describes some basic types of stones that EVPath makes available for data stream construction. Most of the design points we have discussed can be addressed just with these building blocks. Stones in the same process are linked together by directing the output of one stone to the input of another. Bridge stones transport events to specific stones in another process or on another host. Events may be submitted directly to any stone, may be passed from stone to stone locally, or may arrive from a network link. Event processing at a stone does not depend upon how it arrived, but only upon the type of data and attributes it is carrying, making the overlay transport publisher-blind as a result.

12.3.2 Data Type Handling

A key feature of many event systems is the handling of data between heterogeneous systems — whether by adopting a common format like XML or a common run-time like Java. Unlike systems that transport events as undifferentiated blocks of bytes, EVPath data manipulation is based upon fully-typed events processed natively on the hardware. Event data is submitted to EVPath marked up as a C-style structure, in a style similar to ECho [132] and PBIO [133]. Terminal stones pass similar structures to application-level message handlers, allowing different handlers to be specified for different possible incoming data types. Figure 12.5 shows a very simple structure and its EVPath declaration. EVPath events are structures which may contain simple fields of atomic types (as in Figure 12.5), NULL-terminated strings, static and variable-sized multi-dimensional arrays, substructures and pointers to any of those elements. EVPath structures can also describe recursive data structures such as trees, lists and graphs.

EVPath uses the FMField description to capture necessary information about application-level structures, including field names, types, sizes and offsets from the start of the structure. The type field, either "integer" or

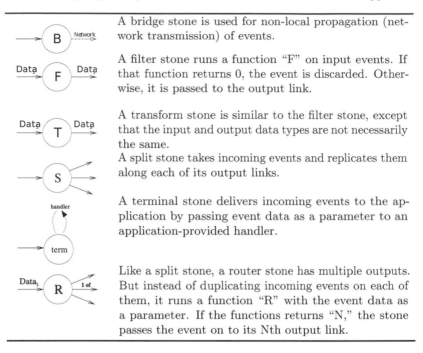

A bridge stone is used for non-local propagation (network transmission) of events.

A filter stone runs a function "F" on input events. If that function returns 0, the event is discarded. Otherwise, it is passed to the output link.

A transform stone is similar to the filter stone, except that the input and output data types are not necessarily the same.

A split stone takes incoming events and replicates them along each of its output links.

A terminal stone delivers incoming events to the application by passing event data as a parameter to an application-provided handler.

Like a split stone, a router stone has multiple outputs. But instead of duplicating incoming events on each of them, it runs a function "R" with the event data as a parameter. If the functions returns "N," the stone passes the event on to its Nth output link.

FIGURE 12.4:   Basic stone types.

"float" in the example, encodes more complex field structures such as static and variable arrays ("integer[1000]", "float[dimen1][dimen2]"), substructures and pointers to other items. In all cases, EVPath can marshall and unmarshall the data structures as necessary for network data transmission, transparently handling differences in machine representation (e.g., endianness) and data layout.

In order to support high-performance applications, EVPath performs no data copies during marshalling. As in PBIO [133], data is represented on the network largely as it is in application memory, except that pointers are converted into integer offsets into the encoded message. On the receiving side, dynamic code generation is used to create customized unmarshalling routines that map each incoming data format to a native memory structure. The performance of the marshalling and unmarshalling routines is critical to satisfying the requirements of high-performance applications.

12.3.3   Mobile Functions and the Cod Language

A critical issue in the implementation of EVPath is the nature of the functions (F, T, *etc.*) used in the implementation of the stones described above. The goal of EVPath is to support a flexible and dynamic computational environment where stones might be created on remote nodes and possibly

```
type def struct {
     int cpu;
     int memory;
     double network;
} Msg, *MsgP;

FMField Msg_field[] = {
     {"load", "integer", sizeof(int), FMOffset(MsgP, load)},
     {"memory", "integer", sizeof(int), FMOffset(MsgP, memory)},
     {"network", "float", sizeof(double), FMOffset(MsgP, network)},
}
```

FIGURE 12.5:   Sample EVPath data structure declaration.

relocated during the course of the computation, as this feature is necessary to scalably realize the complex data processing portion of the design space. In this environment, a pointer to a compiled-in function is obviously insufficient. There are several possible approaches to this problem, including:

- severely restricting F, such as to preselected values or to boolean operators,

- relying on pre-generated shared object files, or

- using interpreted code.

As noted in Section 12.2, having a relatively restricted filter language, such as one limited to combinations of boolean operators, is the approach chosen in other event systems, such as the CORBA Notification Services [317] and in Sienna [80]. This approach facilitates efficient interpretation, but the restricted language may not be able to express the full range of conditions useful to an application, thus limiting its applicability. To avoid this limitation, it is desirable to express F in the form of a more general programming language. One might consider supplying F in the form of a shared object file that could be dynamically linked into the process of the stone that required it; indeed, EVPath does support this mode if a user so chooses. Using shared objects allows F to be a general function but requires F to be available as a native object file everywhere it is required. This is relatively easy in a homogeneous system, but it becomes increasingly difficult as heterogeneity is introduced, particularly if the semblance of type safety is to be maintained.

In order to avoid problems with heterogeneity, one might supply F in an interpreted language, such as a TCL function or Java byte code. This would allow general functions and alleviate the difficulties with heterogeneity, but it potentially impacts efficiency. While Java is attractive in general as a unified solution in this scenario because of its ability to marshall objects, transport byte-code and effectively execute it using just-in-time compilation, its use in event systems can be problematic. In particular, consider that traditional decoupling between event suppliers and consumers might restrict them from

```
{
    if ((input.trade_price < 75.5) ||
        (input.trade_price > 78.5)) {
        return 1; /* pass event over output link */
    }
    return 0; /* discard event */
}
```

FIGURE 12.6:   A specialization filter that passes only stock trades outside a pre-defined range.

effectively sharing events as specific application-defined objects of a common class. Some solutions have been proposed [137], but at this point, Java-based marshalling of unbound objects is not the highest-performing messaging solution.

Because of this, EVPath chose a different approach in order to reach as far as possible into the design space and achieve the highest event rate and throughput possible. In particular, EVPath's approach preserves the expressiveness of a general programming language and the efficiency of shared objects while retaining the generality of interpreted languages. The function F is expressed in Cod (C On Demand), a subset of a general procedural language, and dynamic code generation is used to create a native version of F on the source host. Cod is a subset of C, supporting the C operators, `for` loops, `if` statements and `return` statements.

Cod's dynamic code generation capabilities are based on the Georgia Tech DILL package that provides a virtual RISC instruction set. Cod consists primarily of a lexer, parser, semanticizer, and code generator. The Cod/DILL system generates native machine code directly into the application's memory without reference to an external compiler. Only minimal optimizations are performed, but, because of the simplicity of the marshalling scheme (allowing direct memory-based access to event data), the resulting native code still significantly outperforms interpreted approaches.

Event filters may be quite simple, such as the example in Figure 12.6. Applied in the context of a stock trading example, this filter passes on trade information only when the stock is trading outside of a specified range. This filter function requires 330 microseconds to generate on a 2Ghz x86-64, comprises 40 instructions, and executes in less than a microsecond. Transform stones extend this functionality in a straightforward way. For example, the Cod function defined in Figure 12.7 performs such an average over wind data generated by an atmospheric simulation application, thereby reducing the amount of data to be transmitted by nearly four orders of magnitude.

Event transformations, like the one performed in Figure 12.7, are not achievable in typical enterprise-focused event systems that are mostly concerned

```
{
    int i, j;
    double sum = 0.0;
    for(i = 0; i<37; i= i+1) {
        for(j = 0; j<253; j=j+1) {
            sum = sum + input.wind_velocity[j][i];
        }
    }
    output.average_velocity = sum / (37 * 253);
    return 1;
}
```

FIGURE 12.7: A specialization filter that computes the average of an input array and passes the average to its output.

with smaller-scale and simpler events. However, it is not atypical of the types of data processing concerns that might occur in scientific computing.

12.3.4 Meeting Next Generation Goals

EVPath's architecture is designed to satisfy the next generation goals laid out in Section 12.2.2. It encourages a separation of control and data transport functionality by providing a building block-like set of functional entities for constructing computational overlay networks, but it does not provide any specific higher-level control abstraction that restricts usage of those entities. In order to facilitate the construction of higher-level abstractions without large numbers of opaque objects, EVPath provides insight into such things as low-level resource availability by allowing Cod functions access to system state. Using these facilities, a higher-level control abstraction can, for example, use EVPath to create an event delivery overlay that reconfigures itself in response to changes in computational demands so as to maintain a throughput goal. In this case, EVPath could be used to both implement the data flows as well as the monitoring system that acquires resource data, filters it as necessary and forwards it to a central location for decisions on reconfiguration.

EVPath's type handling and response selection is designed to function efficiently in the common case of homogeneous event types or simple heterogeneity. It also functions effectively where there is less a priori knowledge of datatypes or when there might be multiple generations of clients operating simultaneously, as in an enterprise case, or where there are multiple valid interfaces to a software component, such as a numerical solver in scientific HPC. To support these scenarios, EVPath allows for type flexibility whenever possible. For example, the input type specifications for EVPath filter stones specify the minimum set of data fields that the function requires for operation. When presented with an event whose datatype is a superset of that minimum set, EVPath will generate a customized version of the filter function that accepts the larger event type and conveys it unchanged if it passes the filter criterion. EVPath type and function registration semantics provide sufficient information to allow dynamic morphing of message contents, allowing new senders to seamlessly interact with prior-generation receivers as explored in earlier work [5].

EVPath also combines some of this functionality in novel ways to meet the goals laid out in Section 12.2.2. For example, EVPath allows "congestion handlers" to be associated with bridge stones. These functions are designed to be invoked when network congestion causes events to queue up at bridge stones rather than being immediately transmitted. Because EVPath exposes network-level resource measures such as estimated bandwidth, and because EVPath type semantics allow reflection and views into the application level data in events, congestion handlers have the ability to customize an event stream in an application-specified manner to cope with changing network bandwidth. This capability relies upon the run-time event filtering, open access to network resources, and insight into application-level types that are unique features of EVPath. Because this feature is supported at the data transport level in EVPath, it also allows higher-level control schemes to utilize this mechanism when and as appropriate.

In order to understand the types of performance considerations that might impact the utility of an event infrastructure in the high performance domain, we consider a set of relevant benchmarks.

12.4   Performance Microbenchmarks

There are a number of characteristics that impact the suitability of a system for high-performance applications, including interprocess event communication, intraprocess computation, and communication between local stones. EVPath is built using CM, a point-to-point message passing layer which provides an abstraction for managing multiple physical transports. In this context, the properties of CM are largely irrelevant, except in that it abstracts the point-to-point connection protocol without sacrificing EVPath's high performance goals. All measurements here are performed on a RedHat Linux cluster of dual socket, quad-core Intel-based blades, with both non-blocking Infiniband and gigabit Ethernet networking available. The nodes are provisioned with 1 Gigabyte of RAM per core; a typical if not low ratio for current high performance computing deployments. Specific microbenchmark tests are described below; if networking is needed, the Infiniband network has been used even for TCP traffic.

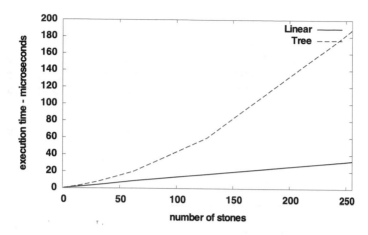

FIGURE 12.8: Local stone transfer times for linear and tree-structured paths.

12.4.1 Local Data Handling

Stones are relatively lightweight entities, and creating a new stone mostly consists of allocating and initializing a few dozen bytes of memory. Stones do accumulate additional state during the course of execution as EVPath caches such things as handler choice decisions for previously seen data types, but the stones' storage demands remain modest. Initial creation requires less than a microsecond.

Local data handling is also quite efficient. Events moving between stones are not copied but are enqueued at each stone for processing. When an action such as a split stone would send an event to multiple destinations, the event is simply multiply-enqueued, with reference counts to indicate when it is to be freed. To evaluate the cost of inter-stone transfer, we create linear sequences of split stones ending in a terminal stone and measure the time required to move an event from the beginning to the end. As these are split stones, no processing is done on the event; it is just dequeued from one stone and re-enqueued on the next until it reaches the terminal stone. Measurements show EVPath is capable of propagating events in this scenario at about .13 microseconds/stone.

However, as Figure 12.8 shows, the time for an event to traverse a local overlay depends upon the structure of the graph. For simple linear graphs, the time increases linearly as expected. For stones structured as binary trees, using split stones to duplicate an event on two outputs at each stage, the total execution time increases more rapidly. The problem here is not in the copying cost, but instead, it is the result of the event tracking algorithm in the EVPath implementation. EVPath can follow a single path through a complex graph; when a split stone enqueues an event on multiple stones, only one path is followed directly. Processing on the other stones is tracked in

TABLE 12.1:   Comparing split stone and filter stone execution times

first execution of a filter stone	350 microseconds
subsequent executions of a filter stone	.24 microseconds
processing time on a split stone	.13 microseconds

their individual queues, but no coherent execution ordering (like a stack-based model) is enforced. As a result EVPath knows that there are additional events "in-play" but not where they are enqueued, and so it performs a linear search through the stones to find them. In a binary graph of split stones, this gives an execution time that increases with the square of the number of stones. A more sophisticated event tracking algorithm might ameliorate this problem, but we do not expect it to be common in real-world scenarios.

As described above, EVPath has significant flexibility with respect to the manner in which filter functions are deployed. Because they are instantiated the first time an event of a matching type arrives, there is additional computational cost for dynamic code generation associated with the arrival of the first event. In order to measure this, we construct linear graph of Filter stones, each with a different filter function of the form `return input.integer_field < 12345;`. We then submit a sequence of events, ensuring that each will pass the filter test (i.e., it returns true) so that no event is discarded. Table 12.1 shows EVPath's performance in this scenario.

The first time an event of a particular type arrives at a filter stone, EVPath essentially compiles the filter function to generate a customized native subroutine that evaluates the function. This accounts for a first-time execution time that is three orders of magnitude higher than that of a simple split stone. Subsequent executions on that filter stone are only slightly more expensive than the computational demands of the split stone. The numbers are obviously dependent upon the actual function being compiled, in terms of its length and complexity for both the time to generate the native version and its actual execution time when operating on the data. However, because application-level data filtering can be performed with such low overheads, EVPath filters can be employed in data paths with high throughput demands.

12.4.2   Network Operation

The local data processing performance explored in the previous section is important for data filtering and transformation, but its ability to perform efficient network data transfers is also key for many of our application scenarios. In order to evaluate EVPath's network operation, we employ a simple stone configuration, submitting fully typed events directly to an output stone on process A where they will be transmitted to process B and delivered to

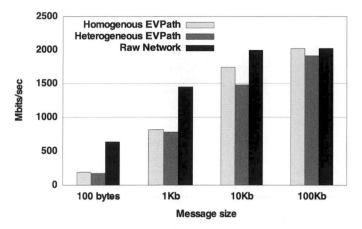

FIGURE 12.9: EVPath throughput for various data sizes.

a terminal stone. Figure 12.9 compares EVPath's delivered bandwidth to available bandwidth (as measured by netperf [307]) over a range of data sizes. Note that while netperf is just delivering a block of bytes, EVPath is delivering fully typed messages. This simple case of a transfer from a 64-bit x86 machine to a similar box with a terminal stone that is expecting the same data type that is sent is the best case for EVPath, and as Figure 12.9 shows, EVPath delivers nearly all of available bandwidth at large message sizes.

As a more complex test, Figure 12.9 also presents results from experiments where the sender's data structure does not perfectly match that required by the receiver. We simulate a byte-endian mismatch and a transfer between 32-bit and 64-bit machines by having the original senders instead transmit data that was pre-encoded on other architectures. Both scenarios require a more complex decoding transformation on the receiving side before the data can be delivered to the application. Figure 12.9 shows that the byteswap decoding operation, handled with custom-generated subroutines in EVPath, has no impact on delivered bandwidth on smaller data sizes and causes less than a 5% degradation (as compared to a homogeneous transfer) at the worst case.

12.5 Usage Scenarios

The microbenchmark results depicted in the previous section show a basic suitability for applications requiring high-performance data transfers and data processing. It is also interesting to examine how the building blocks described in the EVPath architecture are actually used to implement event-based systems

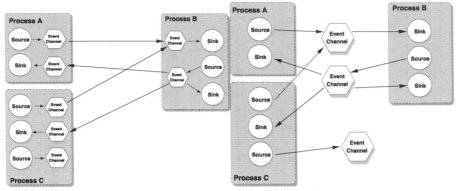

(a) Abstract view of event channels.   (b) ECho realization of event channels.

FIGURE 12.10:   Using event channels for communication.

at various locations in the design space. This section examines the use of this architecture in application scenarios. In particular, we explore the use of EVPath to:

- implement the event transport functionality in a different publish/subscribe system, ECho [132];

- manage enterprise information flows with the IFlow scalable infrastructure; and

- create novel peta-scalable I/O interfaces through in-network staging.

12.5.1   Implementing a Full Publish/Subscribe System

The ECho publish/subscribe system is an efficient middleware designed for high-speed event propagation. Its basic abstraction is an event channel, which serves as a rendezvous point for publishers and subscribers and through which the extent of event notification propagation is controlled. Every event posted to a channel is delivered to all subscribers of that channel, in a manner similar to the CORBA Event Service [318].

ECho event channels, unlike many CORBA event implementations and other event services such as Elvin [391], are not centralized in any way. ECho channels are light-weight virtual entities. Figure 12.10 depicts a set of processes communicating using event channels. The event channels are shown as existing in the space between processes, but in practice they are distributed entities, with bookkeeping data residing in each process where they are referenced as depicted in Figure 12.10b. Channels are *created* once by some process, and *opened* anywhere else they are used. The process that creates the event channel is distinguished in that it is the contact point for other processes that use the channel.

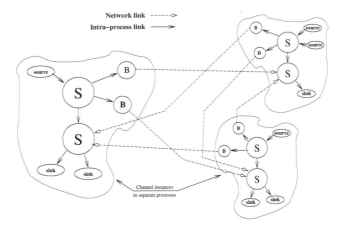

FIGURE 12.11: ECho event channel implementation using EVPath stones.

However, event notification distribution is not centralized, and there are no distinguished processes during notification propagation. Messages are always sent directly from an event source to all sinks, and network traffic for individual channels is multiplexed over shared communications links. One of ECho's novel contributions is the concept of a *derived* event channel, a channel in which the events delivered to subscribers are a filtered or transformed version of the events submitted to another channel [132].

The style of event propagation provided by ECho can be easily implemented through a stylized used of EVPath stones. Each open channel in a process is represented by two split stones as shown in Figure 12.11. The upper split stone is the stone to which events are submitted. Its outputs are directed to bridge stones that transmit the events to other processes where the channel is open and to the lower split stone. The lower split stone's outputs are directed to terminal stones which deliver events to local subscribers on that process. While local events are always submitted to the upper split stone, the bridge stones that transmit events between processes target the lower split stone on remote hosts. This arrangement, maintained by ECho's subscription mechanism, duplicates the normal mechanics of ECho event delivery.

From a control perspective, ECho2 leverages the pre-existing control messaging system that ECho used. An initial enrollment message is sent from a new publisher and/or subscriber to the channel opener. In reply, the opener sends back a complete list of all of the publishers and subscribers and their contact information. In the EVPath implementation with a TCP transport, this contact include the IP address, port number, and lower stone ID number. The joiner then contacts each member on the list with an update to register itself as either a subscriber or publisher.

ECho's derived event channels can be implemented with a simple addition

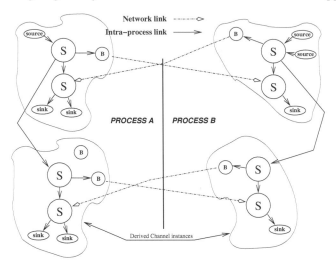

FIGURE 12.12:   Derived event channel implementation using EVPath stones.

to the convention described above. When a derived event channel is created, ECho creates an instance of the new channel on every process in which the original channel is represented. Then, the two-stone representation of the original channel and the derived channel are linked with a filter or transform stone as shown in Figure 12.12. This arrangement duplicates ECho's derived event channel semantics in which the filtering is always performed at the point of event submission in order to avoid unnecessary network overheads.

The earlier work on ECho demonstrated the interesting result that, while abstractions like the derived event channel move more computation to the publisher of events, the total computational demands on the publisher can be lowered even if the filter actually rejects very few of the events it considers. Figure 12.13 depicts the total CPU time incurred by a process sending 2 million relatively small (1000 byte) records as a function of the rejection ratio of the filter. The EVPath implementation of ECho event distribution shows a similar effect, with the filter reducing CPU overheads with less than a 10% rejection ratio.

12.5.2   IFlow

Another application area where distributed messaging plays an important role is in federated database queries, which are here treated as a type of scalable, data subscription event system. IFlow (for Information Flow) is hierarchically-controlled and designed both to provide correct functionality in satisfying complex SQL-like queries over federated databases and to make best-

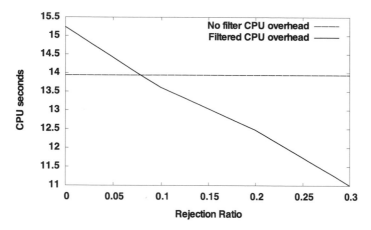

FIGURE 12.13: CPU overhead as a function of filter rejection ratio.

effort autonomic end-to-end performance optimizations. In particular, IFlow explicitly addresses dynamically changing network and server performance — the network graph will undergo continuous modification and rerouting in the event that the SLA on end-to-end performance is not being met.

As a control layer on top of EVPath, IFlow presents an interesting set of challenges. The ability to dynamically route and reroute data requires EVPath to suspend processing of data at certain stones so that downstream connections may be reconfigured in a clean way. In the event that data would be "orphaned" by such a move — left as local state in a stone queue from which there will not be processing routes under the new path structure — IFlow also uses the local control actions in EVPath to dequeue and ship that data to its new location before destroying stones and routing actions.

The details of autonomic routing decisions in overlay networks are worth additional treatment on their own, but those used in IFlow can be found in the literature [243,386]. However, IFlow has two interesting features that arise out of the features of EVPath architecture: IFlow's use of a SOAP RPC layer for distributed control and the management of local caching in the flow network to enable reliability. The SOAP control library for IFlow, called SoapStone, clearly outlines the implications of the separation of the data fast-path from control in EVPath. Given the Enterprise application focus, using an industry-standard format for the control layer so that it would easily interoperate with existing tools is important. The SOAP message bodies correspond to a control language that can invoke the various local functions on a remote node; the handler stubs merely translate the request parameters into the local context.

Another interesting corollary of the EVPath queueing model is that it is possible with very little run-time cost to maintain a record of the flow to support automatic recovery in the event of node or transport link failure. The control process can place logging actors into the flow as dictated by policy and then issue commands to compare logical stamps on upstream and downstream logs in the event that a platform error is detected. Once the correct place in the log is determined, the data can be replayed automatically. This technique does not work under all possible types of data flow processing, of course, particularly with complicated, stateful data fusion applications. However, within the domain space targeted by IFlow, these techniques are sufficient for offering reliability service agreement parameters that are novel in their application [244, 386].

12.5.3   I/OGraph

As a final example in the scientific computing space, we consider an I/O library for scalable petascale applications in the I/OGraph project. The client-side interface registers data buffers as ready to write, and these get turned into formatted messages which are published through an EVPath overlay for sorting, naming, and delivery to the process which commits the data to disk.

A key feature of this work is the idea of delaying global metadata consistency for scalable I/O — for example, the process of creating a particular file and getting a shared file pointer (or agreeing on how to partition into non-overlapping file regions for non-shared pointers) can be prohibitively expensive when scaled to tens of thousands of nodes or more. By enabling this metadata to be generated and associated with the data packet somewhere downstream in the network, the blocking time for the application can be reduced. For example, data can go through partial aggregation as it is gathered into the storage nodes, making for a much faster calculation of the prefix sums for writes into a shared file. Indeed, some application domains may prefer that the data not be gathered into a shared file at all, if there is a mechanism for doing intelligent aggregation; for example, generating files based on the quadrant of simulation space, rather than the particular order of commits from the compute nodes.

Another key feature of this work is that exactly these sorts of domain-specific customizations can be accommodated using an imperative-style data treatment specification. Coupling the queue logging functionality of EVPath that was discussed in the previous section with the arbitrary imperative data parsing abilities of Cod and with the router stone shown in Figure 12.4 leads to a new type of functionality. As messages arrive, they are enqueued and a custom function is invoked to evaluate them. However, that function has full control of its queue — it can publish events, delete events from the queue, create and publish synthetic events (e.g. windowed averages), and provide in-situ coordination.

The I/OGraph infrastructure can leverage this feature of the EVPath data handling layer to implement the delayed metadata assignment in a clean way. Application-specific modules must only determine the thresholding criteria for when certain pieces of data should be written together; these could be physical (queue length $>= 8$) or domain driven (data outside of bounding box B). The I/OGraph library will create the EVPath overlay and place the queue handling function on one or more nodes that are set aside as a data staging partition. Because the data is published in a single act from the queue handler, any relevant file data (such as prefix sums for the individual pieces) can be much more easily determined.

Some ongoing research involves developing a more straightforward user interface for mapping application-driven metadata (e.g., containing bounding box, existence of a particular feature in the data) and file system metadata (e.g., file names or ownership). This will add an additional control layer to the current EVPath infrastructure; a higher-level control language specified in an input language will be parsed and used to automatically generate the Cod function code for the queue handling functions. Autonomic provisioning and deployment of the queue handlers for both local-area and wide-area data consumption complete this application. This vision of a delayed consistency framework for I/O fully exercises the EVPath design targets for event overlay, run-time evolution of data, efficient transformation and filtering, and support of varied network protocols.

12.6   Summary

In this chapter, we have surveyed the design space for high performance event-based publish/subscribe systems and more deeply explored EVPath, an infrastructure aimed at supporting applications and middleware with high performance eventing needs. This is an area of continuing research and many topics are beyond the scope of a single chapter. In particular, as multi-core computing platforms continue to evolve, the importance of messaging in general and event-based approaches in particular would seem to be growing. With deep memory hierarchies and System-on-a-Chip architectures, messaging as a communication approach between "clumps" of cores would seem to be a viable and scalable approach. Additionally, the software trends towards composable, componentized applications are inevitable as well. So both from a platform perspective and from an application evolution perspective, event-based middleware will have a strong role to play in the future of high performance communications.

The work presented here, both generally and the authors' specific research, is intended to spur further thought over the possible development of new techniques that might break down some of the classic design space trade-offs. Future high performance event messaging systems have the potential to become generically useful tools for high performance, alongside other middleware toolkits like MPI.

Chapter 13

High Performance Communication Services on Commodity Multi-Core Platforms: The Case of the Fast Financial Feed

Virat Agarwal, Lin Duan, Lurng-Kuo Liu,
Michaele Perrone, Fabrizio Petrini
  IBM TJ Watson Research Center
Davide Pasetto
  IBM Computational Science Center
David Bader
  Georgia Institute of Technology

13.1   Introduction

Traditional scientific HPC applications running on high-end computing platforms are not the only class of applications dependent on the availability of high-performance communication services. As Chapter 12 points out, many classes of enterprise codes have strict communications requirements — airline operations systems, applications for package delivery monitoring and planning, financial trading platforms, and others. The requirements include the need for low latency and response time guarantees, as well as the ability to handle large data volumes, and are addressed by mechanisms which enable high-speed interpretation, distribution, and manipulation of the applications' message streams, often supported by custom hardware engines.

In this chapter we discuss an alternative approach of leveraging the computational capabilities of modern multicore platforms to support high-speed communications for streaming applications. While the mechanisms described in the chapter are more general, we present them and demonstrate their feasibility in the context of financial trading services, as a class of applications with particularly critical communication requirements. By 2010, the global options and equity markets are expected to produce an average of more than 128 billion messages per day. Trading systems, that form the backbone of the low latency and high frequency business, need fundamental research and innovation to overcome their current processing bottlenecks. With market data rates rapidly growing, the financial community is demanding solutions that are extremely fast, flexible, adaptive, and easy to manage.

Towards this end, in the following sections, we first provide greater detail on the class of financial trading applications and their current and future communication needs. We then discuss our approach to providing high performance decoding and manipulation services for financial data streams on commodity platforms, and experimentally demonstrate the capabilities it provides on a range of modern multicore architectures. In fact, our measurements demonstrate that with some of the current platforms we can sustain future data rates of 15 million messages per second! Many other application domains with streaming data exhibit similar trends of rapid increase in data volumes (though perhaps at reduced rate), and need for lower response time and for efficient support of value added services. Therefore, the approach described in this chapter is of importance to these other contexts, as well.

13.2   Market Data Processing Systems

A typical market data processing system consists of several functional units that receive data from external sources (such as exchanges), publish financial data of interest to their subscribers (such as traders at workstations), and route trade information to various exchanges or other venues. A high-level design of a market data processing system or *ticker plant* is sketched in Figure 13.1. Examples of functional units include feed handlers, services management (such as permission, resolution, arbitration, normalization, etc.), value-added, trading system, data access, and client distribution. The feed handler is the component that directly interacts with the feed sources for handling real time data streams, in either compressed or uncompressed format, and decodes them converting the data streams from source-specific format into an internal format, a process often called *data normalization*. According to the message structure in each data feed, the handler processes each field value with a specified operation, fills in the missing data with value and state of its cached records, and maps it to the format used by the system. The ability of a feed handler to process high volume market data stream with low latency is critical to the success of the market data processing system.

13.2.1   The Ticker Plant

A *ticker plant* is an important component in a market data processing system: it handles much of the processing associated with market data, receives data feeds from external sources, OPRA being one of them, and publishes the

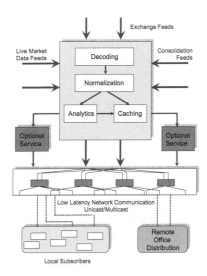

FIGURE 13.1:   High-level overview of a ticker plant.

processed data of interest to its subscribers. Reuters, Bloomberg, Wombat, and Exegy are examples of ticker plants. A typical ticker plant infrastructure is shown in Figure 13.1 and usually provides access to data from a wide range of exchange data feeds, and consolidated third-party feeds. These feeds could be in any one of the many protocols used worldwide. A ticker plant must first parse/decode incoming messages and translate them into a common format; after the decoding, a normalization stage integrates the new pieces of information into complete quotes, usually in a proprietary format. A quote cache is then updated with the integrated data and the new information is distributed in real-time to subscribing clients. There are also several platform services associated with a ticker plant, like entitlement systems for offering controlled access, administrative utilities for operational management, and resolution services for offering single namespace capability.

Each stage of the ticker plant process adds unwanted latency, therefore high throughput and low latency processing at every stage is a critical factor to the success and competitive position of the product. Several approaches and products are available on the market. Solutions range from specialized hardware accelerators to software implementation on commodity hardware. For example Exegy [139] and Active Financial [1] employ reconfigurable hardware (FPGAs) to deliver very high ticker plant performance. Others, like Solace Content Router [409] and Tervela TMX [428], use more traditional architecture based on multiple servers or blades and TCP/IP hardware engines.

13.3   Performance Requirements

U.S. exchanges that allow trading of securities options have been authorized under the Securities Exchange Act of 1934 to agree to a "Plan for Reporting of Consolidated Options Last Sale Reports and Quotation Information." This activity is administered by a committee called Options Price Reporting Authority or OPRA [322]. OPRA is the securities information processor that disseminates, in real-time on a current and continuous basis, information about transactions that occurred on the options markets. OPRA receives options transactions generated by participating U.S. exchanges, calculates and identifies the "Best Bid and Best Offer," and consolidates this information and disseminates it electronically to the financial community in the U.S. and abroad. The protocol, OPRA FAST, is the most widely used protocol in ticker plants.

13.3.1   Skyrocketing Data Rates

Fueled by the growth of algorithmic and electronic trading, the global options and equities markets are expected to produce an average of more than 128 billion messages a day by 2010, rising from an average of more than 7 billion messages a day in 2007, according to estimates from the TABB Group [422]. In the options market, the OPRA consolidated feed for all US derivatives business represents a very significant portion of market data in the national market system. Figure 13.2 shows that the OPRA market data rates, on a one-second basis, have dramatically increased over the course of the past 4 years, approaching 1 million messages per second. The traffic projection for OPRA alone is expected to reach an average of more than 14 billion messages a day in 2010 [322]. As market data rates continue to skyrocket, high speed, low latency, and reliable market data processing system are becoming increasingly critical to the success of the financial institutions.

13.3.2   Low Latency Trading

In today's financial world, the growth of algorithmic and electronic trading has provided opportunities and exposed many challenges to financial institutions. The study of market data to predict market behavior helps these institutions increase profits. The performance of a trading process is critical in electronic trading. How fast a trading system can respond to the market will determine who wins and who loses. Every millisecond and every microsecond count in electronic trading. The first one to identify a trading opportunity generally doesn't leave much behind for the other players in the line. A few milliseconds difference in latency is enough to make difference between a profitable trade and a losing one, which can cause huge losses to

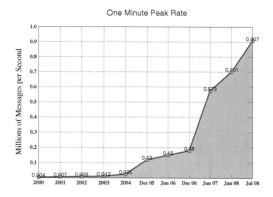

FIGURE 13.2: OPRA market peak data rates.

the financial institutions and their clients. In fact some analysts estimate that 1 millisecond latency difference can impact $100 million dollars worth for an investment institution [368]. With such high stakes, the need for high speed and low latency trading system presents tremendous pressures to the technology division, in financial institutions, to optimize their market data messaging and trading system.

The latency in electronic trading is measured as the time it takes for a trade message to traverse the market data distribution network, starting from an exchange to when a trading application becomes aware of the trade. There are several sources that contribute to the latency in this messaging system. The first one is the physical distance between the trade application and the exchange. The market data cannot travel faster than the speed of light, thus financial institutions suffer increasing latency with increasing distance from the exchange. The switching network introduces another source of latency, latency due to store and forward, switch fabric processing, and frame queuing. The basic task of an Ethernet switch operation is to store the received data in memory until the entire frame is received. The switch then transmits the data frame to an appropriate output port. The latency of the store and forward operation depends on the frame size and the data rate. An Ethernet switch uses queues in conjunction with the store and forward mechanism to eliminate the problem of frame collisions. When the load on a network is very light, latency due to queuing is minimal. However, as mentioned before in Section 13.3.1, the market data rate is skyrocketing. OPRA data feeds are disseminated over 24 lines (48 when counting redundant delivery). Under a heavy network load, the switch will queue frames in buffer memory and introduce frame queuing latency. Typically, market data feeds terminate at ticker plants on the customer's premises. The ticker plants decode/process the market data stream, normalize it into a common format, and republish it for a wide variety of applications (e.g., trading application) through a messaging middleware. The delay introduced by the ticker plants and the messaging middleware also contributes to the latency in the overall trading process.

With high speed and low latency a necessity for trading processes in financial markets, there is continuing pressure for investments by financial institutions in low-latency trading and market data architectures. To increase profits, new applications are being deployed on these low latency architectures. To fight with the latency attributed to the physical distance, companies are offering low latency collocation services to financial institutions [1, 362, 426]. It also saves the financial institutions from building/expanding their own facilities to house the network architecture. Direct exchange feed solutions that connect from a customer's site directly to the exchange are another way to reduce the latency [1, 368]. As opposed to a data consolidator, the direct-exchange feed eliminates data hops and thus reduces the latency. Re-evaluation of ticker plant and messaging middleware is another place to look for latency reduction. Many financial institutions are still using in-house ticker plants to handle the market data. As market data rates continue to increase, we are not just talking about low latency, but low latency at high-throughput rate. This also gives rise to the issues of system scalability, power consumption, and consolidation. Managing in-house ticker plants becomes challenging for the market data technology staff. The traditional software-based ticker plants on commodity white boxes seem to be breaking. There are vendors doing this in hardware to accelerate the ticker plant functionality. For example, FPGA is used to build an acceleration appliance in a reconfigurable way for ticker plant functionality [139].

13.3.3   High Performance Computing in the Data Center

Not surprisingly, many of the technologies that are commonly used in high performance computing are appearing in the data centers of many financial institutions. Advanced communication fabric such as Infiniband or 10 Gigabit Ethernet, low-latency OS-bypass communication protocols, hardware accelerators, such as GPGPUs and FPGAs [230, 274, 312], and the latest generation of multi-core processors are becoming essential building blocks to meet the increasing demands of the financial markets.

Together with the increased performance, data centers are also encountering typical problems in high performance computing: the complexity of developing, testing and validating parallel software, and the need to reduce power consumption in the processing units that are often "co-located" near the data feeds to minimize the communication latency and reduce floor space requirement, which typically comes at a high premium. For these reasons, the financial community is demanding solutions that are extremely fast, and easy to program and configure, so as to adapt to dynamically changing requirements.

13.4   The OPRA Case Study

In the remainder of this chapter we describe our approach to developing high-speed communication support for the OPRA market feeds. OPRA messages are distributed in a compact encoded format, so as to minimize the bandwidth requirements. Therefore, the ability to parse and decode them with minimal latency, while sustaining incoming message rates, is key for addressing the performance needs described in the previous section.

We explore several avenues to implement a fast OPRA decoder and data normalization engine on commodity multi-core processors. We have first attacked the problem by optimizing a publicly available version of an OPRA decoder, and based on this preliminary implementation we have developed two more versions of the decoder, one that has been written from scratch and hand-optimized for multi-core processors and one that uses DotStar [116], a high-level protocol parsing tool that we have recently designed and implemented.

The main contributions of our work are as follows.

1. The implementation of a high-speed hand-optimized OPRA decoder for multi-core processors. The decoder, described in Section 13.4.1 has a simplified control flow that enables a number of low-level optimizations that are typical of scientific applications [21], and has been designed using a bottom-up approach, starting from highly optimized building blocks. In the fastest configuration, this decoder is able to achieve an impressive processing rate of 14.6 (3.4 per thread) million messages per second with a single Intel Xeon E5472 quad-core socket.

2. A second implementation, based on the DotStar protocol processing tool described in Section 13.4.4, that is able to capture the essence of the OPRA structure in a handful of lines of a high-level description language. The DotStar protocol specification is combined with a set of actions (the same basic actions used in the bottom-up version) that are triggered by the protocol scanner. The DotStar protocol parser is only marginally slower than the hand-optimized parser, on average 7% on 5 distinct processor architectures, and provides a very compact and flexible representation of the OPRA protocol. In the fastest single-socket configuration a single Intel Xeon E5472 quad-core is able to achieve a rate of 15 million messages per second.

3. An extensive performance evaluation that exposes important properties of the DotStar OPRA protocol and parser, and analyzes the scalability of five reference systems, two variants of the Intel Xeon processors, and AMD Opteron, the IBM Power6, and the SUN UltraSPARC T2.

4. Finally, we provide insight into the behavior of each processor architecture trying to explain *where* the time is spent by analyzing, for all the

processor architectures under evaluation, each action associated with DotStar events. All the action profiles are combined in a cycle-accurate performance model, that is presented in Section 13.4.6, that helps determine the optimality of the approach, and to evaluate the impact of architectural or algorithmic changes for this type of workload. We believe that this level of accuracy can be useful to both application developers and processor designers to clearly understand how these algorithms map to specific processor architectures, allowing interesting what-if scenarios.

One evident limitation of this work is that it addresses only a part of the feed handler, leaving many important questions unanswered. For example, the network and the network stack are still areas of primary concern. Nevertheless, we believe that the initial results presented in this chapter are ground-breaking because they show that is possible to match or exceed speeds that are typical of specialized devices such as FPGAs, using commodity components and a high-level formalism that allows quick configurability. We believe that the methodology we present can be readily extended to the other parts of the ticker plant, in particular the value-added functionalities, and other data protocols taking full advantage of the multi- and many-core architectures that are expected to become a ubiquitous form of computing within the next few years.

In the remainder of this section we first provide greater detail on the encoding protocol used in OPRA market feeds. Next, we described in greater detail our decoding alternatives, and show experimental data to demonstrate the impressive capability of our approach to handle current and future market requirements.

13.4.1   OPRA Data Encoding

An essential component in ensuring the timely reporting of option equity/index and every other transaction is the OPRA IP multicast data stream. OPRA messages are delivered through the national market system with a UDP-based IP multicast [404]. The options market data generated by each participant is assembled in prescribed message formats and transmitted to the appropriate TCP/IP processor address via the participant's private communications facility. As each message is received, it is merged with messages received from other participants, and the consolidated message stream is transmitted simultaneously to all data recipients via their private communications facilities. Each message is duplicated and delivered to two multicast groups. OPRA messages are divided into 24 data lines (48 when counting redundant delivery) based on their underlying symbol. Multiple OPRA messages are encapsulated in a block and then inserted in an Ethernet frame. The original definition of OPRA messages is based on an ASCII format [402], which uses only string based encoding and contains redundant information. With the growth of data

TABLE 13.1: OPRA message categories with description.

	Description
Category a	Equity and Index Last Sale
Category d	Open Interest
Category f	Equity and Index End of Day Summary
Category k	Index and Stock Quotes
Category C	Administrative
Category F	FCO End of Day Summary
Category H	Control
Category O	FCO Last Sale
Category U	FCO Quote
Category Y	Underlying Value Message
default	Contains Text Value

volume, a more compact representation for messages was introduced: OPRA FAST (FIX Adapted for STreaming) [401, 403].

The OPRA FAST messages belong to one of several categories, shown in Table 13.1. The generic message format has a number of fields (50, see [322] for full list of all fields), however, not all fields are required for messages of each category. For instance, only "End-of-Day-Summary" categories ("f" and "F") have fields indicating OPEN_, HIGH_, LOW_, and LAST_PRICE. On the other hand, certain fields, such as PARTICIPANT_ID are required for all message categories. Furthermore, across a sequence of messages some fields have successive values which don't change frequently, or change in a very predictable manner, i.e., are incremented or modified by a specific "delta." The OPRA protocol allows such fields to be omitted even if they are required for a given message category, by specifying what type of "encode operator" is associated with each field (e.g., copy or increment).

The techniques used in the FAST protocol include implicit tagging, field encoding, stop bit, and binary encoding. Implicit tagging eliminates the overhead of field tags transmission. The order of fields in the FAST message is fixed, and thus the meaning of each field can be determined by its position in the message. The implicit tagging is usually done through XML-based FAST template. The presence map (PMAP) is a bit pattern at the beginning of each message where each bit is used to indicate whether its corresponding field is present. Field encoding defines each field with a specific action, which is specified in a template file. The final value for a field is the outcome of the action taken for the field. Actions such as "copy code," "increment," and "delta" allow FAST to remove redundancy from the messages. A stop bit is used for variable-length coding, by using the most significant bit in each byte as a delimiter. FAST uses binary representation, instead of text string, to represent field values. OPRA is an early adopter of the FAST protocol for reducing the bandwidth needed for OPRA messages.

S O H	L E N	OPRA FAST ENCODED MESSAGE	L E N	OPRA FAST ENCODED MESSAGE	- - - - - - -	E T X

(a) Version 1.03

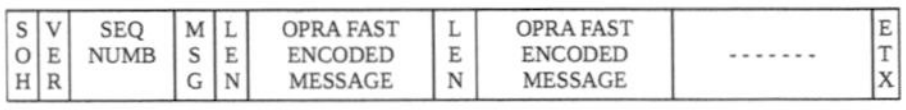

S O H	V E R	SEQ NUMB	M S G	L E N	OPRA FAST ENCODED MESSAGE	L E N	OPRA FAST ENCODED MESSAGE	- - - - - - -	E T X

(b) Version 2.0

FIGURE 13.3:   OPRA FAST encoded packet format.

Figure 13.3a shows the format of an encoded OPRA version 1 packet. Start of Header (SOH) and End of Text (ETX) are two control characters that mark the start and end of the packet. One transmission block can contain multiple messages, where a message is a unit of data that can be independently processed by the receiver. In OPRA FAST version 2 (see Figure 13.3b) there is a header after SOH and before the first message to further reduce redundant information. The first byte of an encoded message contains the length in bytes and is followed by the presence map. For example, presence map 01001010 means field 1, field 4 and field 6 are present. The type of each field is specified by message category and type. Data fields that are not present in an encoded message but that are required by the category will have their value copied from the same field of a previous messages and optionally incremented. OPRA data fields can be either unsigned integer or string.

13.4.2   Decoder Reference Implementation

Figure 13.4a shows a block diagram of the reference OPRA FAST decoder provided along with the standard. The implementation starts by creating a new message and parsing the presence map, computing its length by checking the stop bit of every byte until one is found set, masking the stop bit and copying all the data into temporary storage. The presence map bits are then examined to determine which fields are present and which require dictionary values. The implementation proceeds checking the category of the message and calls a specific decoder function, where the actual decoding for each required field is implemented. There are two basic building blocks: decoding unsigned integers and decoding strings. Both examine PMAP information, input data, and manipulate the last value dictionary. We initially tried to optimize the reference decoder using a *top-down* approach, by speeding up the functions that are more time-consuming. Unfortunately the computational load is distributed across a large number of functions, as shown in Figure 13.4b, and our effort resulted in a very limited performance improvement.

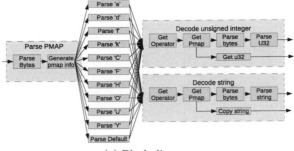

(a) Block diagram

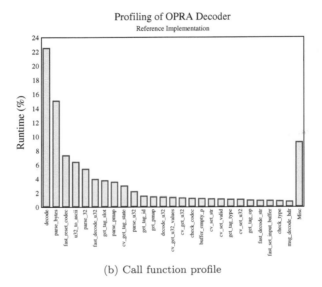

(b) Call function profile

FIGURE 13.4: OPRA reference decoder.

13.4.3 A Streamlined Bottom-Up Implementation

We then designed another version of the OPRA decoder following a bottom-up approach, trying to eliminate the complex control flow structure existing in the reference implementation. In this approach we identify important computationally intensive kernels, optimize these routines independently and analyze the assembly-level code to get the best performance. These optimized routines are then crafted together to perform OPRA feed decoding as shown in Figure 13.5.

To further improve the performance of these kernels, we introduce important novel optimizations to the OPRA FAST decoding algorithm, which we describe below. First, we replace branches and category-specific routines, by a single routine that uses a category-specific bitmap. A category bitmap is a bit stream

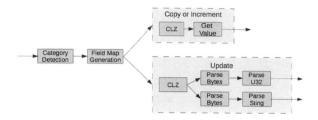

FIGURE 13.5:   Bottom-up reference decoder block diagram.

signifying all the required bits of that category, where the bit is on whenever the corresponding field is needed by the category. Once the category of a message has been determined, this bitmap is passed to the next building block for specifying the fields that are relevant to a message of this category.

The second design choice is related to presence map parsing. The reference code parses the PMAP field bit by bit, testing every presence bit to see if it is set or not. However, since only a subset of fields is required for each message category, it is tedious to check for each corresponding bit. Since PMAP points to the location of all the present fields, it can also be considered as a field map for the data fields that need to be updated. To get the field map for the rest of the data fields that are not present but required, we can simply create another bitmap, through an *xor* of PMAP and the category-specific bitmap. This bitmap gives information about all the fields that need to be copied from the last known values. After computing these bitmaps, we use *clz* (count leading zeroes) operation to determine the field ID of the next field present in the message block. This operation can also be used on the copy bitmap to determine the field IDs of the fields that need to be copied. Figure 13.6 provides an overview of the overall decoding algorithm.

The third optimization deals with the string field processing. If a string field relevant to a category is not in the message block, it is copied from the last known value. Instead of copying the string value multiple times, we create a separate buffer, copy the string once to this buffer and reference that to the decoded messages through an array offset. As shown in Figure 13.5, the bottom up algorithm has a much simpler control flow structure than the reference decoder described in Figures 13.4a and 13.4b.

13.4.4   High-Level Protocol Processing with DotStar

A fundamental step in the semantic analysis of network data is to parse the input stream using a high-level formalism. This process transforms raw bytes into structured, typed, and semantically meaningful data fields that provide a high-level representation of the traffic. Constructing protocol parsers by hand is a tedious process and it is error-prone due to the complexity of the

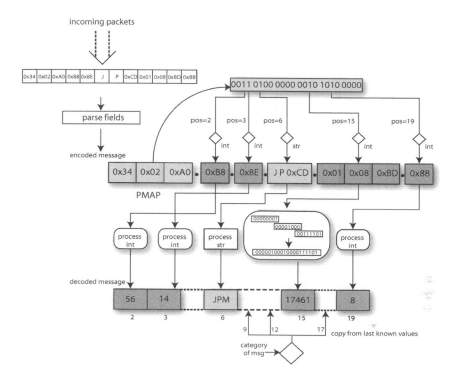

FIGURE 13.6: Presence and field map bit manipulation.

low level details in most protocols. Moreover optimizing the performance of the implementation for one or more target architecture is very complex and can be challenging. Finally, any change to the protocol, like a new version or new fields, may require extensive rewrite of the parser code.

A number of tools that simplify development of protocol parsers are available in literature; these tools are normally based on declarative languages to describe data layout and transfer protocol. Examples are Interface Definition Languages (IDL), like XDR [412], RPC [412] or ASN.1 [261], and regular languages and grammars, like BNF, ABNF [323], lex and yacc, or declarative languages like binpac [327]. These solutions are often tailored for a specific class of protocols, like binary data or text oriented communication. A common characteristic of all these solutions is that they are very high level and "elegant," and provide the user with great expressiveness, but the generated parser performance is often one order of magnitude slower than an equivalent handcrafted one, and may be unable to parse real-time heavy volume applications.

While exploring how to extend our fast keyword scanner automaton in order to handle regular expression sets, we developed DotStar [116], a complete tool-chain that builds a deterministic finite automaton for recognizing the

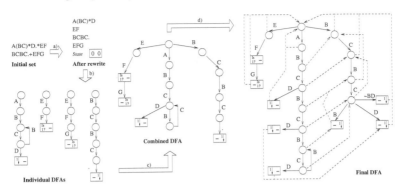

FIGURE 13.7:   A graphical representation of DotStar compiler steps: a) is the pre-processor, which reduces the number of states, b) builds individual automata for each regular expression, c) combines them in a unique automaton and d) computes the failure function.

language of a regular expression set. The generated automaton is an extension of the Aho-Corasick [304], the de facto standard for fast keyword scanning algorithms. The Aho-Corasick algorithm operates on a keyword tree that contains a combined representation of all keywords. The keyword tree is transformed into a Deterministic Finite Automaton (DFA) by creating a failure function $F()$ that is followed when no direct transition is available; $F()$ points to the longest proper suffix already recognized.

Finding every instance of a regular expression pattern, including overlaps, is a complex problem either in space or time [276]. Matching a complete set of regular expressions adds another level of complexity. DotStar employs a novel mechanism for combining several regular expressions into a single engine while keeping the complexity of the problem under control. DotStar is based on sophisticated compile time analysis of the input set of regular expressions and on a large number of automatic transformations and optimizing algorithms. The compilation process proceeds through several stages (see Fig. 13.7): at first each regular expression is simplified in a normal form, rewriting it and splitting it into sub-expressions. Next it is transformed into a Glushkov NFA [164]. We selected the Glushkov representation because it has a number of interesting properties [79]. The Glushkov NFA is then turned into a DFA. These automata are then combined together by an algorithm that operates on their topology. The resulting graph is then extended, as in the Aho-Corasick algorithm, by a "failure" function, whose computation can further modify the graph structure. The result is a single pass deterministic automaton that:

- groups every regular expression in a single automaton,

- reports exhaustive and complete matches, including every overlapping pattern,

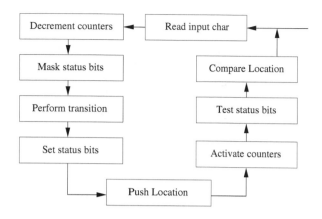

FIGURE 13.8: DotStar runtime.

- in most practical cases it is as memory efficient as an NFA.

During the compilation process DotStar can be tuned and optimized for a specific architecture: for example it is possible to trade the number of states with the size of the state bit word, or it is possible to reduce the size of the automaton using parallel counters if the target architecture supports vector instructions.

The DotStar runtime shares the same execution model of the well known Aho-Corasick algorithm and for each symbol the automaton transitions at most once across a keyword tree edge or a finite number of times across F() edges until either a proper suffix is detected or the root node is reached. A linear time DFA is obtained by pre-computing every F(). DotStar extends this runtime by adding actions to be executed when a state is entered. These actions have been defined in such a way as to be parametric and exploit modern CPU features, thus allowing extensive runtime optimizations targeted at a specific architecture class. Our runtime system, depicted in Fig. 13.8, leverages the available memory hierarchy, pipeline, and multiple cores, using caching, partitioning, data replication and mapping [383, 441, 442]; it also employs automata interleaving, bit-level parallelism and vector instructions to operate on data structures.

The DotStar solution is applicable in several fields, from network intrusion detection to data analysis and also as a front end for a protocol parser, since the first logical step for any protocol parser is recognizing where data fields start and end. For supporting this task we defined a simple declarative language, called DSParser, that describes a data layout as a sequence of regular expression fragments "connected" using standard imperative constructs, like if/then/else/while. Actions can be inserted after a (partial) expression is recognized; these actions are either *system actions*, that perform common

```
MATCH "......................................"
MATCH "\x01\x02"
PUSH
MATCH ".........."
EXECUTE sequence_number
MATCH "..."
LOOP
  WHILECNT "."
    PUSH
    MATCH "[\x00-\x7f]*[\x80-\xff]"
    EXECUTE action_pmap
    PUSH
    WSWITCH
      CASE "[\x80-\xff]"
        EXECUTE action_field
        PUSH
        ENDCASE
      CASE "[\x00-\x7f]"
        MATCH "[\x00-\x7f]*[\x80-\xff]"
        SEND 0x02 0
        EXECUTE action_field
        ENDCASE
    ENDWSWITCH
  ENDWHILECNT
ENDLOOP
```

FIGURE 13.9: DotStar source code.

operations or user defined functions on blocks of input data. A precompiler tool parses the declarative language and builds a suitable regular expression set that is compiled using the DotStar toolchain. A small number of states of the resulting automaton is then annotated with the system and user defined actions specified in the initial protocol definition. The DSParser source code for analyzing OPRA V2 messages is remarkably simple, compact and easy to manage[1] (see Figure 13.9).

Intuitively, the resulting automaton will start by skipping the IP/UDP headers and will match the OPRA v2 start byte and version number. It will then mark the stream position for the successive action and recognize the initial sequence number for the packet, calling the appropriate user defined code. After that it will loop, examining every message in the packet and detecting first the PMAP and then all individual fields. User defined actions operate on blocks of input data recognized by the parser automaton, and the overall system processing model proceeds along the following high level steps:

1. Look for an interesting section of data: the automaton reads input symbol(s) and switches state until a PUSH action is reached.

2. Save the start of interesting data: the current stream position is saved in state machine memory.

---

[1] It is worth noting that the full OPRA FAST protocol is described in a 105-page manual!!

3. Look for the end of a data field: the automaton reads more input symbols until a state with a user defined action is found.

4. Handle the data field: the user defined action is invoked over the block of data from the saved position to the current stream position.

Thanks to DotStar's arsenal of optimization and configuration parameters, we can fine-tune the compiled automaton and we can select how this automaton "connects" with user defined actions. For example, our modular runtime can either invoke user actions when detected, for minimum latency, or can "schedule" their execution at a later moment, eventually by a different thread, for maximum throughput (and this will allow automata interleaving for better CPU pipeline utilization). The individual user actions are exactly the basic building blocks used in the handmade parser for converting integers and strings and parsing PMAP content, as shown in Figure 13.6. The rest of the decoder, which takes care of input handling, data sequencing and field delimitation, is handled by the DSParser compiler and is automatically tuned for the architecture, exploiting optimization that is difficult to implement in the hand-optimized code.

13.4.5   Experimental Results

OPRA market data feeds are transmitted in a set of 24 channels. In this experimental section, we assume that each channel can inject messages at full speed by storing the OPRA feeds in main memory, and therefore is not the bottleneck of the decoder. While this hypothesis may not be realistic in practice, it serves the purpose of pushing to the limit the OPRA decoding algorithm and provides an upper bound.

OPRA packets are 400 bytes on average, each containing multiple messages that are encoded using the FAST protocol. We use real market data feeds, obtained by capturing a few seconds of the network traffic, totalling one gigabyte of version 1 and 2 OPRA format for experimental analysis throughout this section.

Figure 13.10a gives the distribution of the message size for both OPRA version 1 and 2 formats. The messages are typically distributed across the 10-50 byte range, with an average message size of 21 bytes. Thus each packet contains 19 messages on average. Our additional analysis shows that, under the assumption of full injection, the feeds across the various channels have very similar data pattern and distribution and tend to have the same processing rate. Since the performance is insensitive to the OPRA protocol version, we will consider only OPRA version 2 traces in the rest of the evaluation.

Figure 13.10b gives a distribution of the OPRA messages among 11 categories, as described earlier. We observe that 99% messages flowing in the market are category K equity and index quotes. The first field of each message after

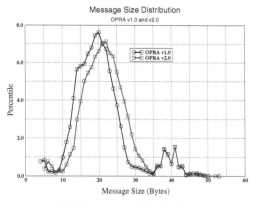

(a) Message size distribution

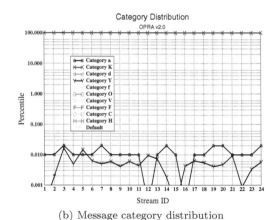

(b) Message category distribution

FIGURE 13.10:   OPRA message distribution.

the *message length* is the PMAP, containing encoded information about the position and type of the data fields that follow. Additional analysis of the message stream indicates that for OPRA version 2, a large majority of the messages contain a 5 byte PMAP (the size of the PMAP may vary from 1 to 8 bytes). The data fields can be either integer or string type, that is given as a part of the protocol specification. Each OPRA message can contain multiple integer fields, with length varying from 1 to 5 bytes. Our analysis shows that more than 80% of the encoded integers are less than 2 bytes.

Next, we present an extensive performance analysis of our algorithm on a variety of multi-core architectures. In our tests, we use Intel Xeon Q6600 (Quad), Intel Xeon E5472 (Quad, a.k.a. Harpertown), AMD Opteron 2352 (Quad), IBM Power 6, and Sun UltraSparc T2 (a.k.a. Niagara-2). Power6 and Niagara-2 are hardware multithreaded with 2 threads and 8 threads per core,

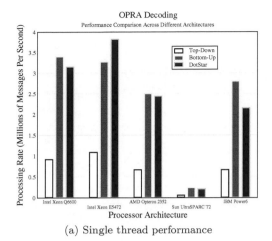

(a) Single thread performance

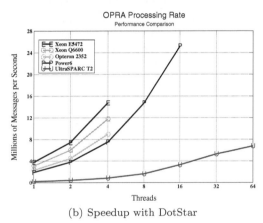

(b) Speedup with DotStar

FIGURE 13.11: Performance comparison on several hardware platforms.

respectively. Table 13.2 gives more information about the CPU speed, number of cores, threads, cache size, and sockets on each architecture.

Here, we discuss the performance of the three approaches for processing OPRA FAST messages, the top-down reference implementation, the hand-optimized bottom-up version, and DotStar as discussed in sections 13.4.2, 13.4.3 and 13.4.4, respectively. Figure 13.11a gives the decoding rate per processing thread in millions of messages/second using the three approaches on the multi-core architectures described in Table 13.2. We observe that our bottom-up and DotStar implementations are consistently 3 to 4 times faster than the reference implementation; the Intel Xeon E5472 gives the maximum processing rate for a given thread, and that our optimized Bottom-Up and DotStar implementations are similar in terms of performance, with the hand-optimized version being only 7% faster, on average. The Sun UltraSparc T2

TABLE 13.2:   Hardware platforms used during experimental evaluation.

	CPU Speed (GHz)	Sockets	Cores per Socket	Threads per Core	Threads	Cache Size (KB)
INTEL Xeon Q6600	2.4	1	4	1	4	4096
INTEL Xeon E5472	3.0	1	4	1	4	6144
AMD Opteron 2352	2.1	1	4	1	4	512
Sun UltraSPARC T2	1.2	1	8	8	64	4096
IBM Power6	4.7	4	2	2	16	4096

processor is designed to operate with multiple threads per core, thus the single thread performance is much slower than other processors. It is also important to note that our single core OPRA decoding rate is much higher than the current needs of the market, as given in Figure 13.2.

Figure 13.11b presents a scalability study of our high-level DotStar implementation across threads/cores on the five multi-core architectures. For the Intel Xeon processors our implementation scales almost linearly and processes on a single socket, 15 million messages/second, on the E5472 and 12 million messages/second, on the Q6600. On Niagara-2, the performance scales linearly up to 16 threads, and reaches 6.8 million messages/second using 64 threads (8 threads/8 cores). We believe we start hitting the memory bandwidth wall beyond 32 threads. For the IBM Power6, our single thread performance is not as impressive as the Intel E5472, but gets a scaling advantage up to 16 threads using 8 cores, giving a performance of 26 million messages/second. The performance scales linearly up to 8 threads (1 thread per core), with a performance benefit of 1.5x up to 16 threads.

13.4.6   Discussion

In this section we provide insight into the performance results, and present a reference table (Tables 13.3 and 13.4) that explains in detail where the time is spent. The OPRA FAST decoding algorithm can be broken down into 5 main components, processing of the PMAP, *integer*, *string* fields, *copying previous values*, and *computing field index from PMAP*. Every OPRA message contains one PMAP field with an average size of 5 bytes, *integer* fields of 13 bytes, *string* fields of 4 bytes, in total 22 bytes.

Decoding a message requires *copying* the last known values into a new data structure, as the encoded message only contains information in a subset of fields. To find the index of each new data fields, we perform a *clz* (count leading zeroes) operation. For each of these actions we calculate the instructions from assembly, Tables 13.3 and 13.4 give the instructions/byte for processing the PMAP, *integer* and *string* fields, on each multicore processor. For nsec and cycles per byte we compute actual performance, by performing multiple iterations of the algorithm with and without the corresponding action and normalizing the difference. Similarly, for the *clz* action, we compute cycles,

TABLE 13.3: Detailed performance analysis (Intel Xeon Q6600, E5472 and AMD Opteron 2352)

	size	%	INTEL Xeon Q6600			INTEL Xeon E5472			AMD Opteron 2352		
			instr	ns	cyc	instr	ns	cyc	instr	ns	cyc
INT	1	62.6	3.0	1.4	3.4	3.0	1.5	4.5	3.0	4.0	8.4
	2	22.1	3.5	1.3	3.2	3.5	1.6	4.9	3.5	4.8	10.1
	3	13.6	3.7	1.4	3.4	3.7	1.7	5.2	3.7	5.0	10.5
	4	2.7	3.7	1.5	3.6	3.7	1.8	5.4	3.7	5.1	10.7
	5	0.0	3.8	1.5	3.8	3.8	1.8	5.5	3.8	5.1	10.8
PMAP	1	0.0	7.0	2.6	6.3	7.0	3.3	9.9	7.0	9.4	19.7
	2	0.0	6.0	2.5	6.0	6.0	3.0	9.0	6.0	8.0	16.9
	3	0.0	5.7	2.5	6.0	5.7	2.8	8.5	5.7	7.6	15.9
	4	8.1	5.5	2.4	5.9	5.5	2.8	8.3	5.5	7.4	15.5
	5	87.1	5.4	2.4	5.8	5.4	2.7	8.1	5.4	7.3	15.4
	6	4.8	5.3	2.3	5.6	5.3	2.7	8.0	5.3	7.1	15.0
	7	0.0	5.2	2.3	5.5	5.2	2.6	7.8	5.2	7.0	14.7
	8	0.0	5.1	2.3	5.5	5.1	2.6	7.7	5.1	6.9	14.4
Dcopy/msg			-	11.0	26.4	-	15.0	45.0	-	45.1	94.7
Scopy/byte			-	1.4	3.4	-	1.9	5.7	-		
CLZ/field			6	2.8	6.7	5	2.0	6	5		

	msgs M/s	ns /b	cyc /b	msgs M/s	ns /b	cyc /b	msgs M/s	ns /b	cyc /b
Basic blocks peak	11.8	3.6	8.5	11.8	3.5	10.6	4.7	9.0	19.0
Actions only	8.8	4.8	11.5	9.5	4.4	20.6	4.3	9.7	20.3
Dotstar rate only	5.8	8.1	19.4	7.5	6.4	30.0	5.7	7.5	15.7
Optimal	3.5	-	-	4.2	-	-	2.5	-	-
Actions w/Dotstar	**3.1**	13.3	32.0	**3.8**	11.0	51.7	**2.4**	17.5	36.8
Optimality ratio	**0.89**	-	-	**0.91**	-	-	**0.96**	-	-

nsec, and instructions per data field, and for the *copy msg* we compute average nsec and cycles per message.

Using the field length distributions, their occurrence probability rates, and individual action performance, we compute the aggregate estimated peak performance in the first row of the last block in the tables. Note that this peak estimate does not take into account any control flow or I/O time. In the second row we give the actual actions-only performance that is computed by taking the difference between the total performance and performance after commenting out the actions, let's call it A. The DotStar rate is the peak performance of our parsing algorithm on OPRA feeds, that does not include any actions on the recognized data fields, let's call it B.

The *optimal* row in that block is the estimated peak performance that we should have gotten by combining the DotStar routine and our optimized-actions routines, that is computed using the formula $(\frac{1}{A} + \frac{1}{B})^{-1}$. We compare

TABLE 13.4: Detailed performance analysis (SUN UltraSparc T2 and IBM Power6)

	size	%	SUN UltraSPARC T2 instr	ns	cyc	IBM Power6 instr	ns	cyc
INT	1	62.6	3.0	6.5	7.8	4	2.3	10.8
	2	22.1	3.0	6.8	8.2	3.5	2.2	10.3
	3	13.6	3.0	5.4	6.5	3.3	2.0	9.4
	4	2.7	3.0	6.6	7.9	3.2	1.9	8.9
	5	0.0	3.0	6.6	7.9	3.2	1.9	8.8
PMAP	1	0.0	8.0	17.3	20.7	7	4.4	20.7
	2	0.0	7.4	16.5	19.8	6.5	4.2	19.7
	3	0.0	7.3	16.0	19.2	6.3	4.1	19.0
	4	8.1	7.5	16.9	20.3	6.2	3.9	18.6
	5	87.1	7.2	15.9	19.1	6.2	3.9	18.1
	6	4.8	7.7	17.5	21.0	6.0	3.8	17.8
	7	0.0	7.9	18.2	21.8	6	3.8	17.9
	8	0.0	7.5	16.9	20.3	6	3.9	18.1
Dcopy/msg			-	706.0	847.2	-	14.0	65.8
Scopy/byte			-	15.2	18.2	-	3.9	18.6
CLZ/field			17	36.9	44.3	6	4	18.8
			msgs M/s	ns /b	cyc /b	msgs M/s	ns /b	cyc /b
Basic blocks peak			0.7	59.0	70.8	7.5	5.6	26.2
Actions only			0.6	69.3	83.2	5.3	7.9	37.2
Dotstar rate only			0.4	109.8	131.8	4.1	11.2	52.6
Optimal			0.24	-	-	2.3	-	-
Actions w/Dotstar			**0.2**	200.0	240.0	2.1	20.0	94.0
Optimality ratio			**0.84**	-	-	**0.92**	-	-

this to our actual performance and compute the *optimality ratio* in the last row of the table.

We notice that:

- On the IBM Power 6, we get similar nsec/byte performance as compared to Intel Xeon processor. We believe that our code is not able to take advantage of the deep pipeline and in-order execution model of this CPU.

- To get maximum performance for the *copy msg* routine, we developed our own vectorized memory copy using SSE intrinsics, and pipelined load and stores to obtain best performance. On the Sun Niagara-2 we used a manual copy, at the granularity of an integer, and unrolled the load and store instructions. With vectorized memory copy we get about 2x performance advantage over the *memcpy* library routine, whereas for the manual scalar copy we get an advantage of 1.2x on Sun Niagara-2.

TABLE 13.5: DotStar latency on different platforms.

	Latency/msg (nsec)
INTEL Xeon Q6600	317
INTEL Xeon E5472	261
AMD Opteron 2352	409
Sun UltraSPARC T2	4545
IBM Power6	476

- On Sun-Niagara, the *copy* action is much slower than on other processors, that leads to a significant performance penalty.

- Table 13.5 gives the latency for processing/decoding an OPRA message on each of the multicore processors discussed in the papers. We decode one message at a time on a single processing thread, thus the average latency to process a message is given by taking the inverse of the processing rate. We observe a latency between 200 and 500 nsec on the Intel/IBM/AMD processors, which accounts for only a negligible latency inside the ticker plant, under a very high throughput.

- Our implementation requires integer/string processing as opposed to floating point computation. This fails to utilize the much improved floating point units on the Sun Niagara-2 processor.

- On Intel Q6600 processor, the cycles/byte performance matches closely with the instructions/byte count.

- Our implementation is close to the optimal performance we could obtain with the high-level approach, with an average optimality ratio of 0.9.

13.5 Chapter Summary

The increasing rate of market data traffic is posing a very serious challenge to the financial industry and to the current capacity of trading systems worldwide. This requires solutions that protect the inherent nature of the business model, i.e., providing low processing latency with ever increasing capacity requirements. Similar requirements are present in other application domains dealing with streaming data, where the increase in data volumes, and the need for streamlined execution of value-added application-processing services, need to be addressed.

Using as a case study the handling of OPRA FAST financial feeds, we describe an approach to address the above requirements on commodity platforms. One component of our approach is DSParser — a high-level descriptive language which allows application developers to easily capture the essence of the underlying OPRA FAST protocol. This tool is the programming interface of our second component, the DotStar protocol parser, which enables efficient realization of the protcol engine. With these tools, our approach results in a solution which is high-performance, yet flexible and adaptive.

We demonstrate impressive processing rates of 15 million messages per second on the fastest single socket Intel Xeon, and over 24 million messages per second using the IBM Power6 on a server with 4 sockets. We present an extensive performance evaluation that helps one understand the intricacies of the decoding algorithm, and expose, many distinct features of the emerging multicore processors, that can be used to estimate performance on these platforms. As a future extension of this work, we plan to extend our approach to design other components of the ticket plant, and evaluate a threaded-model in DotStar, that would help pipeline independent analytics processing kernels seamlessly.

In addition, by describing the details of the OPRA FAST protocol, we highlight the importance of compact and efficient data encoding techniques for streaming applications, so as to reduce the overall data throughput requirements, and to be able to leverage common architectural features in modern hardware platforms.

Chapter 14

Data-Movement Approaches for HPC Storage Systems

Ron A. Oldfield
Sandia National Laboratories[1]
Todd Kordenbrock
Hewlett-Packard
Patrick Widener
University of New Mexico

14.1 Introduction

The efficiency of the storage system for a MPP system depends heavily on the data-movement scheme implemented by the file system. This is primarily because MPP systems, unlike traditional clusters, tend to use a "partitioned architecture" [168] (illustrated in Figure 14.1) where the system is divided into sets of nodes that have different functionality and requirements for hardware and software. For example, the Cray XT-3 and IBM Blue Gene systems have compute, I/O, network, and service nodes. The compute nodes use a "lightweight kernel" [273, 369, 425] operating system with no support for threading, multi-tasking, or memory management. I/O and service nodes use a more "heavyweight" operating system (e.g., Linux) to provide shared services. Recent decisions to use Compute Node Linux on Cray XT-4 compute nodes make available some of these services, but Compute Node Linux still only provides a restricted set of functionality when compared to a standard Linux distribution. These restrictions represent a significant barrier for porting file systems originally designed for clusters of workstations.

System size also has a large impact on file system implementation and data-movement schemes. Current Teraflop systems at DOE laboratories have tens of thousands of multi-core compute nodes, and projected Petaflop systems expect

---
[1]Sandia is a multiprogram laboratory operated by Sandia Corporation, a Lockheed Martin Company, for the United States Department of Energy's National Nuclear Security Administration under contract DE-AC04-94AL85000.

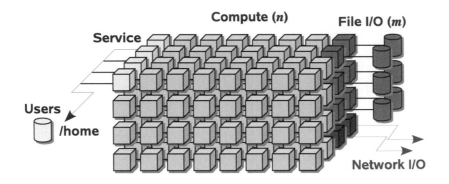

FIGURE 14.1:   Both IBM and Cray use a partitioned architecture with a "lightweight" operating system for compute nodes (limited or no support for threading, multi-tasking, or memory management). I/O and service nodes use a more "heavyweight" operating system (e.g., Linux) to provide shared services.

to have millions of cores in the near future [252]. At these scales, resilience of the file system to compute-node failures becomes a significant issue. This implies that any scheme relying on client-side management of system data, for example file caching, has to be carefully managed or avoided. In addition, as Table 14.1 illustrates, the number of compute nodes is often one to two orders of magnitude greater than the number of I/O nodes. The disparity in the number of I/O and compute nodes, coupled with the fact that compute nodes are diskless, puts a significant burden on the communication network. Because many I/O operations for scientific applications are "bursty," an I/O node may receive tens of thousands of near-simultaneous I/O requests. An MPP I/O system should be able to handle such bursts of I/O in a reliable way. Another often ignored issue is the impact I/O traffic has on application traffic. Cray systems use the same torus mesh network for application and I/O traffic, which can cause network contention between data-intensive and communication-intensive applications. The Blue Gene/L architecture addressed this issue by using five different networks:[2] a torus network for intra-compute node communication; three separate tree networks for I/O, barriers, and global communication; and an N-to-1 network for control and reliability traffic.

Another challenge for file system designers is dealing with the unique characteristics of low-level networking interfaces and protocols for MPP systems (see Chapter 6). In many cases the network interfaces are custom designed for

---

[2]The Blue Gene/L uses five independent networks to avoid contention: a 3-D torus for compute-node communication; three tree networks for global operations, barriers, and I/O traffic; and an N-to-1 backplane for control information. See Chapter 1 for more detail.

TABLE 14.1: Compute and I/O nodes for production DOE MPP systems used since the early 1990s.

Computer	Compute Nodes (core/node)	I/O Nodes	Ratio
Intel Paragon (1990s)	1840 (2)	32	58:1
Intel ASCI Red (1990s)	4510 (2)	73	62:1
Cray Red Storm (2004)	10,368 (2,4)	256	41:1
IBM Blue Gene/L (2005)	65,536 (2)	1024	64:1
Cray Jaguar (2007)	11,590 (2,4)	72	160:1

the particular platform. In general, the requirements for latency, bandwidth, and scalability preclude the use of standard network protocols like TCP/IP due to excessive data copies, CPU-utilization, or persistent connections — all of which limit scalability [168]. In addition, the use of lightweight-kernels with no thread support prevents the types of asynchronous I/O approaches used for Linux clusters. To support asynchronous I/O and limit memory copies and computational overheads of I/O operations, low-level MPP network interfaces often provide special features to support remote direct-memory addressing (RDMA) and operating-system bypass. The protocols used by the MPP storage system should be able to exploit these features for efficient data movement.

In this chapter, we describe the data-movement approaches used by a selection of storage systems used by MPP platforms. In particular, we examine data-movement schemes for Lustre, Panasas, the Parallel Virtual File System (PVFS2), and the Lightweight File Systems (LWFS) project. While this is not a complete survey of parallel file systems, it represents an interesting mix of commercial, open-source, and research-based solutions, as well as a fairly complete set of data-movement options for parallel file systems.

14.2 Lustre

Lustre [95] is a Sun Microsystems-supported object-based file system designed for clusters with tens of thousands of clients. Lustre was originally designed to meet the needs of the MPP systems of the Department of Energy (DOE) laboratories and is the primary parallel file system used by the Cray XT systems at Sandia National Laboratories and Oak Ridge National Laboratory, and the IBM Blue Gene systems at Lawrence Livermore National Laboratory. Lustre obtains scalability on MPP systems by careful attention to the network between the compute-node clients and the storage targets, by using flow-control mechanisms to avoid congestion, and by using distributed lock management and adaptive I/O locks to provide POSIX consistency semantics at scale [335].

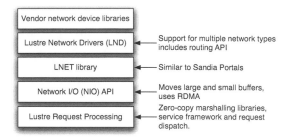

FIGURE 14.2:   Lustre software stack.

14.2.1   Lustre Networking (LNET)

Lustre uses the LNET (Lustre Networking) [54] interface to control data movement between compute-node clients and the Lustre storage and metadata servers. The LNET software stack, illustrated in Figure 14.2, consists of a remote procedure API with interfaces for recovery and bulk transport, a network I/O (NIO) API that provides an RDMA abstraction to move large and small buffers, an LNET library that implements the NIO API interface, a set of drivers for different network types, and the vendor-supplied device libraries.

Lustre derived its message-passing API for the LNET library from the Portals [64] message passing API developed by Sandia National Laboratories and the University of New Mexico (see Chapter 6). The API provides a low-latency connectionless communication model with inherent support for remote-direct memory access (RDMA) on networks that support RDMA.

Lustre supports different network types through pluggable drivers called Lustre Network Drivers (LND). These software modules are similar in concept to the Network Abstraction Layer (NAL) modules in the Portals library in that they provide an abstraction layer that allows access to multiple network types. Lustre currently has LNDs for InfiniBand (IB), TCP, Quadrics, Myricom, and Cray.

A Lustre network consists of a set of nodes, each with a unique node ID (NID), that can communicate directly with any other NID in the Lustre network. The file system consists of three types of nodes: an Object Storage Client (OSC), a Meta-Data Server (MDS), and Object Storage Targets (OST). In a typical cluster, LNET can define the nodes in a Lustre network by enumerating the IP addresses in the network. When more than one network exists, some of the nodes also act as routing nodes. If multiple routers exist between a pair of networks, Lustre balances load between the two networks.

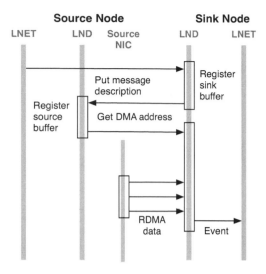

FIGURE 14.3: Lustre uses a server-directed DMA handshake protocol to pull data to the server.

14.2.2 Optimizations for Large-Scale I/O

As with other object-storage file systems, Lustre achieves fine-grained parallel access to storage by decoupling metadata operations from the I/O operations, allowing direct access between the OSCs and the OSTs. Lustre stripes data to multiple OSTs to aggregate bandwidth available from a number of OSTs. The challenge for large systems, where there are at least an order-of-magnitude more OSCs than OSTs, is maintaining a steady flow of data to the OSTs without overwhelming the OSTs and the network with requests. Lustre implements a number of optimizations to address these challenges.

First, Lustre separates the I/O request from the bulk data transfer and uses a server-directed [236] approach to maintain sustained bandwidth of data to the servers. The protocol, illustrated in Figure 14.3, requires a handshake at the operating system to initiate the RDMA, preventing the ability to implement purely asynchronous I/O from the client. However, Lustre allows up to eight concurrent requests from each client, so the server can be pulling data from one request (without OS interference) while the client negotiates the next exchange with the server.

To increase request size, Lustre makes use of read-ahead and write-back caches to aggregate small requests on the OSC [478]. This optimization mitigates the network latency overhead of a large number of small requests and improves disk throughput, which is severely impacted by seek latency. Lustre also modified the block allocator on the local file system (ext3) for the OST to merge requests — converting a sequence of logical extents into large contiguous

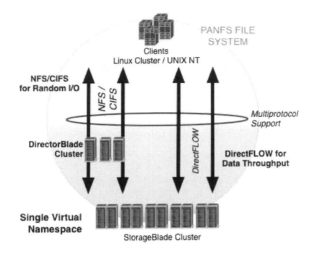

FIGURE 14.4:   Panasas DirectFlow architecture.

extents. This optimization led to significant bandwidth improvement on the server when under heavy loads [335].

To avoid congestion due to the potential number of requests, the Lustre Network Drivers (LNDs) use flow-control credits [335] in their message-passing protocol to avoid congestion. Before a node sends a request, the sender consumes a sender credit from the source and a receiver credit from the destination node. The sender credit from the source guarantees that there is buffer space on the source for the outgoing packet, while the receiver credit for the destination guarantees that the receiver has space for the incoming packet. The credits are returned after the appropriate node frees its buffer. If a credit is not available, the transmission has to wait. In addition, to avoid the overhead of sending and fetching credits, Lustre "piggybacks" credit information with every network packet.

14.3   Panasas

The Panasas PanFS parallel file system [324] also provides scalable parallel storage by separating metadata management from data storage operations. Previous attempts to scale storage systems to the terabyte-petabyte ranges in traditional NFS-style arrangements (with a server interposed between clients and storage devices) had produced bottlenecks in metadata operations. SAN solutions suffer similar problems, in addition to requiring non-standard storage access protocols between clients and storage devices. PanFS was designed to provide clients with high-capacity parallel storage access while still maintaining the interoperability benefits of a standardized solution.

PanFS provides an object interface to clients, but organizes storage in blocks. Metadata services are provided by a dedicated server, but that server is not part of the high-performance data path. The data-movement protocol used by PanFS is called *DirectFlow*.

14.3.1 PanFS Architecture

PanFS objects are made up of storage blocks. In traditional file systems, block management is performed by the file server; this requires all clients to include the file server in each storage operation, making it both a bottleneck and a single point of failure. PanFS addresses this issue by delegating block management to smart storage devices. This allows clients to bypass the file server and write data directly to the storage devices, once the set of blocks comprising each object has been identified. Besides enabling high-performance parallel data access, this architecture allows PanFS to proactively avoid and recover from hardware failures in a transparent manner.

14.3.1.1 Data movement in PanFS

Clients are able to directly access data blocks in PanFS. This is accomplished by using smart storage devices and an Object Storage Device (OSD) interface. OSD devices present a higher-level abstraction than typical mass-storage devices. Traditional device interfaces represent a disk as a large array of blocks and rely on file systems to map a collection of blocks to logical entities such as files. OSD-capable devices move this mapping down to the device, and allow file systems to use an object-based interface. The OSD interface used by PanFS is a set of extensions to the SCSI command set that is very close to the ANSI-standard T10 [407] protocol.

SCSI commands can be issued to directly-connected devices over a hardware bus. PanFS uses an emulation protocol called iSCSI [381] to issue SCSI commands to devices over an TCP/IP network. Specialized network hardware that handles the TCP communication operations is attached to the PanFS storage clusters, allowing remote hosts to use standard interconnection infrastructure while behaving as though the storage devices are actually locally attached.

The DirectFlow protocol requires the use of a Linux kernel module at the client. In order to maximize interoperability, PanFS also allows clients to use standard NFS protocols to access data. Windows clients are also supported using the CIFS protocol.

14.3.1.2 Eliminating the metadata bottleneck

A PanFS metadata server maintains mappings between file names and the collection of storage objects that make up the file. These collections are also known as *layouts*. Layouts may also contain parameters for data striping or security information for particular objects. Once a client knows the layout for a particular file, the metadata server is not needed for data reads or writes.

In order to use a file, a client requests the layout from the metadata server. The client then "owns" the layout and may use it to perform direct I/O to the storage devices (using the iSCSI/OSD interfaces described above). When the client is finished, any changes to the layout are communicated to the metadata server and the client is considered to have returned the layout. The server can revoke access to the file at any time by recalling the layout from the client, which must roll back any changes to the layout.

14.3.1.3   Blade-based architecture and fault-tolerance

PanFS is commercially available [326] in storage clusters marketed under the ActiveStor name. These clusters are arranged as collections of StorageBlades (which host the OSD-capable storage devices) and DirectorBlades (which provide metadata management services). By varying the deployment of Director and StorageBlades, different clustering solutions can be created to address scalability or fault-tolerance requirements.

The flexibility provided by moving block-level mappings into storage devices allows ActiveStor clusters to implement advanced behaviors like "hot spot" management. Hot spots form when a single StorageBlade is serving disproportionately more requests than other StorageBlades. ActiveStor clusters detect these potential I/O bottlenecks and automatically relocate or replicate any involved data objects to less-utilized StorageBlades. Clients are minimally affected, as only the object layouts on the metadata server have to be changed.

Similarly, objects can be relocated from potentially failing disks before any actual fault occurs, eliminating the need to perform RAID rebuilds. If a RAID reconstruction is necessary, all object layouts in the storage cluster are updated in parallel, making the rebuild more efficient [454].

14.3.2   Parallel NFS (pNFS)

The data movement strategy used by Panasas is a specialization of proposed extensions to the Network File System version 4 standard, known as pNFS or NFS version 4.1. Complete specifications of the entire set of pNFS proposals are still under review and revision at the time of this writing. However, it is illustrative to look briefly at how pNFS expands on what is present in Panasas.

14.3.2.1   Differences from Panasas

The most important difference between Panasas and the emerging pNFS standard is in the nature of the data movement between clients and storage devices. Where PanFS uses OSD interfaces over an iSCSI transport, pNFS broadens the available choices. pNFS clients may also use SCSI block com-

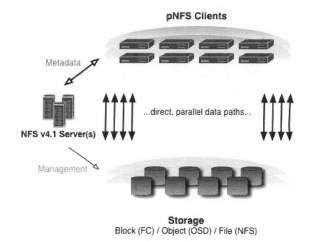

FIGURE 14.5: Parallel NFS Architecture.

mands (SBC) over a FibreChannel interface, for example. This alternative can provide higher performance than using iSCSI over TCP, but the relatively high cost of FibreChannel hardware and the ubiquity of TCP interconnects make SBC/FC more of a niche solution. pNFS also allows normal NFS data movement (via XDR-encoded RPC) between clients and storage systems. pNFS is primarily concerned with removing metadata operations from the parallel fast data storage path, and so its design is deliberately somewhat agnostic and open with respect to I/O control and storage protocols.

14.3.2.2 Standard adoption

The original IETF RFC documents proposing the pNFS standard were introduced in 2005. As of this writing, these original documents have been superseded by more detailed specifications that are still under consideration by the IETF Network File System Version 4 Working Group (see [204] for more information).

14.4 Parallel Virtual File System 2 (PVFS2)

The Parallel Virtual File System 2 [348] (PVFS2) is a distributed parallel file system that emphasizes a modular adaptable design. Modularity is provided both vertically in the design of the software stack (Figure 14.6a) and across components by separating PVFS2 functionality into separate server and client libraries. Additional separation between server components is achieved by

Software Stack

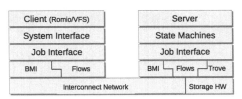

(a) Software stack

Client Server

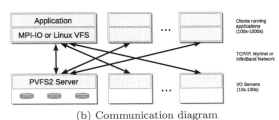

(b) Communication diagram

FIGURE 14.6:   PVFS2.

minimizing inter-server communication (Figure 14.6b). When a client creates or opens a file, the client receives an opaque handle that represents all the information needed to directly contact servers and distribute the data amongst them.

In the PVFS2 software stack, the lowest level is the Buffered Message Interface (BMI) network abstraction. BMI was specifically designed with the I/O requirements of parallel file systems in mind. BMI strikes a balance between the flexibility of an interconnect API and the restrictiveness of a structured I/O library.

14.4.1   BMI Design

During the PVFS2 design cycle, the PVFS2 team identified eight critical design criteria for a data movement library in the context of a parallel file system [78]:

- efficiency
- parallel I/O access pattern support (scatter/gather)
- thread safety
- explicit buffer management for RDMA
- dynamic client/server model

- fault tolerance
- minimal exposed state
- simple high level API

The team surveyed existing data movement libraries in search of an solution, but found none that met all the requirements. The Buffered Message Interface (BMI) was designed to specifically addresses the needs of a parallel file system.

BMI defines two programming APIs — a user API and a module API. The user API is the public API to which clients and servers program. It allows servers to establish themselves at a well known location on the network and allows clients to find peers, to manage data movement operations and to manage special memory regions.

The BMI user API employs a simple non-blocking post/test model. The client posts operations followed by a test for completion. Because posts are non-blocking, the client (a single thread or multiple threads) can post multiple (related or unrelated) concurrent operations. As each operation is posted, BMI assigns it a unique ID that is later used for completion testing. Unlike some network protocols, the BMI user API does not require the user to post a dangling receive for unexpected messages. Instead, the BMI user can specifically test for unexpected messages. In the context of a parallel file system, unexpected messages usually occur on the server side when a client initiates a message/data exchange.

The BMI module API defines a common network device interface that isolates BMI and the user from the complexities of different network models, which simplifies the process of porting BMI to new network infrastructures. The network modules can be started and stopped dynamically with multiple modules active simultaneously. Modules exist for a variety of common network protocols — TCP/IP, Myrinet GM, Myrinet MX, InfiniBand, and Portals.

14.4.2 BMI Simplifies the Client

Servicing independent application components (most commonly threads) without confusing them keeps the user code simple. BMI allows the user to define execution contexts that separate operations into groups. In its simplest form, execution contexts group operations by thread, but arbitrary groups can be created to suit the needs of the user. When threads use execution contexts, they can safely test for completion without fear of receiving completion events for operations belonging to another thread.

Network complexity can add greatly to the overall complexity of a parallel file system design. BMI has gracefully hidden the most significant complexity by presenting a single view of the network to the user. The BMI user API is connectionless, locates peers using simple URIs, transparently manages flow control and does not require dangling receives for unexpected messages. BMI minimizes the amount of state exposed to the user by holding network specific state in the network module. The user only needs three pieces of information: URI of the peer, execution context, and the IDs of incomplete operations.

After examining available approaches to fault tolerance, the PVFS2 team decided that there would be no specific fault tolerance mechanism in BMI. Determining retransmission points is tricky business, particularly when the context of the data movement is unknown. BMI has chosen a notification only approach, in which failures are delegated to higher levels. The PVFS2 Job Interface takes responsibility for managing retransmission in the case of a fault. If a PVFS2 Job is to be aborted, outstanding BMI operations are individually canceled.

14.4.3   BMI Efficiency/Performance

The BMI design identifies and addresses specific system areas (network, memory copies, scatter/gather) that can either boost or hinder the efficiency of a parallel file system.

Testing for the completion of operations can be an expensive endeavor. Queues must be maintained and polling must be done. Polling can be quite expensive. All BMI post functions may be immediately completed, which means the operation is complete when the post returns. The user can test for immediate completion and avoid the need to set up a test. Immediate completion occurs most commonly when posting short sends or when posting receives for data that has already been received and buffered.

Another technique for minimizing completion testing is aggregating many small operations into a single operation that manipulates a list of buffers. PVFS2 supports user-defined I/O access patterns that allow the user to specify the manner in which memory regions map to on disk storage. This is commonly used to select regions of discontiguous memory to create a more compact contiguous structure on disk. Instead of individual operations for each memory region, the operations can be aggregated into a single operation. All BMI post operations have a corresponding list version that allows a set of buffers to be sent or received in a single operation. If the network supports them, the network module can use scatter/gather hardware to increase the performance of such operations.

Kernel buffer copies are expensive operations. With many types of network hardware, as the data streams in, it is copied into temporary kernel buffers, then into user buffers when the receive operation is posted. Many modern cluster interconnects offer hardware assisted Remote Direct Memory Addressing (RDMA). After the user allocates a memory region and pins it, the NIC can stream the data directly out of the source memory buffer onto the wire and then off the wire into the destination memory region. To maximize performance on RDMA networks, the BMI user API has memory management functions to allocate and free pinned memory buffers. The BMI memory functions enable the user to manage pinned memory regions if the network module supports them. If the client doesn't allocate any memory for the RDMA operations, the network module will do it for the user at possibly reduced performance.

14.4.4 BMI Scalability

Busy servers cannot afford to spend too much time performing lookups. To keep lookups to a minimum, BMI provides a feature called User Pointers. When an operation is posted, the user supplies a pointer to some data structure. In the case of PVFS2, the User Pointer refers to a state machine that is keeping track of some higher level operation. Regardless of the number of BMI operations in flight, the user data structure (state machine) can be found in O(1).

Another time consuming/resource intensive operation performed by busy servers is maintaining lists of outstanding operation IDs. BMI provides the two common test operations users would expect, `test()` and `testsome()`. `test()` checks for the completion of a specific operation, while `testsome()` checks for the completion of any operation in a list. In addition, BMI provides `testcontext()` that tests for any operation previously submitted within a particular context. This performance booster makes use of internally maintained lists optimized for each network infrastructure, so the user doesn't have to recreate lists of operation IDs for each polling cycle.

14.4.5 BMI Portability

When a common interface of any sort is created, dissimilar APIs and semantics must be molded to the common interface. The BMI network module API is connectionless and message-based, so network modules must emulate connectionless and message-based communication for networks that do not support that natively.

TCP/IP was the first module to be implemented. The TCP/IP module overcomes particular challenges, including its connection-based packet-oriented nature. The module caches open connections for quick easy reuse and automatically opens new connections as needed. To ensure all connections make progress, connections are polled regularly for incoming messages. Small messages are buffered on the receive side, while large messages require a posted receive to ensure memory is available. TCP guarantees the in-order delivery of a stream of bytes, but it is not aware of messages or message boundaries. The TCP/IP module takes responsibility for the proper delivery of messages.

A higher performance BMI module is the InfiniBand [19] module. The InfiniBand interconnect is very popular with medium-sized clusters as well as in the data center. InfiniBand offers a couple of challenges that separate it from the TCP/IP module. The InfiniBand protocol requires that a receive be posted for every message. To maximize the performance of large transfers, an RDMA buffer must be used. The RDMA buffer is a pinned memory region

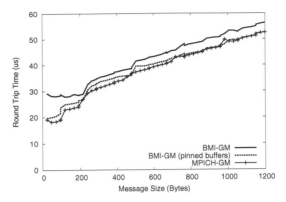

FIGURE 14.7: Round trip latency.

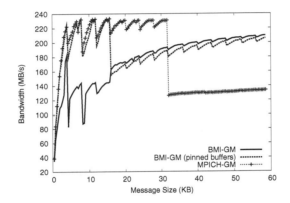

FIGURE 14.8: Point-to-point bandwidth for 120MB transfer.

into which the source process can directly deposit data. Pinning memory can be an expensive operation, because before memory can be pinned each page must be touched to ensure it is in core. For long running applications, the most efficient way to do this is to allocate a ring of memory buffers at startup using them as necessary.

14.4.6   Experimental Results

An analysis of BMI's performance [78] is presented in this section. None of the existing network benchmarks were able to properly exercise BMI, because they are meant to evaluate raw network performance. BMI's intent as a parallel file system data movement library required a new set of benchmarks that evaluate it in that context.

The first two tests are simple round trip and point-to-point tests that will be use to determine BMI overhead. The third test emulates parallel file system

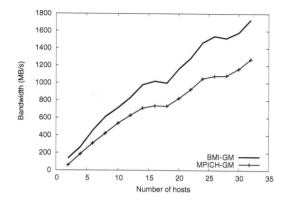

FIGURE 14.9: Aggregate read pattern, 10MB per client/server pair.

activity using I/O access patterns and a series of client/server configurations.

Although TCP/IP is the most mature network module, the Myrinet GM [205] protocol is more representative of a cluster interconnect, so the Myrinet GM module (BMI-GM) was chosen for testing. Myrinet GM also has a mature vendor-supplied MPI implementation (MPICH-GM [205]) derived from MPICH [171], which makes an effective reference point for evaluating test results.

The first test is a round trip latency test. BMI-GM is tested in two different modes — with and without pinned memory buffers. The MPI standard and, therefore, MPICH-GM, doesn't offer any explicit method for allocated pinned memory buffers, so data is supplied in regular user buffers.

The results in Figure 14.7 show that BMI-GM without pinned memory buffers is 3 to 10μs slower than the MPICH-GM implementation. This difference can be attributed to MPICH-GM's use of a memory registration cache. MPICH-GM's memory registration cache is implemented by overloading the `malloc` library. Since only a single library can overload the `malloc` library without symbol clashes at link time, BMI-GM cannot implement a similar registration cache. Unlike MPICH-GM, which doesn't provide explicit memory management, BMI-GM does give the user the ability to explicitly manage memory buffers. PVFS2 makes use of these pinned buffers by encoding into network format directly into the pinned buffers. The same BMI-GM round trip experiment using pinned buffers brings the results in line with MPICH-GM. Beyond a message size of 700 bytes, the results are nearly identical.

The second point-to-point test measures the bandwidth between a pair of clients using both BMI-GM and MPICH-GM. This test transfers 120MB of data from one host to the other using an increasing message size from 400 bytes to 60000 bytes. The bandwidth is computed as the total number of bytes transferred divided by the total time to transfer the data. The first thing to notice in Figure 14.8 is that the BMI-GM performance drops off significantly at 16KB, while MPICH-GM drops off at 32KB. The difference can be attributed to the point at which the two libraries switch from eager mode to rendezvous mode. Above 32KB when both libraries are using rendezvous mode, BMI-GM shows a 50% performance improvement over MPICH-GM even without pinned buffers. It is also interesting to note that in eager mode, BMI-GM with pinned buffers and MPICH-GM track very closely, which is similar to the round trip results.

While the point-to-point and bandwidth benchmarks are interesting when trying to determine baseline performance of the two libraries, they are simple benchmarks that do not properly emulate the real usage of a parallel file system. To better test the behaviors of parallel file systems, a third benchmark was created. This benchmark simulates a large contiguous parallel read by dividing an arbitrary number of hosts into N servers and M clients, where $N = M$. Each of the clients initiates a data transfer by sending a request and receiving an ACK from each server. Each client then receives a stream of bytes from each server. This data stream is divided into messages, which allows the server to overlap network communication with disk I/O. The complete simulation results in $N * M * 2$ small messages in phase 1 plus a large number of data messages in phase 2. In this simulation, the servers do not perform any real disk I/O.

This test simulates PVFS2 request and response messages using messages of size 25 bytes and 400 bytes, respectively. For the data transfer portion of the test, each client receives 10MB from each server. The total number of data bytes transferred per test run is $N * M * 10MB$. Figure 14.9 illustrates the test results. The number of nodes ranges from 2 to 32 with the nodes divided evenly between clients and servers. The bandwidth is calculated as the total bytes transferred ($N * M * 10MB$) divided by the data transfer time of the slowest client.

The BMI API is optimized for the types of parallel file system operations simulated in this test. On the other hand, the MPI API implemented by MPICH-GM is a more generic communication API. For this reason, the implementation of the MPICH-GM test client is a bit more cumbersome. The convenient `BMI_testunexpected()` is emulated using a combination of `MPI_Iprobe()` and `MPI_Recv()`. The BMI list I/O operations are emulated with MPI hvector datatypes. This test uses asynchronous operations, which must be tested for completion. The BMI-GM test implementation uses the optimized `BMI_testcontext()`, but the MPICH-GM test implementation relies on the more expensive `MPI_Testsome()`.

The results of this test clearly show the advantages of the BMI design. For a single server with a single client, the performance difference is dramatic. BMI-GM is more than twice as fast as MPICH-GM. As the number of hosts increases, BMI-GM continues to outpace MPICH-GM by 35% to 42%. BMI-GM's performance continues to scale all the way to the 16x16 case peaking at 1728 MB/s. Analysis of the results by the PVFS2 team revealed that BMI-GM's better performance can be attributed to the lightweight testing mechanism, memory management optimized for the network infrastructure, efficient discontiguous memory region handling and explicit efficient unexpected message handling.

14.5   Lightweight File Systems

The Lightweight File Systems (LWFS) project [314] investigates the applicability of lightweight solutions in storage systems. This work is patterned after research into lightweight kernels for MPP systems, in which kernels are customized for individual applications by including only services that are necessary (the canonical example being the irrelevance of the line-printer daemon *lpd* to a MPP compute node). In the case of LWFS, traditional file system semantics such as atomicity and naming are not provided by the storage system. Instead, LWFS emphasizes secure and direct access to storage devices. The LWFS architecture is extensible through the use of libraries, which may be combined according to application needs to provide exactly the services required. For example, the LWFS code base includes LWFS extensions for naming and metadata services as well as transactional semantics for applications that need them.

14.5.1   Design of the LWFS RPC Mechanism

Data movement in LWFS [315] relies on an asynchronous remote procedure call (RPC) interface based on the Sun RPC interface [293,410]. This interface is responsible for both access to remote services and efficient transport of bulk data.

In contrast to the Sun RPC interface, the LWFS RPC interface is completely asynchronous. This allows clients to overlap computation and I/O — a feature particularly important given that I/O operations are remote for most MPP architectures.

Like Sun RPC, LWFS relies on client and server stub functions to encode/decode (i.e., *marshal*) procedure call parameters to/from a machine-independent format. This approach is portable because it allows access to services on heterogeneous systems, but it is not efficient for I/O requests that contain raw buffers that do not need encoding. It also employs a "push" model for data transport that puts tremendous stress on servers when the requests are large and unexpected, as is the case for most I/O requests.

To address the issue of efficient transport for bulk data, the LWFS uses separate communication channels for control and data messages. In this

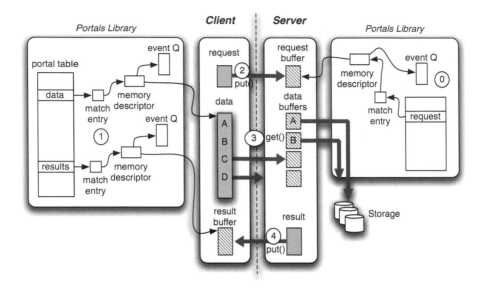

FIGURE 14.10: The figure illustrates the required Portals data structures and network protocol used by the LWFS_WRITE() function.

model, a *control message* is typically small. It identifies the operation to be performed, where to get arguments, the structure of the arguments, and so forth. In contrast, a *data message* is typically large and consists of "raw" bytes that, in most cases, do not need to be encoded/decoded by the server. The LWFS client uses the RPC-like interface to push control messages to the servers, but the LWFS server uses a different, one-sided API to push or pull data to/from the client. This protocol allows interactions with heterogeneous servers, but also benefits from allowing the server to control the transport of bulk data [236, 390]. The server can thus manage large volumes of requests with minimal resource requirements. Furthermore, since servers are expected to be a critical bottleneck in the system (recall the high proportion of compute nodes to I/O nodes in MPPs), a server-directed approach allows the server to optimize the processing of requests for efficient use of underlying network and storage devices — for example, reordering requests to a storage device [236].

14.5.2 LWFS RPC Implementation

On the Cray XT3 Red Storm system at Sandia National Laboratories, the LWFS RPC is layered on top of the Portals Message Passing Interface [64], the standard protocol used on Sandia MPP machines. Portals is a particularly good choice for for MPP storage systems because it is connectionless, it has one-sided communication APIs that enable the exploitation of remote DMA and operating-system bypass to avoid memory copies in the kernel-managed

```
struct data_t {
        int int_val;              /* 4 bytes */
        float float_val;          /* 4 bytes */
        double double_val;        /* 8 bytes */
};
```

FIGURE 14.11: The 16-byte data structure used for each of the experiments.

protocol stack, and it is designed for lightweight operating systems such as the Catamount OS for the Cray XT3 [75].

When a remote service starts, before it can accept remote requests, it allocates a buffer for incoming requests and creates all the necessary Portals data structures required to direct an incoming request to the right location in the buffer. The Portals data structures, shown in Figure 14.10 include a memory descriptor to describe the server's buffer for incoming requests, a match list used to identify appropriate messages, a Portal table to index match lists in the Portals library, and an event queue that contains a log of successfully matched messages.

To illustrate how the RPC protocols work, Figure 14.10 shows the network protocol and Portals data structures used for the LWFS_WRITE() function. The client initiates the protocol by encoding/marshaling an RPC request buffer and putting the request on the server's incoming-request buffer. The request buffer includes the memory descriptor of the source data buffer, the memory descriptor of the result buffer, the operation code of the request, and the arguments required for that operation. When the PUT() completes, the server gets a notification (i.e., Portals event) that a new request has arrived. The server then decodes the request, identifies the request as a WRITE, and calls the local write function with the decoded arguments. The write function then begins a sequence of one-sided GET() calls to pull the data from the client into pre-allocated buffers on the server. While the data is being pulled from the client, the server is writing data from the data buffer to the back-end storage, overlapping the I/O to disk with the network I/O. When the server receives all the data from the client, it puts a message on the client's result buffer notifying the client of completion.

It is important to note that LWFS_WRITE() is an asynchronous operation. The client does not have to sit idle waiting for the remote operation to complete. When the client is ready for the result, the LWFS_WAIT() function blocks the client until the specified remote request is either complete (possibly with an error) or has timed out. When the LWFS_WAIT() function returns, the client may release or re-use all buffers reserved for the remote operation.

14.5.3   Performance Analysis

To investigate the performance of the LWFS data-movement protocols, the following experiments compare the throughput (measured in MB/sec) of the LWFS RPC mechanism to three other approaches: TCP/IP, Sun's RPC, and Sandia's reference implementation of Portals 3.3. Each experiment consists of two processes: a client that sends an integer followed by an array of 16-byte data structures to the server, and a server that receives the array and returns a single data structure to the client. Each 16-byte data structure (defined in Figure 14.11) consists of a 4-byte integer, a 4-byte float, and an 8-byte double.

All experiments were conducted on a cluster of 32 IA32 compute nodes, each with 2-way SMP Dell PE1550 processors (1.1 GHz PIII Xeons) with a Myrinet 2000 interconnect. The LWFS and Portals experiments used the reference implementation of Portals 3.3 that is not optimized to use the Myrinet GM library. In all cases, the lowest-level transport was TCP/IP over Myrinet.

For the raw TCP/IP version, the client makes a socket connection to the server, writes the length of the buffer, writes the buffer, then reads the result. When the server receives the size of the buffer, it allocates space for the incoming array, reads the buffer, then writes back the first data structure in the buffer.

For the raw Portals version, the client and server perform the same steps as the TCP version, except that a send from the client involves a PtlPut() operation from the client buffer into the remote buffer on the server. Like the TCP/IP version, when the server receives the length of the incoming buffer, it allocates space for the array, waits for the buffer to arrive (an event gets logged to the receive buffer's event queue), then sends the first value of the array back to the client.

In contrast to raw TCP and Portals, the Sun RPC version encodes and transfers the entire array in one transaction. The decoding mechanism on the server side allocates the necessary space for the incoming array and reads and decodes the data as it is transferred to the server.

The LWFS version works much like the Portals version except it encodes/decodes a small request that includes the size of the incoming array. The server then fetches the data from the client-side array using the one-sided PtlGet() function.

Figure 14.12 shows the throughput in MB/Sec of the various data-transport schemes as we increase the size of the data array. The worst performer is Sun RPC because it has to encode/decode all messages sent across the network. This is particularly inefficient because the XDR encoding scheme used by Sun RPC has to visit every data-structure in the array, adding a substantial processing burden on both the client and the server. The plot flattens out around 32KB — at which point the transfer becomes bound by the CPU processing required for the XDR encoding.

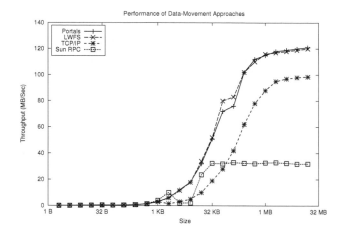

FIGURE 14.12: Comparison of LWFS RPC to various other mechanisms.

The LWFS implementation also uses XDR to encode requests, but it transfers the array (the data portion) in "raw" binary format (the assumption is that the server knows the format of the data and does not need to do a conversion), allowing for larger buffer transfers and server-control of the transport. The LWFS scheme achieves almost exactly the same throughput as the raw Portals version, notable in light of the fact that LWFS uses Portals as a middleware layer, and demonstrating that the overhead of encoding/decoding the request structure is minimal for large requests.

Just below the performance of the raw Portals and LWFS, but following the same general shape, is the raw TCP/IP implementation. This may seem a bit strange since our Portals (and thus the LWFS) implementation is layered on top of TCP/IP. We believe that the raw TCP (that uses default settings) performs worse due to TCP/IP optimizations implemented in the Portals library. Such optimizations are common among HPC data-transport protocols that use TCP/IP [431].

14.6 Other MPP File Systems

A number of parallel file systems are not represented in this survey. Perhaps the most notable is IBM's General Parallel File System (GPFS) [385]. GPFS is a block-based shared-disk parallel file system that achieves scalability by implementing the striping algorithms in the file system clients, instead of sending block storage commands through a separate logical volume manager. While the Schmuck et al. paper has information about GPFS, we could not find sufficient information about the data movement protocols to justify using GPFS in our survey.

Also missing from our survey is the Ceph [453] object-based file system from the University of California — Santa Cruz. Ceph's use of client-side hash functions to discover data placement, distributed metadata servers with adaptive load balancing, custom object-based storage servers, and their particular emphasis on resilience make Ceph attractive for large-scale systems. Space limitations as well as a general lack of information on the data-movement protocols used by Ceph prevented a more detailed description.

14.7   Chapter Summary

This chapter described the data-movement challenges for MPP storage systems and presented a survey of four commercial and research-based solutions currently in use. The systems we have described are possibly more notable for what they have in common rather than the differences in their approaches to data movement. Each separates metadata operations from the data operations, enabling direct access to the storage devices for high-volume data transfer; each takes advantage of RDMA and OS-bypass when it is available in the network; and each uses a server-directed approach for the transfer of bulk data.

This separate handling of metadata and data reflects a convergence in design attitudes toward the differences between metadata and so-called POD ("plain old data"). Metadata has different, and almost always more stringent, consistency, quality and timeliness requirements. Metadata volumes are orders of magnitude smaller, and its transfers tend to be burstier and less predictable. The designs of the parallel storage systems represented in this chapter reflect a growing appreciation and understanding of the differences between metadata and POD. These differences are, to varying degrees, exposed and exploited in order to provide the extremely high levels of scalability required by emerging parallel applications.

The differences in how MPP storage systems handle data movement are in the details of the design and implementation. One difference is in the use of standard data-movement and storage protocols. Lustre, LWFS, and PVFS2 use custom data-movement protocols, Panasas uses the pNFS standard, which supports a number of different storage protocols, including iSCSI, standard NFS, and others.

Another differentiating design choice has to do with managing storage system state in the client. PVFS2 and LWFS use stateless clients that access the storage system through user-space libraries, while Panasas and Lustre require kernel-access on the client OS. There are performance and security advantages to running in the kernel. For example, client-side caching can provide significant performance advantages (especially for writing), but the system also has to worry about coherency and consistency on failures — an especially important problem as systems increase in size.

These systems also chose different designs for data distribution. PVFS2 and Lustre use a standard striping scheme [179] to distribute data across storage servers, where Panasas uses a block-based scheme with a layout map, and LWFS relies on libraries and a flexible naming scheme that maps a name to a generic metadata object that provides details about how the data is mapped to storage servers. Design choices about data distribution and block size can have a significant impact on data-movement performance and reliability. Even among the striped schemes, there is a fair amount of flexibility. For example, Lustre allows the user to specify stripe size, count, and even specify OSTs for the data.

One final notable difference is support for asynchronous I/O. Asynchronous I/O is an important feature that, because of lightweight kernels, relies on the I/O system's ability to take advantage of RDMA and OS-bypass when it is provided by the underlying network. While all of the systems discussed provide this support, Lustre's handshake protocol prevents the ability to perform truly asynchronous I/O; however, the next version of LNET is reported to overcome that issue.

14.8   Acknowledgements

We would like to thank Rob Ross and Phil Carns for providing BMI performance details and graphs.

Chapter 15

Network Simulation

George Riley
Georgia Institute of Technology

15.1 Introduction

The communications infrastructure between individual computing elements in a high-performance cluster of systems is often the limiting factor in the overall performance of the distributed application. Thus, the ability to carefully study and measure the performance of the communications network becomes a critical part of the overall system design. Further, it is often necessary to evaluate performance of *proposed* communications infrastructure or protocols, which may or may not yet have been designed or deployed.

In order to analyze network performance, both for existing networks and new designs, the use of *network simulation* provides the ability to perform experiments on the simulated network to measure various performance metrics such as throughput, delay, jitter, packet loss, and link utilization, just to name a few. There are a number of network simulation tools available, with varying degrees of cost, performance, capabilities, and accuracy. This chapter will discuss the overall design and approach to creating the simulator, rather than any one particular simulation tool.

15.2 Discrete Event Simulation

Virtually all network simulation tools use *discrete event simulation* in order to manage the state of the simulated objects. Discrete event simulations maintain a local notion of the *simulation time*, which simply maintains a logical time value, usually in units of seconds, and almost always relative to the start of the simulation. In other words, the simulation time generally starts at the value of zero, and then advances as the state of the network elements changes due to the advancement of time. There are two important concepts to consider when understanding the notion of simulation time.

353

1. The simulation time always advances (in other words we never back up time and process actions out of order), and advances at a rate that is unrelated to the actual wall-clock time advancement. For a simulation that is modeling a small number of network elements such as routers and desktop systems, the simulation time will often advance much faster than the wall clock time.

2. The simulation time advances in discrete jumps, rather than continuously. In other words, the current time might jump from 1.234 seconds to 1.400 seconds. In this particular example, this would occur if there were nothing "interesting" happening in the simulation between the times of 1.234 and 1.400 seconds. This will be explained in more detail in the subsequent sections.

15.3   Maintaining the Event List

Fundamental to any discrete event simulation is the creation and maintenance of the *event list*. The event list is a collection of state change events that are pending, to be processed at some time in the future. The event contains a timestamp, indicating *when* the event will occur, and the *action*, indicating *what* will occur. An example of an event might be the arrival of a packet at a router, or the initiation of a TCP connection from a web browser model. This list is typically maintained in sorted order of ascending timestamp, with the earliest future event at the head of the list and the latest at the tail of the list.

When the event list is maintained in sorted order as described, then the *main loop* of the simulator can be implemented very simply. The code snippet below shows a simple example of how this might be implemented.

```
void Simulator::Run()
{
  while(!eventList.empty())
    {
      Event* ev = eventList.head();    // Get the earliest event
      eventList.deleteHead();          // Remove the earlist event
      simulationTime = ev->time;       // Advance the simulation time
      ProcessEvent(ev);                // Process the event
    }
}
```

The above code snippet is notional and not intended to show any particular implementation. Rather, it simply illustrates the steps needed to implement the main event processing loop. In this example, the main loop is simply four simple actions:

1. Get the earliest pending event from the event list. Since the list is sorted in ascending timestamp order, this event is by definition the next thing of interest that happens in the simulation.

2. Remove this event from the list.

3. Advance the simulation time to the timestamp of the event just obtained. From this step, it is easy to see how the simulation time advances in discrete steps. For example, if the current simulation time is 1.234 and the timestamp of the event is 1.400, this indicates that there is *nothing of interest* happening in the simulation between those two times. In other words, while in fact the state of the real-world system being modeled does change during this interval (the state of real-world networks of course changes continuously in time), the level of detail of the network model does not takes these state changes into account. Suppose that a network packet is transmitted from router A to router B at some time T. In the real network, the packet advances from router A to B continuously in time. In the simulation, we only change the state of the system when the packet arrives at router B. In this example (assuming there is not other activity being modeled), the simulation time would advance discretely from time T to $T + \Delta T$, where ΔT is the time it takes for the packet to travel from A to B.

4. Finally we *process the event*. In the case of a packet arrival event at a router, this would result in the examination the protocol headers in the received packet, examining the routing tables at the receiving router, and placing the packet in the transmit queue for transmission to the next hop.

15.4 Modeling Routers, Links, and End Systems

The previous sections have discussed the design of discrete event simulation engines, without regard to the application domain. This section begins the discussion of the use of discrete event simulation in the particular problem domain of simulating computer networks.

The first step in constructing a network simulation is to define and create a model of the *network topology*. This simply means informing the network simulation tool of the desired communication infrastructure for the system being modeled. Typically this consists of a number of *routers* and *communications links* connecting the routers together. Further, the topology model usually contains one or more *end–systems* which will act as the data demand generator for the simulated network, and optionally the data sink for the generated data. Each of these will be discussed in a bit more detail.

1. **Network Routers.** When modeling a network router, a number of
   specific details regarding the capabilities and configuration of the router
   are needed in order to construct a realistic behavior model.

   (a) How many network interfaces does the router have? Typical routers
       have as few as three or four interfaces to other routers, but often
       as many as several dozen to a hundred or more.

   (b) What are the characteristic physical communications media used by
       each interface? This determines the type of media access protocol
       to be used to access the medium on that interface. For example, if
       the medium is the well-known *Ethernet*, the protocol must listen
       for a quiet network before trying to transmit a packet, and it must
       listen for and react to *collisions* when two routers are transmitting
       at the same time on the same medium.

   (c) What type of *queuing* is used at each output interface? Virtually
       all router interfaces have queues that are used to temporarily store
       packets to be transmitted when the packet cannot be transmitted
       immediately. There are many different types of queuing methods,
       that vary significantly in the action taken when the queue is full
       or nearly full, or when one packet has a higher priority than other
       packets.

   (d) What type of *routing protocol* is used by each interface on the
       router? Routers make routing decisions by looking up the desti-
       nation network address in a *routing table*, which specifies which
       directly connected routing neighbor is the next hop on the best
       path to the destination. These routing tables must be constructed
       by the routers during the normal operation of the network. There
       are many different routing protocols that are used to construct
       and maintain these routing tables, and these different choices have
       different amounts of overhead and responsiveness, both of which
       affect the overall performance of the network being modeled.

   (e) Does the router use *input queuing* or *output queuing*? The design
       of high-performance routers typically involves queuing the packets
       both at the ingress point to the router (the input queue) and the
       egress point (the output queue). The choice of input queuing and
       how the packets are moved through the router can make a difference
       in the overall behavior of the router.

2. **Communication Links.** The communications links are the medium
   used to transmit data (packets) between routers or between end-systems
   and routers. When defining links in the simulation, there are a number
   of characteristics of the links that must be specified.

   (a) What is the transmission rate of the link? This determines how fast
       the bits can be transmitted on the medium, which affect how long it

takes to transmit a packet. This rate is usually specified in *bits per second*, with typical values in the range of 10Mbps (Mega-bits per second) up to 10Gbps (Giga-bits per second). It should be noted that the actual data rate available on a medium is determined both by the medium and the network interface used to connect to the medium. For example, a 100Mbps Ethernet network is capable of 100 million bits per second, but some Ethernet interface cards are only capable of 10Mbps.

(b) What is the propagation delay on the medium, which indicates how long it takes for a single bit to travel from one end of the medium to the other end. This is a function of the speed-of-light delay on the medium and the distance between the endpoints of the medium.

(c) What happens on the medium when more than one network interface tries to transmit packets at the same time? A simple point-to-point link is almost always *full duplex*, meaning that both endpoints can transmit at the same time without interfering with each other. Other medium types, such as wireless channels, do not work properly when two nearby systems transmit simultaneously.

3. **End-Systems.** Finally, the network topology consists of one or more *end-systems*, which represent end user desktop workstations or laptops. These end-systems act as the data demand on the network, making connection requests, data generation actions, and data sink actions. This is typically done by modeling network applications, which is discussed in more detail below. However, there are a number of capabilities and configurations that must be considered when defining end-systems.

(a) What *protocols* are supported by the end-system? Nearly all end-systems have support for IPV4, TCP, and UDP, but what other protocols can be used and supported? Does the end-system support other less common protocols such as RTP and IPV6?

(b) What is the delay experienced by packets as they progress up and down the protocol stack? Most simulator instances assume these delays are negligible, but for high-speed and closely-coupled networks, these delays can be a performance bottleneck.

(c) What does the end-system do in response to illegal or unexpected packets? For many experiments the end-systems will frequently received unintended or malicious data packets and must respond to these in some way.

A moderately complex network topology is shown in Figure 15.1 . This topology consists of a number of end-systems, routers, encryption devices, and satellites.

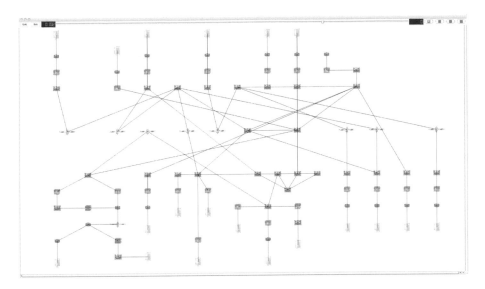

FIGURE 15.1:   Moderately complex network topology.

15.5   Modeling Network Packets

Another important requirement for any network simulation environment is the ability to represent and model the *packets* of data that are transmitted through the simulated network. In general, a packet is a collection of *protocol headers* and *application data*. The protocol headers are typically fairly small, and contain information that allows the protocol endpoints to communicate with each other. For example, the TCP protocol exchanges information between the endpoints regarding the sequence numbers of the data in the packet and the sequence being acknowledged, as well as some other flags and housekeeping information.

Additionally, the packets normally contain applications data that is the useful data being exchanged. For example, in a web browser application, the application data would be the http *GET* request, or the actual web page response from a server.

In network simulation packages, it is frequently the case that the actual packet contents (the application data) are not relevant to the performance metrics being collected. An example might be the contents of a web page. The metric being collected is the web browser response time, which is a function only of the size of the web objects, and is independent of the actual contents of the web objects. In this case, most simulation tools have the ability to contain *abstract* data, without actually modeling packet contents. As long as the logical size of the packet is correct, the actual packet representation can be much smaller, thus saving memory.

In representing packets, there are two basic approaches that are typically used.

1. The *packed* format, in which the packet in the simulator is represented as a simple array of bytes, exactly as is done in real network protocol stacks. Each protocol header uses the exact format and the exact size of the corresponding header in actual networks. This means the bit and byte layout of the headers must follow the protocol specifications found in the appropriate *Request for Comments* (*RFC*), and thus requires a bit of bit manipulation in the simulation model. However, this results in a fairly memory efficient representation and easily can be extended to send and receive packet from actual network devices.

2. The *unpacked* format uses a collection of *objects* to represent the headers, without regard to the actual size of the headers. These objects can have virtual functions (for printing for example) and can contain extra information not typically found in the actual protocol header. As long as the simulated size of the header is assumed to match the actual protocol header size, the time to transmit the packet (with the unpacked headers) will be the same as an actual packet. This approach provides more flexibility in analyzing extensions to existing protocols and ease of debugging.

15.6   Modeling the Network Applications

Clearly, the simulated network being modeled exists to move data from some point in the network to some other point. In the case of web browsing for example, the browser sends a relatively small *request* to the server, which responds with a set of web objects representing the web page being requested. There are hundreds of network applications in use in the Internet today, and no one simulation tool has models for every one. Typically, a network model will choose a few applications that are of particular interest (web browsing or peer-to-peer file transfer) and create a behavior model for the application. The web browser model, for example, contains a set of probability distributions for the size of the request, size of the response, and the amount of *think-time* between requests. Peer to peer models might have distributions for object popularity and size, and model the searching and downloading of music or videos.

However, it is rarely the case that the application being studied is the only data demand on the network. Clearly, there is always *competing* or *background*

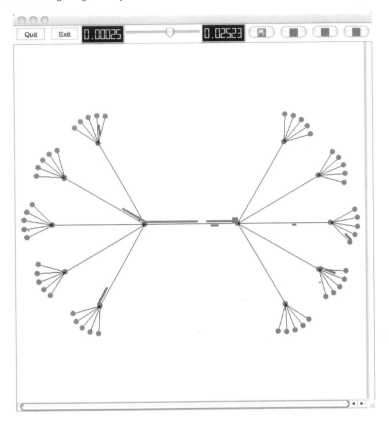

FIGURE 15.2:   Network simulation animation.

traffic on the network that uses of varying amounts of network bandwidth and queue space in the simulated network. This is frequently modeled by a simple *on–off application* that sends data at some rate for a random amount of time and then goes quiet (not sending data) for another random amount of time. This provides a *bursty* traffic model, that is somewhat representative of actual background network traffic.

15.7   Visualizing the Simulation

Network simulation tools normally include the ability to create a *visualization* of the state of the network being simulated. This is normally used by the simulation model developer to insure that the topology created is as planned, and that the data flows are between the correct sets of endpoints. A typical

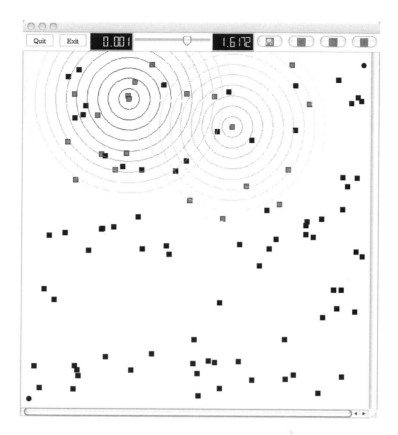

FIGURE 15.3: Network simulation visualization.

animation is shown in figure 15.2 below. The elements represented in this particular visualization are:

1. Circles, representing the end systems and routers in the simulated topology. In this particular example, the end systems are on the left and right side of the figure, and the routers are in the middle.

2. Lines, representing the communications links between the various end systems and routers.

3. Rectangles, representing the individual packets being transmitted through the network. The location and size of the packet on the line connecting two systems or routers is calculated by taking into account the speed of light delay on the link, as well as the length of time taken to transmit all of the bits represented in the packet.

4. Vertical rectangles, representing the queue of packets stored at a congested link.

For wireless simulations, the needed information is significantly different from that of wired simulations. Figure 15.3 shows a representative visualization of a wireless network. This consists of:

1. Small circles and squares, representing the mobile, wireless devices in the network.

2. Concentric circles, representing any packet being transmitted. The circles show the 2-dimensional propagation of the wireless signal from the transmitting element.

Creating and displaying the visualization is almost always a significant CPU overhead in the simulation. In many cases, the simulation can slow down by a factor of 10 or 100 when the visualization is created. For this reason, the visualization and animations are normally used only during the debugging phase of the experiment, and bypassed when actually running data collecting experiments.

15.8   Distributed Simulation

The memory and CPU requirements for any non-trivial network topology with non-trivial data flow models can grow very quickly, and can easily exhaust the resources on any single computing platform. For this reason, *distributed simulation* is often employed when constructing larger-scale models of computer networks.

The most common approach used to construct a distributed network simulation is the *space–parallel* method. In this method, the entire simulated topology model is sub-divided into N sub-models, and each sub-model is independently simulated using a separate instance of the simulation software. An example of this approach is shown in figure 15.4.

In this simple example, the four network elements on the left are assigned to simulator A, and the remaining four nodes are assigned to simulator B. Clearly, simulators A and B cannot simply simulate their own small piece of the overall network, as it is frequently the case that a packet originating on A is destined for some element in simulator B. Further, we must insure that the simulation time representation in each of two simulators stays more or less the same value, within some small variation. The issues that must be addressed to correctly execute a distributed network simulation are:

1. Time synchronization, which insures that the local representation of the current simulation time between the various simulators does not diverge too far. There are a number of well known approaches to this problem,

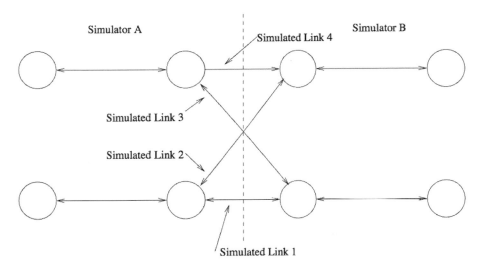

FIGURE 15.4: The space-parallel method for distributed network simulation.

but none have good performance as the number of simulation instances gets large.

2. Message passing between the simulators, which is used to notify simulator B that a packet has been transmitted from A and will arrive at B at some future time. Depending on the type of packet representation chosen, this can result in *serialization* and *de-serialization* of the packet data. This serialization can be a source of significant overhead.

3. Determining routing information between simulation instances, which allows each simulator to independently determine the *best* path for packets, particularly when the packet path traverses two or more simulator instances. Since each simulator only has global knowledge of a smaller sub–model of the topology, it might not have sufficient information to make correct routing decisions. One approach to solve this problem is the use of *ghost nodes*. A ghost node is a memory efficient representation of a network node (router or end system), that contains only enough information for routing decisions, but none of the other more memory intensive objects (such as applications, queues, protocols, etc). Each simulator will then have a full-state representation of its local nodes, and a ghost representation of the non-local nodes assigned to other simulator instances. This results in a complete topology picture at every simulator instance, and thus a correct and common decision for packet routing.

It is clear that the overhead discussed for time synchronization and message passing actions are detrimental to the overall performance of the network simulation, particularly if the number of cooperating simulator instances becomes large. In an extreme case, a distributed simulation using a large number simulator processes can sometime execute more slowly as additional CPU resources are added to the simulation. The actual performance of a large scale distributed simulation is dependent on a large number of variables, many of which are beyond the control of the simulation developer.

However, any case using distributed simulation will result in the ability to support larger models, which is often the desired result, regardless of overall execution time. If the goal is to simulate a significant fraction of the Internet, for example, the only reasonable approach is large-scale distributed simulation on a large number of CPUs with a high speed efficient interconnect network.

Using the newly developed *Georgia Tech Network Simulator (GTNetS)*, the network simulation research group at Georgia Tech set out to construct the largest network simulation experiments ever conducted. The *GTNetS* simulation tool was designed from the outset for large-scale distributed execution. The computing platform used was the *Lemieux* supercomputer at the Pittsburgh Supercomputer Center. The *Lemieux* system consists of 750 instances of quad-CPU HP Alpha systems, each with 4Gb of main memory. The 750 computing platforms are connected with a specialized *Quadrics* high-speed interconnect network. Using a subset of these available computing platforms, the experiment successfully modeled an extremely large network as follows:

1. Total topology size was 1,928,192 nodes.

2. Total number of data flows was 1,820,672 flows.

3. Total simulation event count was 18,650,757,866 events.

4. Total simulation time modeled was 1,000 seconds.

5. Total wall clock time was 1,289 seconds.

This experiment successfuly demonstrated that, with sufficient hardware to execute the simulation, and with a simulation environment specifically designed for large-scale distributed simulation, topology models of more than 1 million network elements are certainly possible. As supercomputers become more capable, with more memory and faster interconnects, even larger simulation experiments will be within reach.

15.9   Summary

The field of network simulation design and construction is an ongoing research avenue. Many of the principles and examples discussed here are from the *Georgia Tech Network Simulator*, which is a simulation tool designed at Georgia Tech for the specific purpose of simulator research and distributed simulation. There are many other network simulation tools available, including both open-source, free tools, as well as moderately expensive commercial tools.

References

[1] ACTIV Financial Systems. www.activfinancial.com.

[2] N. R. Adiga, G. Almasi, et al. An Overview of the BlueGene/L Supercomputer. In *Proceedings of the 2002 ACM/IEEE Conference on Supercomputing*, Baltimore, Maryland, November 16–22, 2002. IEEE Computer Society Press.

[3] Advanced Micro Devices, Inc. HyperTransport Technology I/O Link: A High-Bandwidth I/O Architecture. White Paper 25012A, Advanced Micro Devices, Inc., Sunnyvale, California, July 20, 2001.

[4] Asmara Afework, Michael Beynon, et al. Digital Dynamic Telepathology - the Virtual Microscope. In *Proceedings of the 1998 AMIA Annual Fall Symposium. American Medical Informatics Association*, 1998.

[5] Sandip Agarwala, Greg Eisenhauer, and Karsten Schwan. Lightweight Morphing Support for Evolving Data Exchanges in Distributed Applications. In *Proc. of the 25th International Conference on Distributed Computer Systems (ICDCS-25)*, June 2005.

[6] Agere Systems. *The Case for a Classification Language. White Paper*, February 2003.

[7] Sheldon B. Akers, Dov Harel, and Balakrishnan Krishnamurthy. The Star Graph: An Attractive Alternative to the *n*-Cube. In *Proceedings of the 1987 International Conference on Parallel Processing (ICPP)*, pages 393–400, University Park, Pennsylvania, August 1987. Pennsylvania State University Press.

[8] Sheldon B. Akers and Balakrishnan Krishnamurthy. A Group-Theoretic Model for Symmetric Interconnection Networks. *IEEE Transactions on Computers*, 38(4):555–566, April 1989.

[9] Mark Allman. RFC3465: TCP Congestion Control with Appropriate Byte Counting (ABC). In *RFC3465*, 2003.

[10] George Almási, Charles Archer, et al. Design and Implementation of Message-Passing Services for the Blue Gene/L Supercomputer. *IBM Journal of Research and Development*, 49(2/3):393–406, March–May 2005.

[11] George Almasi, Gabor Dozsa, C. Chris Erway, and Burkhardt Steinmacher-Burow. Efficient Implementation of Allreduce on Blue-

Gene/L Collective Network. In *Proceedings of EuroPVM/MPI*, September 2005.

[12] Gheorghe Almási, Sameh Asaad, et al. Overview of the IBM Blue Gene/P Project. *IBM Journal of Research and Development*, 52(1/2):199–220, January–March 2008.

[13] Robert Alverson. Red Storm. In *Hot Chips 15, Invited Talk*, August 2003.

[14] Applied Micro Circuits Corporation. www.amcc.com.

[15] Glenn Ammons, Jonathan Appavoo, Maria Butrico, Dilma Da Silva, David Grove, Kiyokuni Kawachiya, Orran Krieger, Bryan Rosenburg, Eric Van Hensbergen, and Robert W. Wisniewski. Libra: a Library Operating System for a JVM in a Virtualized Execution Environment. In *VEE '07: Proceedings of the 3rd International Conference on Virtual Execution Environments*, pages 44–54, 2007.

[16] George Apostolopoulos, David Aubespin, Vinod Peris, Prashant Pradhan, and Debanjan Saha. Design, Implementation and Performance of a Content-Based Switch. In *Proc. of IEEE INFOCOM 2000*, volume 3, pages 1117–1126, Tel Aviv, Israel, March 2000.

[17] Leon Arber and Scott Pakin. The Impact of Message-Buffer Alignment on Communication Performance. *Parallel Processing Letters*, 15(1/2):49–65, March–June 2005.

[18] Bruce W. Arden and Hikyu Lee. Analysis of Chordal Ring Network. *IEEE Transactions on Computers*, C-30(4):291–295, April 1981.

[19] InfiniBand Trade Association. InfiniBand Architecture Specification, Release 1.2, October 2004.

[20] ATM Forum. *ATM User-Network Interface Specification, Version 3.1*, September 1994.

[21] David Bader and Virat Agarwal. Computing Transforms on the IBM Cell Broadband Engine. *Parallel Computing*, 2008.

[22] David A. Bader and Kamesh Madduri. Design and Implementation of the HPCS Graph Analysis Benchmark on Symmetric Multiprocessors. In *HiPC*, pages 465–476, 2005.

[23] Stephen Bailey and Tom Talpey. Remote Direct Data Placement (RDDP), April 2005.

[24] Randal S. Baker. A Block Adaptive Mesh Refinement Algorithm for the Neutral Particle Transport Equation. *Nuclear Science and Engineering*, 141(1):1–12, May 15, 2002.

[25] Pavan Balaji, Wu-chun Feng, and Dhabaleswar K. Panda. Bridging the Ethernet-Ethernot Performance Gap. *IEEE Micro Journal Special Issue on High-Performance Interconnects*, (Issue 3):24–40, May/June 2006.

[26] Pavan Balaji, Karthikeyan Vaidyanathan, Sundeep Narravula, Hyun-Wook Jin, and Dhabaleswar K. Panda. Designing Next-Generation Data-Centers with Advanced Communication Protocols and Systems Services. In *Proceedings of Workshop on NSF Next Generation Software (NGS), held in conjunction with Int'l Parallel and Distributed Processing Symposium (IPDPS 2006)*, 2006.

[27] Die Shot of AMD's Quadcore Barcelona. elnexus.com/mail/07-09-barcelona.jpg.

[28] Kevin Barker, Kei Davis, Adolfy Hoisie, Darren Kerbyson, Michael Lang, Scott Pakin, and José Carlos Sancho. Entering the Petaflop Era: The Architecture and Performance of Roadrunner. In *Proceedings of the ACM/IEEE SC2008 Conference*, Austin, Texas, November 15–21, 2008. ACM.

[29] Kevin J. Barker, Alan Benner, et al. On the Feasibility of Optical Circuit Switching for High Performance Computing Systems. In *Proceedings of the ACM/IEEE SC 2005 Conference on Supercomputing*, Seattle, Washington, November 12–18, 2005.

[30] Kevin J. Barker, Kei Davis, Darren J. Kerbyson, Mike Lang, Scott Pakin, and José Carlos Sancho. An Early Performance Evaluation of the SiCortex SC648. In *Proceedings of the 2008 IEEE International Symposium on Performance Analysis of Systems and Software (ISPASS), Workshop on Unique Chips and Systems (UCAS)*, Austin, Texas, April 20, 2008.

[31] Ray Barriuso and Allan Knies. *SHMEM User's Guide for C*. Cray Research, Inc., June 20, 1994. Revision 2.1.

[32] Kenneth E. Batcher. The flip network in STARAN. In *Proceedings of the 1976 International Conference on Parallel Processing (ICPP)*, pages 65–71. IEEE Computer Society, August 24–27, 1976.

[33] Andrew Baumann, Paul Barham, Pierre-Evariste Dagand, Tim Harris, Rebecca Isaacs, Simon Peter, Timothy Roscoe, Adrian Schpbach, and Akhilesh Singhania. The Multikernel: A New OS Architecture for Scalable Multicore Systems. In *Proceedings of the 22nd ACM Symposium on OS Principles*, Big Sky, MT, 2009.

[34] Bay Microsystems. http://www.baymicrosystems.com/.

[35] Donald J. Becker, Thomas Sterling, Daniel Savarese, John E. Dorband, Udaya A. Ranawak, and Charles V. Packer. Beowulf: A Parallel Workstation for Scientific Computation. In *International Parallel Processing Symposium*, 1996.

[36] Jon Beecroft, David Addison, David Hewson, Moray McLaren, Duncan Roweth, Fabrizio Petrini, and Jarek Nieplocha. QsNet^{II}: Defining High-Performance Network Design. *IEEE Micro*, 25(4):34–47, July–August 2005.

[37] Jon Beecroft, Fred Homewood, Duncan Roweth, and Ed Truner. The Elan5 Network Processor. In *Proceedings of the International Supercomputing Conference (ISC) 2007*, Dresden, Germany, June 2007.

[38] Muli Ben-Yehuda and Jimi Xenidis others. The Price of Safety: Evaluating IOMMU Performance. In *OLS '07: The 2007 Ottawa Linux Symposium*, pages 9–20, 2007.

[39] Vaclav E. Beneš. Optimal Rearrangeable Multistage Connecting Networks. *Bell System Technical Journal*, 43:1641–1656, July 1964.

[40] Jean Claude Bermond, Francesc Comellas, and D. Frank Hsu. Distributed Loop Computer Networks: A Survey. *Journal of Parallel and Distributed Computing*, 24(1):2–10, January 1995.

[41] Rudolf Berrendorf, Heribert C. Burg, Ulrich Detert, Rüdiger Esser, Michael Gerndt, and Renate Knecht. Intel Paragon XP/S—Architecture, Software Environment, and Performance. Internal Report KFA-ZAM-IB-9409, Central Institute for Applied Mathematics, Jülich Research Centre, Jülich, Germany, May 16, 1994.

[42] Caitlin Bestler and Lode Coene. Applicability of Remote Direct Memory Access Protocol (RDMA) and Direct Data Placement Protocol (DDP). Technical report, IETF RFC 5045, October 2007.

[43] Caitlin Bestler and Randall Stewart. Stream Control Transmission Protocol (SCTP) Direct Data Placement (DDP) Adaptation. Technical report, IETF RFC 5043, October 2007.

[44] Raoul Bhoedjang, Tim Ruhl, and Henri E. Bal. User-Level Network Interface Protocols. *IEEE Computer*, 31(11), November 1998.

[45] Laxmi N. Bhuyan and Dharma P. Agrawal. Generalized Hypercube and Hyperbus Structures for a Computer Network. *IEEE Transactions on Computers*, C-33(4):323–333, April 1984.

[46] Kenneth Birman, AndreSchiper, and Pat Stephenson. Lightweight Causal and Atomic Group Multicast. *ACM Transactions on Computer Systems*, 9(3), August 1991.

[47] Matthias A. Blumrich, Cezary Dubnicki, Edward W. Felten, and Kai Li. Protected, User-level DMA for the SHRIMP Network Interface. In *IEEE International Symposium on High Performance Computer Architecture (HPCA-2)*, pages 154–165, 1996.

[48] Nanette J. Boden, Danny Cohen, Robert E. Felderman Alan E. Kulawik, Charles L. Seitz, Jakov N. Seizovic, and Wen-King Su. Myrinet: A gigabit-per-second local area network. *IEEE Micro*, 15(1):29–36, February 1995.

[49] Taisuke Boku, Hiroshi Nakamura, Kisaburo Nakazawa, and Yoichi Iwasaki. The Architecture of Massively Parallel Processor CP-PACS. In *Proceedings of the Second Aizu International Symposium on Parallel Algorithms/Architecture Synthesis*, pages 31–40, Aizu-Wakamatsu, Japan, March 17–21, 1997. IEEE.

[50] Dan Bonachea. Gasnet specification, v1.1. Technical report, Berkeley, CA, USA, 2002.

[51] Shekhar Borkar, Robert Cohn, et al. Supporting Systolic and Memory Communication in iWarp. In *Proceedings of the 17th Annual International Symposium on Computer Architecture (ISCA)*, pages 70–81, Seattle, Washington, May 28–31, 1990. ACM.

[52] Don Box, Luis Felipe Cabrera, et al. Web Services Eventing (WS-Eventing). *W3C Member Submission*, 2006.

[53] Silas Boyd-Wickizer, Haibo Chen, Rong Chen, Yandong Mao, Frans Kaashoek, Robert Morris, Aleksey Pesterev, Lex Stein, Ming Wu, Yuehua Dai, Yang Zhang, and Zheng Zhang. Corey: An Operating System for Many Cores. In *Proceedings of the 8th USENIX Symposium on Operating Systems Design and Implementation (OSDI'08)*, 2008.

[54] Peter Braam. Lustre Networking: High-Performance Features and Flexible Support for a Wide Array of Networks. Whitepaper, Sun Microsystems, January 2008.

[55] Lawrence Brakmo, Sean O'Malley, and Larry Peterson. TCP Vegas: New Techniques for Congestion Detection and Avoidance. In *Proceedings of the SIGCOMM '94 Symposium*, 1994.

[56] Tim Brecht, G. Janakiraman, Brian Lynn, Vikram Saletore, and Yoshio Turner. Evaluating Network Processing Efficiency with Processor Partitioning and Asynchronous I/O. In *Proceedings of EuroSys*, 2006.

[57] Lars Brenna, Alan Demers, Johannes Gehrke, Mingsheng Hong, Joel Ossher, Biswanath Panda, Mirek Riedewald, Mohit Thatte, and Walker White. Cayuga: A High-Performance Event Processing Engine. In *SIGMOD '07: Proceedings of the 2007 ACM SIGMOD international conference on Management of data*, pages 1100–1102, New York, NY, USA, 2007. ACM.

[58] Ron Brightwell, William Camp, Benjamin Cole, Erik DeBenedictis, Robert Leland, James Tomkins, and Arthur B. Maccabe. Architectural Specification for Massively Parallel Computers: An Experience and

Measurement-based Approach. *Concurrency and Computation: Practice and Experience*, 17(10):1271–1316, March 22, 2005.

[59] Ron Brightwell, Doug Doerfler, and Keith D. Underwood. A Comparison of 4X InfiniBand and Quadrics Elan-4 Technologies. In *Proceedings of the 2004 International Conference on Cluster Computing*, September 2004.

[60] Ron Brightwell, Doug Doerfler, and Keith D. Underwood. A Preliminary Analysis of the InfiniPath and XD1 Network Interfaces. In *Proceedings of the 20th International Parallel and Distributed Processing Symposium (IPDPS), Workshop on Communication Architecture for Clusters (CAC)*, Rhodes Island, Greece, April 25–29, 2006. IEEE.

[61] Ron Brightwell, Sue Goudy, and Keith D. Underwood. A Preliminary Analysis of the MPI Queue Characteristics of Several Applications . In *Proceedings of the 2005 International Conference on Parallel Processing*, June 2005.

[62] Ron Brightwell, Trammell Hudson, Kevin Pedretti, Rolf Riesen, and Keith Underwood. Implementation and performance of Portals 3.3 on the Cray XT3. In *Proceedings of the 2005 IEEE International Conference on Cluster Computing*, September 2005.

[63] Ron Brightwell, Trammell Hudson, Kevin T. Pedretti, and Keith D. Underwood. SeaStar Interconnect: Balanced Bandwidth for Scalable Performance. *IEEE Micro*, 26(3), May/June 2006.

[64] Ron Brightwell, Trammell B. Hudson, Arthur B. Maccabe, and Rolf E. Riesen. The Portals 3.0 Message Passing Interface Revision 2.0. Technical Report SAND2006-0420, Sandia National Laboratories, January 2006.

[65] Ron Brightwell, William Lawry, Arthur B. Maccabe, and Rolf Riesen. Portals 3.0: Protocol Building Blocks for Low Overhead Communication. In *Proceedings of the 2002 Workshop on Communication Architecture for Clusters*, April 2002.

[66] Ron Brightwell, Arthur B. Maccabe, and Rolf Riesen. Design, Implementation, and Performance of MPI on Portals 3.0 . *International Journal of High Performance Computing Applications*, 17(1):7–20, Spring 2003.

[67] Ron Brightwell and P. Lance Shuler. Design and Implementation of MPI on Puma Portals. In *Proceedings of the Second MPI Developer's Conference*, pages 18–25, July 1996.

[68] Walter Brooks, Michael Aftosmis, et al. Impact of the Columbia supercomputer on NASA science and engineering applications. In Ajit Pal, Ajay D. Kshemkalyani, Rajeev Kumar, and Arobinda Gupta, editors, *Proceedings of the 7th International Workshop on Distributed Computing (IWDC)*, volume 3741 of *Lecture Notes in Computer Science*, pages

293–305, Kharagpur, India, December 27–30, 2005. Springer. Keynote Talk.

[69] Ravi Budruk, Don Anderson, and Tom Shanley. *PCI Express System Architecture*. Addison-Wesley Developers Press, 2003.

[70] Darius Buntinas and Dhabaleswar K. Panda. NIC-Based Reduction in Myrinet Clusters: Is it Beneficial? In *Proceedings of the SAN-02 Workshop (in conjunction with HPCA)*, February 2002.

[71] Rajkumar Buyya, Toni Cortes, and Hai Jin, editors. *High Performance Mass Storage and Parallel I/O: Technologies and Applications*. Wiley-IEEE Press, 2001.

[72] Quadrics Has Been Selected by Bull to Provide QsNet for Europe's Fastest Computer. Press release in *Business Wire*, December 23, 2004.

[73] Zhongtang Cai, Greg Eisenhauer, Qi He, Vibhore Kumar, Karsten Schwan, and Matthew Wolf. IQ-Services: Network-Aware Middleware for Interactive Large-Data Applications. *Concurrency & Computation. Practice and Experience Journal*, 2005.

[74] Brent Callaghan, Theresa Lingutla-Raj, Alex Chiu, Peter Staubach, and Omer Asad. NFS over RDMA. In *NICELI '03: Proceedings of the ACM SIGCOMM workshop on Network-I/O convergence*, 2003.

[75] William J. Camp and James L. Tomkins. The Red Storm Computer Architecture and its Implementation. In *The Conference on High-Speed Computing: LANL/LLNL/SNL*, Salishan Lodge, Glenedon Beach, Oregon, April 2003.

[76] Jason F. Cantin, James E. Smith, Mikko H. Lipasti, Andreas Moshovos, and Babak Falsafi. Coarse-Grain Coherence Tracking: RegionScout and Region Coherence Arrays. *IEEE Micro*, 26(1):70–79, 2006.

[77] Willaim W. Carlson, Jesse M. Draper, David E. Culler, Kathy Yelick, Eugene Brooks, and Karen Warren. Introduction to UPC and language specification. Technical Report CCS-TR-99-157, May 1999.

[78] Philip H. Carns, Walter B. Ligon III, Robert Ross, and Pete Wyckoff. BMI: a Network Abstraction Layer for Parallel I/O. In *Proceedings of the International Parallel and Distributed Processing Symposium*, April 2005.

[79] Pascal Caron and Djelloul Ziadi. Characterization of Glushkov Automata. *Theoretical Computer Science*, 233(1-2):75–90, 2000.

[80] Antonio Carzaniga, David S. Rosenblum, and Alex L. Wolf. Challenges for Distributed Event Services: Scalability vs. Expressiveness. In *Proc. of Engineering Distributed Objects (EDO '99), ICSE 99 Workshop*, May 1999.

[81] Miguel Castro, Atul Adya, Barbara Liskov, and Andrew Myers. Hybrid Adaptive Caching. In *Proceedings of Symposium of Operating Systems Principles*, 1997.

[82] Umit Catalyurek, Michael D. Beynon, Chialin Chang, Tahsin Kurc, Alan Sussman, and Joel Saltz. The Virtual Microscope. *IEEE Transactions on Information Technology in Biomedicine*, 7(4):230–248, 2003.

[83] Cavium Networks. www.caviumnetworks.com.

[84] Vint Cerf, Yogen Dalal, and Carl Sunshine. RFC 675: Specification of Internet Transmission Program, 1974.

[85] Vinton Cerf and Robert Kahn. A Protocol for Packet Network Inter-communication. *IEEE Transations on Communication*, 22(5):637–648, 1974.

[86] Sudip Chalal and Todd Glasgow. Memory Sizing for Server Virtualization. *Intel Information Technology*.

[87] Roger D. Chamberlain, Eric J. Tyson, Saurabh Gayen, Mark A. Franklin, Jeremy Buhler, Patrick Crowley, and James Buckley. Application Development on Hybrid Systems. In *Proceedings of the 2007 Int'l Conf. for High Performance Computing, Networking, Storage, and Analytics (Supercomputing)*, Reno, NV, November 2007.

[88] Ken Chang, Sudhakar Pamarti, et al. Clocking and Circuit Design for a Parallel I/O on a First-Generation CELL Processor. In *2005 IEEE International Solid-State Circuits Conference (ISSCC), Digest of Technical Papers*, pages 526–527 and 615, San Francisco, California, February 6–10, 2005. IEEE.

[89] Philippe Charles, Christian Grothoff, Vijay Saraswat, Christopher Donawa, Allan Kielstra, Kemal Ebcioglu, Christoph von Praun, and Vivek Sarkar. X10: An object-oriented approach to non-uniform cluster computing. In *OOPSLA '05*, pages 519–538, New York, NY, USA, 2005. ACM.

[90] Chelsio Communications. *The Terminator Architecture. White Paper*, 2004.

[91] Ludmila Cherkasova and Rob Gardner. Measuring CPU Overhead for I/O Processing in the Xen Virtual Machine Monitor. In *Proc. of USENIX ATC*, 2005.

[92] Imrich Chlamtac, Aura Ganz, and Martin G. Kienzle. An HIPPI Interconnection System. *IEEE Transactions of Computing*, 42(2):138–150, 1993.

[93] Chris Clark, Wenke Lee, David Schimmel, Didier Contis, Mohamed Kone, and Ashley Thomas. A Hardware Platform for Network Intrusion

Detection and Prevention. In *Proceedings of The 3rd Workshop on Network Processors and Applications (NP3)*, Madrid, Spain, 2004.

[94] Charles Clos. A Study of Non-blocking Switching Networks. *Bell System Technical Journal*, 32(5):406–424, March 1953.

[95] Cluster File Systems, Inc. Lustre: A Scalable High-Performance File System. Technical report, Cluster File Systems, November 2002. http://www.lustre.org/docs/whitepaper.pdf.

[96] Compaq, Intel, and Microsoft Corporation. *Virtual Interface Architecture Specification*, 1997.

[97] Pat Conway and Bill Hughes. The AMD Opteron Northbridge architecture. *IEEE Micro*, 27(2):10–21, March–April 2007.

[98] Cray Research, Inc. *SHMEM Technical Note for C, SG-2516 2.3*, October 1994.

[99] Crucial Memory Pricing for Proliant DL785 G5, 2008. www.crucial.com.

[100] David Culler, Richard Karp, et al. LogP: Towards a Realistic Model of Parallel Computation. In *Proceedings of the Fourth ACM SIGPLAN Symposium on Principles and Practice of Parallel Programming (PPoPP)*, pages 1–12, San Diego, California, May 19–22, 1993. ACM Press. Also appears in SIGPLAN Notices 28(7), July 1993.

[101] P. Culley, U. Elzur, et al. Marker PDU Aligned Framing for TCP Specification. Technical report, IETF RFC 5044, October 2007.

[102] Dennis Dalessandro, Ananth Devulapalli, and Pete Wyckoff. Design and Implementation of the iWarp Protocol in Software. In *Proceedings of PDCS'05*, Phoenix, AZ, November 2005.

[103] Dennis Dalessandro, Ananth Devulapalli, and Pete Wyckoff. iWarp Protocol Kernel Space Software Implementation. In *Proceedings of IPDPS'06, CAC Workshop*, Rhodes Island, Greece, April 2006.

[104] Dennis Dalessandro, Ananth Devulapalli, and Pete Wyckoff. iSER Storage Target for Object-based Storage Devices. In *Proceedings of MSST'07, SNAPI Workshop*, San Diego, CA, September 2007.

[105] Dennis Dalessandro and Pete Wyckoff. A Performance Analysis of the Ammasso RDMA Enabled Ethernet Adapter and its iWarp API. In *Proceedings of Cluster'05*, Boston, MA, September 2005.

[106] Dennis Dalessandro and Pete Wyckoff. Accelerating Web Protocols Using RDMA. In *Proceedings of NCA'07*, Cambridge, MA, July 2007.

[107] Dennis Dalessandro and Pete Wyckoff. Memory Management Strategies for Data Serving with RDMA. In *Proceedings of HotI'07*, Palo Alto, CA, August 2007.

[108] Dennis Dalessandro, Pete Wyckoff, and Gary Montry. Initial Performance Evaluation of the NetEffect 10 Gigabit iWARP Adapter. In *Proceedings of IEEE Cluster'06, RAIT Workshop*, Barcelona, Spain, September 2006.

[109] William Dally and Brian Towles. *Principles and Practices of Interconnection Networks*. Morgan Kaufmann Publishers Inc., 2003.

[110] William J. Dally. *A VLSI Architecture for Concurrent Data Structures*. PhD thesis, California Institute of Technology, Pasadena, California, March 3, 1986. Also appears as Computer Science Technical Report 5209:TR:86.

[111] William J. Dally. Performance Analysis of k-ary n-cube Interconnection Networks. *IEEE Transactions on Computers*, 39(6):775–785, June 1990.

[112] Peter B. Danzig. Finite buffers and fast multicast. In *Proceedings of the 1989 ACM SIGMETRICS International Conference on Measurement and Modeling of Computer Systems*, pages 108–117, Berkeley, California, May 23–26, 1989. ACM.

[113] DAT Collaborative. http://www.datcollaborative.org.

[114] DAT Collaborative. *uDAPL: User Direct Access Programming Library*, May 2003.

[115] DAT Collaborative. *DAT collaborative: Direct Access Transport*, 2007.

[116] Davide Pasetto and Fabrizio Petrini . DotStar: Breaking the Scalability and Performance Barriers in Regular Expression Set Matching, 2009. In preparation.

[117] Alan Demers, Johannes Gehrke, Mingsheng Hong, Mirek Riedewald, and Walker White. A General Algebra and Implementation for Monitoring Event Streams. Technical Report TR2005-1997, Cornell University, 2005.

[118] Alan Demers, Johannes Gehrke, Mingsheng Hong, Mirek Riedewald, and Walker White. Towards Expressive Publish/Subscribe Systems. *Lecture Notes in Computer Science*, 3896:627, 2006.

[119] Alvin M. Despain and David A. Patterson. X-Tree: A Tree Structured Multi-processor Computer Architecture. In *Proceedings of the 5th Annual Symposium on Computer Architecture (ISCA)*, pages 144–151, Palo Alto, California, April 1978. ACM.

[120] Wael W. Diab, Howard M. Frazier, and Gerry Pesavento. Ethernet Provides the Solution for Broadband Subscriber Access. In *Ethernet Alliance*, April 2006.

[121] Lloyd Dickman, Greg Lindahl, Dave Olson, Jeff Rubin, and Jeff Broughton. PathScale InfiniPath: A first look. In *Proceedings of the*

13th Symposium on High Performance Interconnects (HOTI'05), August 2005.

[122] DIS Stressmark Suite, 2001. Updated by UC Irvine, www.ics.uci.edu/-amrm/hdu/DIS_Stressmark/DIS_stressmark.html.

[123] Jack J. Dongarra, Piotr Luszczek, and Antoine Petitet. The LINPACK Benchmark: Past, Present and Future. *Concurrency and Computation: Practice and Experience*, 15(9):803–820, August 10, 2003.

[124] Boris Dragovic, Keir Fraser, Steven Hand, Tim Harris, Alex Ho, Ian Pratt, Andrew Warfield, Paul Barham, and Rolf Neugebauer. Xen and the Art of Virtualization. In *Proc. of SOSP*, 2003.

[125] Robert J. Drost, Robert David Hopkins, Ron Ho, and Ivan E. Sutherland. Proximity Communication. *IEEE Journal of Solid-State Circuits*, 39(9):1529–1535, September 2004.

[126] Jose Duato. A Necessary and Sufficient Condition for Deadlock-Free Adaptive Routing in Wormhole Networks. *IEEE Transactions on Distributed Systems*, 6(10):1055–1067, October 1995.

[127] Jose Duato, Federico Silla, et al. Extending HyperTransport Protocol for Improved Scalability. In *First International Workshop on Hyper-Transport Research and Applications*, 2008.

[128] Jose Duato, Sudhakar Yalamanchili, and Lionel Ni. *Interconnection Networks: An Engineering Approach*. Morgan Kaufmann Inc., San Francisco, CA, 2002.

[129] Thomas H. Dunigan, Jr., Jeffrey S. Vetter, and Patrick H. Worley. Performance Evaluation of the SGI Altix 3700. In *Proceedings of the 2005 International Conference on Parallel Processing (ICPP)*, pages 231–240, Oslo, Norway, June 14–17, 2005. IEEE.

[130] Will Eatherton. The Push of Network Processing to the Top of the Pyramid. In *Proceedings of the Symposium on Architectures for Networking and Communications Systems, Keynote Talk*, 2005.

[131] Hans Eberle, Pedro J. Garcia, José Flich, José Duato, Robert Drost, Nils Gura, David Hopkins, and Wladek Olesinski. High-Radix Crossbar Switches Enabled by Proximity Communication. In *Proceedings of the ACM/IEEE SC2008 Conference*, Austin, Texas, November 15–21, 2008. ACM.

[132] G. Eisenhauer, F. Bustamante, and K. Schwan. Publish-Subscribe for High-Performance Computing. *IEEE Internet Computing - Asynchronous Middleware and Services*, 10(1):8–25, January 2006.

[133] Greg Eisenhauer and Lynn K. Daley. Fast Heterogenous Binary Data Interchange. In *Proc. of the Heterogeneous Computing Workshop (HCW2000)*, May 3-5 2000.

[134] Bernard Elspas, William H. Kautz, and James Turner. *Theory of Cellular Logic Networks and Machines, Section III(B): Bounds on Directed (d, k) Graphs*, pages 20–28. Stanford Research Institute, Menlo Park, California, December 1968. Research Report AFCRL-68-0668. Final Report for SRI Project 7258.

[135] IEEE Working Group - 802.1Qau: Congestion Notification, 2008. www.ieee802.org.

[136] IEEE Standard for Local and Aetropolitan Area Networks: Virtual Bridged Local Area Networks, 2006. standards.ieee.org.

[137] Patrick Eugster. Type-based Publish/Subscribe: Concepts and Experiences. *ACM Trans. Program. Lang. Syst.*, 29(1):6, 2007.

[138] Patrick Eugster, Pascal Felber, Rachid Guerraoui, and A.-M. Kerrmarec. The many faces of publish/subscribe. Tech. Report DSC-ID:200104, École Polythechnique Fédérale de Lausanne, Lausanne, France, January 2001.

[139] Exegy Ticker Plant. www.exegy.com.

[140] Kevin Fall and Sally Floyd. Simulation-based comparisons of Tahoe, Reno and SACK TCP. *SIGCOMM Comput. Commun. Rev.*, 26(3):5–21, 1996.

[141] Xiaobo Fan, Wolf-Dietrich Weber, and Luiz Andre Barroso. Power Provisioning for a Warehouse-Sized Computer. In *Proceedings of the 34th International Symposium on Computer Architecture*, pages 13–23, New York, NY, USA, 2007. ACM.

[142] Alexandra Fedorova, David Vengerov, and Daniel Doucette. Operating System Scheduling On Heterogeneous Core Systems. In *Proceedings of the Operating System support for Heterogeneous Multicore Architectures (OSHMA) workshop at the Sixteenth International Conference on Parallel Architectures and Compilation Techniques*, Brasov, Romania, 2007.

[143] Wu-chun Feng, Pavan Balaji, et al. Performance Characterization of a 10-Gigabit Ethernet TOE. In *Proceedings of the IEEE International Symposium on High-Performance Interconnects (HotI)*, Palo Alto, CA, Aug 17-19 2005.

[144] Wu-chun Feng and Kirk W. Cameron. The Green500 List: Encouraging Sustainable Supercomputing. *Computer*, 40(12):50–55, December 2007.

[145] Wu-chun Feng, Justin Hurwitz, Harvey Newman, Sylvain Ravot, R. Les Cottrell, Oliver Martin, Fabrizio Coccetti, Cheng Jin, David Wei, and Steven Low. Optimizing 10-Gigabit Ethernet for Networks of Workstations, Clusters and Grids: A Case Study. In *Proceedings of SuperComputing*, 2003.

[146] Sally Floyd and Thomas Henderson. The NewReno Modification to TCP's Fast Recovery Algorithm. www.ietf.org/rfc/rfc2582.txt.

[147] Sally Floyd, Thomas Henderson, and Andrei Gurtov. The New-Reno Modification to TCP's Fast Recovery Algorithm. www.ietf.org/-rfc/rfc3782.txt.

[148] Michael J. Flynn. Very High-Speed Computing Systems. *Proceedings of the IEEE*, 54(12):1901–1909, December 1966.

[149] The Message Passing Interface Forum. MPI: A message-passing interface standard. Version 2.1, June 2008.

[150] Ian Foster, Karl Czajkowski, et al. Modeling and Managing State in Distributed Systems: The Role of OGSI and WSRF. *Proceedings of the IEEE*, 93(3):604–612, 2005.

[151] Ian Foster, Carl Kesselman, and Steven Tuecke. The anatomy of the Grid: Enabling scalable virtual organizations. *International Journal of High Performance Computing Applications*, 15(3):200–222, Fall 2001.

[152] Doug Freimuth, Elbert Hu, Jason LaVoie, Ronald Mraz, Erich Nahum, Prashant Pradhan, and John Tracey. Server Network Scalability and TCP Offload. In *USENIX Annual Technical Conference*, 2005.

[153] Hiroaki Fujii, Yoshiko Yasuda, et al. Architecture and performance of the Hitachi SR2201 massively parallel processor system. In *Proceedings of the 11th International Parallel Processing Symposium (IPPS)*, pages 233–241, Geneva, Switzerland, April 1–5, 1997. IEEE.

[154] Edgar Gabriel, Graham E. Fagg, George Bosilca, Thara Angskun, Jack J. Dongarra, Jeffrey M. Squyres, Vishal Sahay, Prabhanjan Kambadur, Brian Barrett, Andrew Lumsdaine, Ralph H. Castain, David J. Daniel, Richard L. Graham, and Timothy S. Woodall. Open MPI: Goals, concept, and design of a next generation MPI implementation. In *Proceedings, 11th European PVM/MPI Users' Group Meeting*, pages 97–104, Budapest, Hungary, September 2004.

[155] Mark K. Gardner, Wu-chun Feng, and Michael E. Fisk. Dynamic Right-Sizing in FTP (drsFTP): An Automatic Technique for Enhancing Grid Performance. In *Proceedings of the IEEE Symposium on High-Performance Distributed Computing (HPDC'02)*, Edinburgh, Scotland, 2002.

[156] Tal Garfinkel and Mendel Rosenblum. A Virtual Machine Introspection Based Architecture for Intrusion Detection. In *Proceedings of the 2003 Network and Distributed System Symposium*, 2003.

[157] Ada Gavrilovska, Sanjay Kumar, Himanshu Raj, Karsten Schwan, Vishakha Gupta, Ripal Nathuji, Radhika Niranjan, Adit Ranadive,

and Purav Saraiya. High Performance Hypervisor Architectures: Virtualization in HPC Systems. In *Proceedings of the 1st Workshop on System-level Virtualization for High Performance Computing (HPCVirt), in conjunction with EuroSys 2007*, 2007.

[158] Ada Gavrilovska, Kenneth Mackenzie, Karsten Schwan, and Austen McDonald. Stream Handlers: Application-specific Message Services on Attached Network Processors. In *Proc. of Hot Interconnects 10*, Stanford, CA, August 2002.

[159] Ada Gavrilovska, Karsten Schwan, Ola Nordstrom, and Hailemelekot Seifu. Network Processors as Building Blocks in Overlay Networks. In *Proc. of Hot Interconnects 11*, Stanford, CA, August 2003.

[160] Al Geist, William Gropp, et al. MPI-2: Extending the Message-Passing Interface. In *Euro-Par '96 Parallel Processing*, pages 128–135. Springer Verlag, 1996.

[161] Global Environment for Network Innovations. www.geni.net.

[162] Giganet Corporations. http://www.giganet.com.

[163] The Globus Alliance. www.globus.org.

[164] Victor. M. Glushkov. The abstract theory of automata. *Russian Mathematical Survey. v16. 1-53*.

[165] Brice Goglin. What is Myrinet Express over Ethernet? bgoglin.livejournal.com/1870.html.

[166] Brice Goglin. Design and Implementation of Open-MX: High-Performance Message Passing over Generic Ethernet Hardware. In *Workshop on Communication Architecture for Clusters*, Miami, FL, April 2008.

[167] L. Rodney Goke and G. Jack Lipovski. Banyan Networks for Partitioning Multiprocessor Systems. In *Proceedings of the 1st Annual Symposium on Computer Architecture (ISCA)*, pages 21–28, Gainesville, Florida, December 1973.

[168] David S. Greenberg, Ron Brightwell, Lee Ann Fisk, Arthur B. Maccabe, and Rolf Riesen. A system software architecture for high-end computing. In *Proceedings of SC97: High Performance Networking and Computing*, pages 1–15, San Jose, California, November 1997. ACM Press.

[169] Andrew Grimshaw, Mark Morgan, Duane Merrill, Hiro Kishimoto, Andreas Savva, David Snelling, Chris Smith, and Dave Berry. An open grid services architecture primer. *Computer*, 42(2):27–34, 2009.

[170] William Gropp, Steven Huss-Lederman, Andrew Lumsdaine, Ewing Lusk, Bill Nitzberg, William Saphir, and Marc Snir. *MPI — The Complete Reference: Volume 2, the MPI-2 Extensions*. MIT Press, 1998.

[171] William Gropp, Ewing Lusk, Nathan Doss, and Anthony Skjellum. A high-performance, portable implementation of the MPI message-passing interface standard. *Parallel Computing*, 22(6):789–828, September 1996.

[172] Thomas Gross and David R. O'Hallaron. *iWarp: Anatomy of a Parallel Computing System*. MIT Press, March 1998.

[173] Robert L. Grossman, Yunhong Gu, Michal Sabala, David Hanley, Shirley Connelly, and David Turkington. Towards Global Scale Cloud Computing: Using Sector and Sphere on the Open Cloud Testbed. In *SC08 Bandwidth Challenge Award*, November 2008.

[174] Huaxi Gu, Qiming Xie, Kun Wang, Jie Zhang, and Yunsong Li. X-torus: A variation of torus topology with lower diameter and larger bisection width. In *Proceedings of the International Conference on Computational Science and Its Applications (ICCSA), Workshop on Parallel and Distributed Computing (PDC)*, volume 3984 of *Lecture Notes in Computer Science*, pages 149–157, Glasgow, Scotland, May 8–11, 2006. Springer.

[175] Jiani Guo, Jingnan Yao, and Laxmi Bhuyan. An Efficient Packet Scheduling Algorithm in Network Processors. In *Proceedings of IEEE Infocom*, 2005.

[176] Srabani Sen Gupta, Debasish Das, and Bhabani P. Sinha. The generalized hypercube-connected-cycle: An efficient network topology. In *Proceedings of the 3rd International Conference on High Performance Computing (HiPC)*, pages 182–187, Trivandrum, India, December 19–22, 1996. IEEE.

[177] Vishakha Gupta and Jimi Xenidis. Cellule: Virtualizing Cell for Lightweight Execution. In *Georgia Tech, STI Cell Workshop*, 2007.

[178] Sangtae Ha, Injong Rhee, and Lisong Xu. CUBIC: A New TCP-Friendly High-Speed TCP Variant. *ACM SIGOPS Operating System Review*, 42(5):64–74, July 2008.

[179] John H. Hartman and John K. Ousterhout. The Zebra striped network file system. In Hai Jin, Toni Cortes, and Rajkumar Buyya, editors, *High Performance Mass Storage and Parallel I/O: Technologies and Applications*, chapter 21, pages 309–329. IEEE Computer Society Press and Wiley, New York, NY, 2001.

[180] P. Harvey, R. Mandrekar, et al. Packaging the Cell Broadband Engine Microprocessor for Supercomputer Applications. In *Proceedings of the 58th Electronic Components and Technology Conference (ECTC)*, pages 1368–1371, Lake Buena Vista, Florida, May 27–30, 2008. IEEE.

[181] Mahbub Hassan and Raj Jain. *High Performance TCP/IP Networking: Concepts, Issues, and Solutions*. Prentice Hall, 2004.

[182] Eric He, Jason Leigh, Oliver Yu, and Thomas A. DeFanti. Reliable Blast UDP: Predictable High Performance Bulk Data Transfer. In *Proceedings of IEEE International Conference on Cluster Computing*, 2002.

[183] Eric He, Pascale V. Primet, and Michael Welzl. A Survey of Transport Protocols other than Standard TCP. In *Global Grid Forim, GFD-I.055*, 2005.

[184] Qi He and Karsten Schwan. IQ-RUDP: Coordinating Application Adaptation with Network Transport. In *Proc. of High Performance Distributed Computing (HPDC-11)*, July 2002.

[185] K. Scott Hemmert, Keith D. Underwood, and Arun Rodrigues. An architecture to perform NIC based MPI matching. In *Proceedings of the 2007 IEEE International Conference on Cluster Computing*, September 2007.

[186] John L. Henning. SPEC CPU2006 Benchmark Descriptions. *SIGARCH Comput. Archit. News*, 34(4):1–17, 2006.

[187] Adolfy Hoisie, Gregory Johnson, Darren J. Kerbyson, Mike Lang, and Scott Pakin. A Performance Comparison through Benchmarking and Modeling of Three Leading Supercomputers: Blue Gene/L, Red Storm, and Purple . In *Proceedings of the IEEE/ACM International Conference on High-Performance Computing, Networking, Storage, and Analysis (SC'06)*, November 2006.

[188] Ellis Horowitz and Alessandro Zorat. The binary tree as an interconnection network: Applications to multiprocessor systems and VLSI. *IEEE Transactions on Computers*, C-30(4):247–253, April 1981.

[189] Robert Horst. ServerNet deadlock avoidance and fractahedral topologies. In *Proceedings of the 10th International Parallel Processing Symposium (IPPS)*, pages 274–280, Honolulu, Hawaii, April 15–19, 1996. IEEE.

[190] HP Proliant DL servers - Cost Specifications, 2008. www.hp.com.

[191] HP Labs Award Will Lay Groundwork for Exascale Computers. HPCWire, www.hpcwire.com/offthewire/285158079.html.

[192] HP Power Calculator Utility: A Tool for Estimating Power Requirements for HP ProLiant Rack-Mounted Systems, 2008. www.hp.com.

[193] HyperTransport Consortium. www.hypertransport.org.

[194] Hypertransport specification, 3.00c, 2007. www.hypertransport.org.

[195] Wei Huang, Qi Gao, Jiuxing Liu, and Dhabaleshwar K. Panda. High Performance Virtual Machine Migration with RDMA over Modern Interconnects. In *Cluster 2007: IEEE International Conference on Cluster Computing, 2007*, pages 11–20, 2007.

[196] Wei Huang, Jiuxing Liu, Matthew Koop, Bulent Abali, and Dhabaleswar K. Panda. Nomad: Migrating OS-bypass Networks in Virtual Machines. In *Proc. of VEE'07*, 2007.

[197] Wei Huang, Gopal Santhanaraman, Dhabaleswar K. Panda, and Qi Gao. Design and Implementation of High Performance MVAPICH2:MPI2 over InfiniBand. In *International Symposium on Cluster Computing and the Grid*, 2006.

[198] Yi Huang, Aleksander Slominski, Chathura Herath, and Dennis Gannon. WS-Messenger: A Web Services-based Messaging System for Service-oriented Grid Computing. In *6th IEEE International Symposium on Cluster Computing and the Grid (CCGrid06)*, 2006.

[199] Ram Huggahalli, Ravi Iyer, and Scott Tetrick. Direct cache access for high bandwidth network i/o. In *32nd International Symposium on Computer Architecture*, pages 50–59, Madison, WI, June 2006.

[200] Justin Hurwitz and Wu-chun Feng. End-to-End Performance of 10-Gigabit Ethernet on Commodity Systems. *IEEE Micro*, 24(1):10–22, 2004.

[201] David Husak and Robert Gohn. *Network Processor Programming Models: The Key to Achieving Faster Time-to-Market and Extending Product Life. White Paper.* C-Port. A Motorola Company, 2001.

[202] IBM. http://www.ibm.com/.

[203] IEEE. *Proceedings of the International Conference on Massively Parallel Processing Using Optical Interconnections (MPPOI)*, 1994–1998.

[204] IETF NFSv4 Working Group. NFS Version 4 Minor Version 1. tools.ietf.org/wg/nfsv4/draft-ietf-nfsv4-minorversion1. Work in progress.

[205] Myricom Inc. The GM Message Passing System. http://www.myri.com/.

[206] InfiniBand Trade Association. *http://www.infinibandta.org*, 1999.

[207] InfiniBand Trade Association. *InfiniBand Architecture Specification Release 1.2*, October 2004.

[208] Intel Corporation. *Intel Network Processor Family.* http://developer.intel.com/design/network/producs/npfamily/.

[209] Intel 10 Gigabit Server Adapters. www.intel.com/go/10GbE.

[210] Intel 82541ER Gigabit Ethernet Controller. download.intel.com.

[211] International Organization for Standardization. Open systems interconnection basic reference model. Technical Report 7498-1, ISO/IEC standard, 1994.

[212] http://www.ipv6.org, 2008.

[213] Bruce Jacob. The Memory System and You: A Love/Hate Relationship. In *13th Workshop on Distributed Supercomputing*, 2009.

[214] Van Jacobson and Michael Karels. Congestion Avoidance and Control. In *Proceedings of ACM SIGCOMM*, 1988.

[215] Andrzej Jajszczyk. Nonblocking, repackable, and rearrangeable Clos networks: Fifty years of the theory evolution. *IEEE Communications Magazine*, 41(10):28–33, October 2003.

[216] Nanyan Jiang, Andres Quiroz, Cristina Schmidt, and Manish Parashar. Meteor: a middleware infrastructure for content-based decoupled interactions in pervasive grid environments. *Concurr. Comput. : Pract. Exper.*, 20(12):1455–1484, 2008.

[217] Hai Jin, Toni Cortes, and Rajkumar Buyya, editors. *High Performance Mass Storage and Parallel I/O: Technologies and Applications*, chapter 42: An Introduction to the InfiniBand Architecture, pages 617–632. Wiley Press and IEEE Press, 1st edition, November 26, 2001. Chapter author: Gregory F. Pfister.

[218] Jiuxing Liu and Ahmit Mamidala and Dhabaleswar K. Panda. High Performance MPI-Level Broadcast on InfiniBand with Hardware Multicast Support. In *IPDPS*, 2004.

[219] Jiuxing Liu and Weihang Jiang and others. High Performance Implementation of MPICH2 over InfiniBand with RDMA Support. In *IPDPS*, 2004.

[220] Rajgopal Kannan. The KR-Benes network: A control-optimal rearrangeable permutation network. *IEEE Transactions on Computers*, 54(5):534–544, May 2005.

[221] David Kanter. The Common System Interface: Intel's Future Interconnect. www.realworldtech.com.

[222] Vijay Karamcheti and Andrew A. Chien. Software Overhead in Messaging Layers: Where Does the Time Go? In *Proceedings of the Sixth International Conference on Architectural Support for Programming Languages and Operating Systems (ASPLOS)*, pages 51–60, San Jose, California, October 4–7, 1994. ACM.

[223] Dina Katabi, Mark Handley, and Charlie Rohrs. Congestion Control for High Bandwidth-Delay Product Networks. In *Proceedings of ACM Sigcomm*, 2002.

[224] Darren J. Kerbyson. A Look at Application Performance Sensitivity to the Bandwidth and Latency of InfiniBand Networks. In *Proceedings of the 20th International Parallel and Distributed Processing Symposium (IPDPS), Workshop on Communication Architecture for Clusters (CAC)*, Rhodes Island, Greece, April 25, 2006.

[225] Mukil Kesavan, Adit Ranadive, Ada Gavrilovska, and Karsten Schwan. Active CoordinaTion (ACT) - Towards Effectively Managing Virtualized Multicore Clouds. In *Cluster 2008*, 2008.

[226] Hyong-Youb Kim and Scott Rixner. Connection Handoff Policies for TCP Offload Network Interfaces. In *Proceedings of the Symposium on Operating Systems Design and Implementation (OSDI)*, 2006.

[227] John Kim, Wiliam J. Dally, Steve Scott, and Dennis Abts. Technology-Driven, Highly-Scalable Dragonfly Topology. In *Proceedings of the 35th International Symposium on Computer Architecture (ISCA)*, pages 77–88, Beijing, China, June 21–25, 2008. IEEE Computer Society.

[228] John Kim, William J. Dally, and Dennis Abts. Flattened Butterfly: A Cost-Efficient Topology for High-Radix Networks. In *Proceedings of the 34th Annual International Symposium on Computer Architecture (ISCA)*, pages 126–137, San Diego, California, June 9–13, 2007.

[229] John Kim, J. Dally William, Brian Towles, and Amit Gupta. Microarchitecture of a High-Radix Router. In *Proceedings of the 32nd International Symposium on Computer Architecture (ISCA)*, pages 420–431, Madison, Wisconsin, June 4–8, 2005. IEEE.

[230] Mike Kistler, Michael Perrone, and Fabrizio Petrini. Cell Processor Interconnection Network: Built for Speed. *IEEE Micro*, 25(3), May/June 2006.

[231] Dean Klein. Memory Update. In *13th Workshop on Distributed Super-computing*, 2009.

[232] Mike Ko, Mallikarjun Chadalapaka, et al. iSCSI extensions for RDMA. Technical report, IETF RFC 5046, October 2007.

[233] Kenneth R. Koch, Randal S. Baker, and Raymond E. Alcouffe. Solution of the first-order form of the 3-D discrete ordinates equation on a massively parallel processor. *Transactions of the American Nuclear Society*, 65(108):198–199, 1992.

[234] Peter Kogge et al. ExaScale Computing Study: Technology Challenges in Achieving Exascale Systems. 2008. users.ece.gatech.edu/ mrichard/-ExascaleComputingStudyReports/ECS_reports.htm.

[235] M. Koop, T. Jones, and D. K. Panda. MVAPICH-Aptus: Scalable High-Performance Multi-Transport MPI over InfiniBand. In *International Parallel and Distributed Processing Symposium (IPDPS)*, 2008.

[236] David Kotz. Disk-directed I/O for MIMD multiprocessors. In Hai Jin, Toni Cortes, and Rajkumar Buyya, editors, *High Performance Mass Storage and Parallel I/O: Technologies and Applications*, chapter 35, pages 513–535. IEEE Computer Society Press and John Wiley & Sons, 2001.

[237] Arvind Krishnamurthy, David E. Culler, and Katherine Yelick. Empirical Evaluation of Global Memory Support on the Cray-T3D and Cray-T3E. Technical Report UCB/CSD-98-991, EECS Department, University of California, Berkeley, Aug 1998.

[238] J. Mohan Kumar and L. M. Patnaik. Extended hypercube: A hierarchical interconnection network of hypercubes. *IEEE Transactions on Parallel and Distributed Systems*, 3(1):45–57, January 1992.

[239] Rakesh Kumar, Dean M. Tullsen, Parthasarathy Ranganathan, Norman Jouppi, and Keith Farkas. Single-ISA Heterogeneous Multicore Architectures for Multithreaded Workload Performance. In *Proceedings of the 31st Annual International Symposium on Computer Architecture*, Munich, Germany, 2004.

[240] Sameer Kumar, Gabor Dozsa, Gheorghe Almasi, Philip Heidelberger, Dong Chen, Mark E. Giampapa, Michael Blocksome, Ahmad Faraj, Jeff Parker, Joseph Ratterman, Brian Smith, and Charles J. Archer. The deep computing messaging framework: generalized scalable message passing on the blue gene/P supercomputer. In *ICS '08: Proceedings of the 22nd annual international conference on Supercomputing*, pages 94–103, New York, NY, USA, 2008. ACM.

[241] Sanjay Kumar, Himanshu Raj, Ivan Ganev, and Karsten Schwan. Re-architecting VMMs for Multicore Systems: The Sidecore Approach. In *Proceedings of Workshop on the Interanction between Operating Systems and Computer Architecture (WIOSCA)*, 2007.

[242] Sanjay Kumar, Himanshu Raj, Karsten Schwan, and Ivan Ganev. Re-Architecting VMMs for Multi-core Systems: The Sidecore Approach. In *Workshop on Interaction between Operating Systems and Computer Architecture (WIOSCA), in conjunction with ISCA'07*, San Diego, CA, June 2007.

[243] Vibhore Kumar, Brian F Cooper, Zhongtang Cai, Greg Eisenhauer, and Karsten Schwan. Resource-Aware Distributed Stream Management using Dynamic Overlays. In *Proceedings of the 25th IEEE International Conference on Distributed Computing Systems (ICDCS-2005)*, 2005.

[244] Vibhore Kumar, Brian F. Cooper, Zhongtang Cai, Greg Eisenhauer, and Karsten Schwan. Middleware for Enterprise Scale Data Stream Management using Utility-Driven Self-Adaptive Information Flows. *Cluster Computing Journal*, 10(4), 2007.

[245] James Laudon and Daniel Lenoski. The sgi origin 2000: A cc-numa highly scalable server. In *International Symposium on Computer Architecture*. IEEE/ACM, 1997.

[246] James Laudon and Daniel Lenoski. System overview of the SGI Origin 200/2000 product line. In *Proceedings of IEEE COMPCON '97*, pages 150–156, San Jose, California, February 23–26, 1997. IEEE.

[247] Greg Law. The Convergence of Storage and Server Virtualization. Presented at Xen summit, November 14-16, 2007.

[248] Lawrence Livermore National Laboratory. Facts on ASCI Purple. Technical Report UCRL-TB-150327, Lawrence Livermore National Laboratory, Baltimore, Maryland, November 16–22, 2002. Flyer distributed at SC2002.

[249] Duncan H. Lawrie. Access and Alignment of Data in an Array Processor. *IEEE Transactions on Computers*, C-24(12):1145–1155, December 1975.

[250] Charles Lefurgy, Karthick Rajamani, Freeman Rawson, Wes Felter, Michael Kistler, and Tom W. Keller. Energy management for commercial servers. *IEEE Computer*, 36(12):39–48, 2003.

[251] F. Thomas Leighton. *Introduction to Parallel Algorithms and Architectures: Arrays, Trees, Hypercubes*, chapter 3.2, pages 439–472. Morgan Kaufmann Inc, San Mateo, California, first edition, 1992.

[252] Matt Leininger and Mark Seagar. Sequoia architectural requirements. In *Salishan Conference on High-Speed Computing*, number LLNL-PRES-403472, April 2008.

[253] Charles E. Leiserson. Fat-trees: Universal networks for hardware-efficient supercomputing. *IEEE Transactions on Computers*, C-34(10):892–901, October 1985.

[254] Charles E. Leiserson, Zahi S. Abuhamdeh, et al. The network architecture of the Connection Machine CM-5. In *Proceedings of the Fourth Annual ACM Symposium on Parallel Algorithms and Architectures (SPAA '92)*, pages 272–285, San Diego, California, June 29–July 1, 1992. ACM Press.

[255] Douglas Leith, Lachlan L. H. Andrew, Tom Quetchenbach, and Robert N. Shorten. Experimental Evaluation of Delay/Loss-based TCP Congestion Control Algorithms. In *Proc. Workshop on Protocols for Fast Long Distance Networks*, 2008.

[256] Will E. Leland and Marvin H. Solomon. Dense trivalent graphs for processor interconnection. *IEEE Transactions on Computers*, C-31(3):219–222, March 1982.

[257] Daniel Lenoski, James Laudon, et al. The Stanford Dash multiprocessor. *Computer*, 25(3):63–79, March 1992.

[258] Edgar Leon, Rolf Riesen, Arthur B. Maccabe, and Patrick Bridges. Instruction-Level Simulation of a Cluster at Scale. In *Proceedings of the 2009 ACM/IEEE conference on Supercomputing*, Portland, OR, 2009.

[259] Edgar A. Leon, Kurt B. Ferreira, and Arthur B. Maccabe. Reducing the Impact of the Memory Wall for I/O Using Cache Injection. In *Seventeenth IEEE Symposium on High-Performance Interconnects (HotI'07)*, Palo Alto, CA, August 2007.

[260] Tong Li, Paul Brett, Barbara Hohlt, Rob Knauerhase, Sean D. McElderry, and Scott Hahn. Operating System Support for Shared-ISA Asymmetric Multi-core Architectures. In *Proceedings of the Workshop on the Interaction between Operating Systems and Computer Architecture (WIOSCA'08)*, Beijing, China, 2008.

[261] Huai-An (Paul) Lin. Estimation of the Optimal Performance of ASN.1/BER Transfer Syntax. *SIGCOMM Comput. Commun. Rev.*, 23(3):45–58, 1993.

[262] Ying-Dar Lin, Yi-Neng Lin, Shun-Chin Yang, and Yu-Sheng Lin. DiffServ over Network Processors: Implementation and Evaluation. In *Proc. of Hot Interconnects 10*, Stanford, CA, August 2002.

[263] Heiner Litz, Holger Fröning, Mondrian Nüssle, and Ulrich Brüning. A HyperTranpsort Interface Controller for Ultra-Low Latency Messages, 2008. www.hypertransport.org/docs/wp/UoM_HTX_NIC_02-25-08.pdf.

[264] Jiuxing Liu, Wei Huang, Bulent Abali, and Dhabaleshwar K. Panda. High Performance VMM-Bypass I/O in Virtual Machines. In *ATC'06: USENIX 2006 Annual Technical Conference on Annual Technical Conference*, 2006.

[265] Jiuxing Liu, Weihang Jiang, et al. Design and Implementation of MPICH2 over InfiniBand with RDMA Support. In *Proceedings of Int'l Parallel and Distributed Processing Symposium (IPDPS '04)*, April 2004.

[266] Jiuxing Liu, Ahmit Mamidala, and Dhabaleswar K. Panda. Fast and Scalable MPI-Level Broadcast using InfiniBand's Hardware Multicast Support. In *Proceedings of Int'l Parallel and Distributed Processing Symposium (IPDPS 04)*, April 2004.

[267] Jiuxing Liu, Abhinav Vishnu, and Dhabaleswar K. Panda. Building Multirail InfiniBand Clusters: MPI-Level Design and Performance Evaluation. In *SuperComputing Conference*, 2004.

[268] Los Alamos National Laboratory. The ASCI Blue Mountain 3-TOps system: On the road to 100 TeraOPS. Technical Report LALP-98-127, Los Alamos National Laboratory, Orlando, Florida, November 7–13, 1998. Flyer distributed at SC1998.

[269] Los Alamos National Laboratory. The ASCI Q system: 30 TeraOPS capability at Los Alamos National Laboratory. Technical Report LALP-02-0230, Los Alamos National Laboratory, Baltimore, Maryland, November 16–22, 2002. Flyer distributed at SC2002.

[270] David B. Loveman. High Performance Fortran. *IEEE Parallel & Distributed Technology: Systems & Applications*, 1(1):25–42, February 1993.

[271] Piotr Luszczek, Jack Dongarra, et al. Introduction to the HPC Challenge Benchmark Suite, March 2005. icl.cs.utk.edu/hpcc/pubs/index.html.

[272] Arthur B. Maccabe, Kevin S. McCurley, Rolf Riesen, and Stephen R. Wheat. SUNMOS for the Intel Paragon: A Brief User's Guide. In *Proceedings of the Intel Supercomputer Users' Group. 1994 Annual North America Users' Conference*, pages 245–251, June 1994.

[273] Arthur B. Maccabe and Stephen R. Wheat. Message passing in PUMA. Technical Report SAND-93-0935C, Sandia National Labs, 1993.

[274] Dinesh Manocha. General-Purpose Computations Using Graphics Procecssors. *IEEE Computer*, 38(8):85–88, Aug. 2005.

[275] John Marshall. Cisco Systems – Toaster2. In Patrick Crowley, Mark Franklin, Haldun Hadimioglu, and Peter Onufryk, editors, *Network Processor Design*. Morgan Kaufmann Inc., 2003.

[276] Wim Martens, Frank Neven, and Thomas Schwentick. Complexity of Decision Problems for Simple Regular Expressions. In *MFCS*, pages 889–900, 2004.

[277] Richard P. Martin, Amin M. Vahdat, David E. Culler, and Thomas E. Anderson. Effects Of Communication Latency, Overhead, And Bandwidth In A Cluster Architecture. In *Proceedings of The 24th Annual International Symposium on Computer Architecture (ISCA)*, pages 85–97, Denver, Colorado, June 2–4, 1997.

[278] Timothy G. Mattson, Rob Van der Wijngaart, and Michael Frumkin. Programming the Intel 80-core network-on-a-chip terascale processor. In *Proceedings of the 2008 ACM/IEEE conference on Supercomputing*, Austin, TX, 2008.

[279] Timothy G. Mattson and Greg Henry. An overview of the Intel TFLOPS supercomputer. *Intel Technology Journal*, 2(1), 1st Quarter 1998.

[280] John M. May. *Parallel I/O for High Performance Computing*. Morgan Kaufmann Inc., San Francisco, CA, 2000.

[281] David Mayhew and Venkata Krishnan. PCI Express and Advanced Switching: Evolutionary Path to Building Next Generation Interconnects. In *Proceedings of the 11th Symposium on High Performance Interconnects (HotI)*, pages 21–29, Palo Alto, California, August 20–22, 2003. IEEE.

[282] Mark Meiss. Tsunami: A High-Speed Rate-Controlled Protocol for File Transfer. steinbeck.ucs.indiana.edu/ mmeiss/papers/tsunami.pdf.

[283] Mellanox Technologies. http://www.mellanox.com.

[284] Mellanox Technologies.   I/O Virtualization Using Mellanox InfiniBand And Channel I/O Virtualization (CIOV) Technology. www.mellanox.com/pdf/whitepapers/WP_Virtualize_with_IB.pdf.

[285] Mellanox ConnectX IB Specification Sheet, 2008. www.mellanox.com.

[286] Aravind Menon, Alan L. Cox, and Willy Zwaenepoel. Optimizing Network Virtualization in Xen. In *Proc. of USENIX Annual Technical Conference*, 2006.

[287] Aravind Menon, Jose Renato Santos, et al. Diagnosing Performance Overheads in the Xen Virtual Machine Environment. In *Proc. of VEE*, 2005.

[288] Message Passing Interface Forum. MPI: A Message Passing Interface. In *Proc. of Supercomputing '93*, pages 878–883. IEEE Computer Society Press, November 1993.

[289] Message Passing Interface Forum. *MPI: A Message-Passing Interface Standard*, March 1994.

[290] Message Passing Interface Forum. *MPI-2: Extensions to the Message-Passing Interface*, July 1997.

[291] Cade Metz. Ethernet — A Name for the Ages. *The Register*, 13 March 2009.

[292] Dutch T. Meyer, Gitika Aggarwal, Brendan Cully, Geoffrey Lefebvre, Michael J. Feeley, Norman C. Hutchinson, and Andrew Warfield. Parallax: Virtual Disks for Virtual Machines. In *Eurosys '08: Proceedings of the 3rd ACM SIGOPS/EuroSys European Conference on Computer Systems 2008*, pages 41–54, 2008.

[293] Sun Microsystems. RPC: remote procedure call protocol specification, version 2. Technical Report RFC 1057, Sun Microsystems, Inc., June 1988.

[294] Frank Mietke, Robert Rex, Robert Baumgartl, et al. Analysis of the memory registration process in the Mellanox InfiniBand software stack. In *Proceedings of EuroPar'06*, 2006.

[295] Russ Miller and Quentin Stout. Data movement techniques for the pyramid computer. *SIAM Journal on Computing*, 16(1):38–60, February 1987.

[296] Jeffrey Mogul. TCP offload is a Dumb Idea Whose Time Has Come. In *Proceedings of Hot Topics on Operating Systems - HotOS*, May 2003.

[297] Adam Moody, Juan Fernandez, Fabrizio Petrini, and Dhabaleswar K. Panda. Scalable NIC-based reduction on large-scale clusters. In *Proceedings of the ACM/IEEE SC2003 Conference*, November 2003.

[298] Gordon E. Moore. Cramming more components onto integrated circuits. *Electronics*, 38(8), April 19, 1965.

[299] Myricom, Inc. The GM Message Passing System. Technical report, Myricom, Inc., 1997.

[300] Myricom, Inc. LANai 9, June 26, 2000.

[301] Myricom, Inc. Myrinet Express (MX): A High-Performance, Low-Level, Message Passing Interface for Myrinet. Technical report, July 2003.

[302] Makato Nakamura, Mary Inaba, and Kei Hiraki. Fast Ethernet is Sometimes Faster than Gigabit Ethernet on LFN - Observation of Congestion Control of TCP Streams. In *Proceedings of Symposium on Parallel and Distributed Computing Systems*, 2003.

[303] Sundeep Narravula, Pavan Balaji, et al. Supporting Strong Cache Coherency for Active Caches in Multi-Tier Data-Centers over InfiniBand. In *SAN-3 held in conjunction with HPCA 2004*, 2004.

[304] Gonzalo Navarro and Mathieu Raffinot. *Flexible pattern matching in strings: practical on-line search algorithms for texts and biological sequences*. Cambridge University Press, New York, NY, USA, 2002.

[305] Exploring Science Frontiers at Petascale, 2008. Oak Ridge National Laboratory, Petascale Brochure.

[306] Die Shot of Intel's Quadcore Nehalem. chip-architect.com/news/-Nehalem_at_1st_glance_.jpg.

[307] The Netperf Network Performance Benchmark. http://www.netperf.org.

[308] Netronome's Network Flow Processors: High-performance Solutions at 10Gbps and Beyond. http://www.netronome.com/pages/network-flow-processors.

[309] Network-Based Computing Laboratory. MVAPICH: MPI for InfiniBand on VAPI Layer. http://nowlab.cse.ohio-state.edu/projects/mpi-iba/index.html.

[310] J. Nieplocha, V. Tipparaju, M. Krishnan, and D. K. Panda. High performance remote memory access communication: The ARMCI approach. *International Journal of High Performance Computing Applications*, 20(2):233–253, 2006.

[311] Robert W. Numrich and John Reid. Co-array Fortran for parallel programming. *SIGPLAN Fortran Forum*, 17(2):1–31, 1998.

[312] NVIDIA. GeForce 8800 GPU Architecture Overview – Technical Brief. www.nvidia.com/object/IO_37100.html.

[313] Obsidian Research Corp. http://www.obsidianresearch.com/.

[314] Ron A. Oldfield, Arthur B. Maccabe, Sarala Arunagiri, Todd Korden-brock, Rolf Riesen, Lee Ward, and Patrick Widener. Lightweight I/O for scientific applications. In *Proceedings of the IEEE International Conference on Cluster Computing*, Barcelona, Spain, September 2006.

[315] Ron A. Oldfield, Patrick Widener, Arthur B. Maccabe, Lee Ward, and Todd Kordenbrock. Efficient data-movement for lightweight I/O. In *Proceedings of the 2006 International Workshop on High Performance I/O Techniques and Deployment of Very Large Scale I/O Systems*, Barcelona, Spain, September 2006.

[316] Van Oleson, Karsten Schwan, Greg Eisenhauer, Beth Plale, Calton Pu, and Dick Amin. Operational Information Systems - An Example from the Airline Industry. In *First Workshop on Industrial Experiences with Systems Software (WIESS)*, October 2000.

[317] Object Management Group (OMG). Notification Service Specification 1.0. www.omg.org/pub/doc/formal/00-06-20.pdf, June 2000.

[318] Object Management Group (OMG). Event Service Specification 1.1. www.omg.org/pub/docs/formal/01-03-01.pdf, March 2001.

[319] Diego Ongaro, Alan L. Cox, and Scott Rixner. Scheduling I/O in Virtual Machine Monitors. In *VEE '08: Proceedings of the fourth ACM SIGPLAN/SIGOPS international conference on Virtual execution environments*, pages 1–10, 2008.

[320] Open Cirrus. opencirrus.org.

[321] OpenFabrics Alliance. OpenFabrics. http://www.openfabrics.org/.

[322] OPRA. Option Price Reporting Authority.

[323] Paul Overell. Augmented BNF for Syntax Specifications: ABNF, 1997.

[324] Object-based Storage Architecture: Defining a New Generation of Storage Systems Built on Distributed, Intelligent Storage Devices. Panasas Inc. white paper, version 1.0, October 2003. http://www.panasas.com/docs/.

[325] Scott Pakin, Mario Lauria, and Andrew Chien. High Performance Messaging on Workstations: Illinois Fast Messages (FM) for Myrinet. In *Proceedings of the 1995 ACM/IEEE Supercomputing Conference*, volume 2, pages 1528–1557, San Diego, California, December 1995.

[326] Panasas, Inc. www.panasas.com.

[327] Ruoming Pang, Vern Paxson, Robin Sommer, and Larry Peterson. Bin-pac: A Yacc for Writing Application Protocol Parsers. In *IMC '06: Proceedings of the 6th ACM SIGCOMM conference on Internet measurement*, pages 289–300, New York, NY, USA, 2006. ACM.

[328] Flavio Pardo, Vladimir A. Aksyuk, et al. Optical MEMS devices for telecom systems. In Jung-Chih Chiao, Vijay K. Varadan, and Carles Cane, editors, *Proceedings of the SPIE: Smart Sensors, Actuators, and MEMS II*, volume 5116, pages 435–444, Maspalomas, Gran Canaria, Spain, May 19–21, 2003. SPIE.

[329] Jung-Heum Park and Kyung-Yong Chwa. Recursive Circulant: A New Topology for Multicomputer Networks (Extended Abstract). In *Proceedings of the International Symposium on Parallel Architectures, Algorithms and Networks (ISPAN)*, pages 73–80, Kanazawa, Japan, December 14–16, 1994. IEEE.

[330] Janak H. Patel. Processor-Memory Interconnections for Multiprocessors. In *Proceedings of the 6th Annual Symposium on Computer Architecture (ISCA)*, pages 168–177, Philadelphia, Pennsylvania, April 23–25, 1979. ACM.

[331] Professional Digital Video Gateways for the Broadcaster and Multi-Service Operator: Delivered by Path 1 Network Technologies and Intel Network Processors, 2002. White Paper. www.intel.com/design/-network/casestudies/path1.htm.

[332] Bryan Payne, Martim Carbone, and Wenke Lee. Secure and Flexible Monitoring of Virtual Machines. In *Proceedings of ACSAC*, 2007.

[333] Marshall C. Pease, III. The indirect binary *n*-cube microprocessor array. *IEEE Transactions on Computers*, C-26(5):458–473, May 1977.

[334] Kevin T. Pedretti and Trammell Hudson. Developing Custom Firmware for the Red Storm SeaStar Network Interface. In *Cray User Group Annual Technical Conference*, May 2005.

[335] Peta-Scale I/O with the Lustre File System. Sun Microsystems, Oak Ridge National Laboratory/Lustre Center of Excellence, February, 2008.

[336] Larry Peterson, Scott Shenker, and Jon Turner. Overcoming the Internet Impasse through Virtualization. In *Proc. of HotNets-III*, Cambridge, MA, November 2004.

[337] Fabrizio Petrini, Salvador Coll, Eitan Frachtenberg, and Adolfy Hoisie. Hardware- and Software-Based Collective Communication on the Quadrics Network. In *Proceedings of the 2001 IEEE International Symposium on Network Computing and Applications (NCA)*, pages 24–35, Cambridge, Massachusetts, October 8–10, 2001. IEEE.

[338] Fabrizio Petrini, Wu-chun Feng, Adolfy Hoisie, Salvador Coll, and Eitan Frachtenberg. The Quadrics Network: High Performance Clustering Technology. *IEEE Micro*, 22(1):46–57, January-February 2002.

[339] Fabrizio Petrini, Daniele Scarpazza, and Oreste Villa. High Speed String Searching against Large Dictionaries on the Cell/B.E. Processor. In

Proceedings of the 22nd IEEE International Parallel and Distributed Processing Symposium (IPDPS), Miami, FL, April 2008.

[340] Ian Philp and Yin-Ling Liong. The Scheduled Transfer (ST) Protocol. In *Proceedings of the Third International Workshop on Network-Based Parallel Computing: Communication, Architecture, and Applications*, January 1999.

[341] James Pinkerton and Ellen Deleganes. Direct Data Placement Protocol (DDP)/ Remote Direct Memory Access Protocol (RDMAP) Security. Technical report, IETF RFC 5042, October 2007.

[342] Nicholas Pippenger. On Crossbar Switching Networks. *IEEE Transactions on Communications*, 23(6):646–659, June 1975.

[343] Steve Plimpton, Ron Brightwell, Courtenay Vaughan, Keith Underwood, and Mike Davis. A Simple Synchronous Distributed-Memory Algorithm for the HPCC RandomAccess Benchmark. In *2006 IEEE International Conference on Cluster Computing*, September 2006.

[344] Jon Postel. Transmission Control Protocol: DARPA Internet program protocol specification. Request for Comments 793, Information Sciences Institute, University of Southern California, Marina del Rey, California, September 1981.

[345] IBM PowerNP NP4GS3 Datasheet. IBM Corporation.

[346] Ian Pratt, Keir Fraser, et al. Xen 3.0 and the Art of Virtualization. In *Proc. of the Ottawa Linux Symposium*, 2005.

[347] Franco P. Preparata and Jean Vuillemin. The cube-connected cycles: A versatile network for parallel computation. *Communications of the ACM*, 24(5):300–309, May 1981.

[348] Parallel Virtual File System 2. http://www.pvfs.org/.

[349] Intel quickassist technology community. qatc.intel.com.

[350] Wei Qin, Subramanian Rajagopalan, and Sharad Malik. A Formal Concurrency Model-based Architecture Description Language for Synthesis of Software Development Tools. *SIGPLAN Notices*, 39(7):47 – 56, 2004/07/.

[351] QLogic. http://www.qlogic.com/.

[352] An Introduction to the Intel QuickPath Interconnect. www.intel.com/technology/quickpath/introduction.pdf.

[353] Quadrics Supercomputers World Ltd. http://www.quadrics.com/.

[354] Quadrics Qs 10G for HPC Interconnect Product Family, 2008. www.quadrics.com.

[355] The Cisco QuantumFlow Processor: Cisco's Next Generation Network Processor. http://www.cisco.com/en/US/prod/collateral/routers/ps9343/solution_overview_c22-448936.pdf.

[356] Intel QuickAssist Techcnology Accelerator Abstraction Layer: Enabling Consistent Platform-Level Services for Tightly Coupled Accelerators. Intel White Paper, www.intel.com/go/QuickAssist.

[357] Himanshu Raj and Karsten Schwan. High Performance and Scalable I/O Virtualization via Self-virtualized Devices. In *HPDC*, 2007.

[358] Kaushik Kumar Ram, Jose Renato Santos, Yoshio Turner, Alan L. Cox, and Scott Rixner. Achieving 10 Gb/s using Safe and Transparent Network Interface Virtualization. In *Proc. of VEE*, 2009.

[359] Adit Ranadive, Ada Gavrilovska, and Karsten Schwan. IBMon: Monitoring VMM-Bypass Capable InfiniBand Devices using Memory introspection. In *Workshop on System-level Virtualization for High Performance Computing (HPCVirt) 2009*, Nuremberg, Germany, March 31, 2009.

[360] Adit Ranadive, Mukil Kesavan, Ada Gavrilovska, and Karsten Schwan. Performance Implications of Virtualizing Multicore Cluster Machines. In *Workshop on System-level Virtualization for High Performance Computing (HPCVirt) 2008*, Glasgow, Scotland, March 31, 2008.

[361] Nagaswara S. V. Rao, Weikunan Yu, William R. Wing, Stephen W. Poole, and Jeffrey S. Vetter. Wide-Area Performance Profiling of 10GigE and InfiniBand Technologies. In *Proceedings of Supercomputing'08*, 2008.

[362] RapidSwitch. At http://www.rapidswitch.com/.

[363] IETF RDDP Working Group , 2003. www.ietf.org/html.charters/rddp-charter.html.

[364] RDMA Consortium. rdmaconsortium.org, 2008.

[365] Renato Recio, Bernard Metzler, et al. A Remote Direct Memory Access Protocol Specification. Technical report, IETF RFC 5040, October 2007.

[366] Greg Regnier, Srihari Makineni, Ramesh Illikkal, Ravi Iyer, Dave Minturn, Ram Huggahalli, Don Newell, Linda Cline, and Annie Foong. TCP Onloading for Data Center Servers. *IEEE Computer*, 37(11), 2004.

[367] Greg Regnier, Dave Minturn, Gary McAlpine, Vikram Saletore, and Annie Foong. ETA: Experience with an Intel Xeon Processor as a Packet Processing Engine. In *Proc. of Symposium of Hot Interconnects*, Stanford, CA, 2003.

[368] Thompson Reuters Press Release. Thomson Reuters Launches Ultra Low Latency Direct Fee Connectivity to Turquoise, 2008.

[369] Rolf Riesen, Ron Brightwell, Patrick Bridges, Trammell Hudson, Arthur Maccabe, Patrick Widener, and Kurt Ferreira. Designing and Implementing Lightweight Kernels for Capability Computing. *Concurrency and Computation: Practice and Experience.* Accepted for publication.

[370] Allyn Romanow and Stephen Bailey. An Overview of RDMA over IP. In *Proceedings of International Workshop on Protocols for Long-Distance Networks (PFLDnet2003)*, 2003.

[371] R. B. Ross. Parallel I/O Benchmarking Consortium. www.mcs.anl.gov/-rross/pio-benchmark.

[372] Marcel-Catalin Rosu, Karsten Schwan, and Richard Fujimoto. Supporting Parallel Applications on Clusters of Workstations: The Virtual Communication Machine-based Architecture. *Cluster Computing, Special Issue on High Performance Distributed Computing*, May 1998.

[373] Duncan Roweth and Ashley Pittman. Optimised global reduction on QsNet-II. In *Thirteenth IEEE Symposium on High-Performance Interconnects (HotI'05)*, August 2005.

[374] Paul Royal, Mitch Halpin, Ada Gavrilovska, and Karsten Schwan. Utilizing Network Processors in Distributed Enterprise Environments. In *Proceedings of the 5th International Conference on Network Computing and Applications (IEEE NCA06)*, Cambridge, MA, July 2006.

[375] Martin W. Sachs and Anujan Varma. Fibre Channel. *IEEE Communications*, pages 40–49, Aug 1996.

[376] Bratin Saha, Ali-Reza Adl-Tabatabai, Anwar Ghuloum, Mohan Rajagopalan, Richard L. Hudson, Leaf Petersen, Vijay Menon, Brian Murphy, Tatiana Sphiesman, Eric Sprangle, Anwar Rohillah, Doug Carmean, and Jesse Fang. Enabling Scalability and Performance in a Large-scale CMP Environment. In *Proceedings of EuroSys'07*, Lisbon, Portugal, March 2007.

[377] Maheswara R. Samatham and Dhiraj K. Pradhan. The de Bruijn multiprocessor network: A versatile parallel processing and sorting network for VLSI. *IEEE Transactions on Computers*, 38(4):567–581, April 1989.

[378] Gopalakrishnan Santhanaraman, Sundeep Narravula, Abhinav Mamidala, and Dhabaleswar K. Panda. MPI-2 One Sided Usage and Implementation for Read Modify Write operations: A case study with HPCC. In *Proceedings of EuroPVM/MPI 2007*, 2007.

[379] Jose Renato Santos, John Janakiraman, Yoshio Turner, and Ian Pratt. Netchannel 2: Optimizing Network Performance. Presented at Xen Summit, November 14-16, 2007.

[380] Jose Renato Santos, Yoshio Turner, G. Janakiraman, and Ian Pratt. Bridging the Gap between Software and Hardware Techniques for I/O Virtualization. In *ATC'08: USENIX 2008 Annual Technical Conference on Annual Technical Conference*, pages 29–42, 2008.

[381] J. Satran, K. Meth, C. Sapuntzakis, M. Chadalapaka, and E. Zeidner. Internet Small Computer Systems Interface (iSCSI), 2004.

[382] Daniele Paolo Scarpazza, Patrick Mullaney, Oreste Villa, Fabrizio Petrini, Vinod Tipparaju, and Jarek Nieplocha. Transparent System-level Migration of PGAS Applications using Xen on InfiniBand. In *Cluster 2007: IEEE International Conference on Cluster Computing, 2007*, pages 74–83, 2007.

[383] Daniele Paolo Scarpazza, Oreste Villa, and Fabrizio Petrini. Peak-Performance DFA-based String Matching on the Cell Processor. In *In Proc. SMTPS '07*, 2007.

[384] Michael Schlansker, Nagabhushan Chitlur, Erwin Oertli, Paul M. Stillwell Jr, Linda Rankin, Dennis Bradford, Richard J. Carter, Jayaram Mudigonda, Nathan Binkert, and Norman P. Jouppi. High-Performance Ethernet-Based Communications for Future Multi-Core Processors. In *Proceedings of SuperComputing'07*, 2007.

[385] Frank Schmuck and Roger Haskin. GPFS: A shared-disk file system for large computing clusters. In *Proceedings of the USENIX FAST '02 Conference on File and Storage Technologies*, pages 231–244, Monterey, CA, January 2002. USENIX Association.

[386] Karsten Schwan and Brian F. Cooper et. al. AutoFlow: Autonomic Information Flows for Critical Information Systems. In Manish Parashar and Salim Hariri, editors, *Autonomic Computing: Concepts, Infrastructure and Applications*. CRC Press, 2006.

[387] Thomas Scogland, Pavan Balaji, Wu-chun Feng, and Ganesh Narayanaswamy. Asymmetric Interactions in Symmetric Multi-core Systems: Analysis, Enhancements and Evaluation. In *Proceedings of the International Conference on High-Performance Computing, Networking, Storage, and Analysis, Supercomputing'08*, Austin, Texas, 2008.

[388] Steven L. Scott. Synchronization and communication in the T3E multi-processor. In *Proceedings of the Seventh International Conference on Architectural Support for Programming Languages and Operating Systems (ASPLOS)*, pages 26–36, Cambridge, Massachusetts, October 1–5, 1996. ACM.

[389] SDP Specification. http://www.rdmaconsortium.org/home.

[390] K. E. Seamons, Y. Chen, P. Jones, J. Jozwiak, and M. Winslett. Server-directed collective I/O in Panda. In *Proceedings of Supercomputing '95*, San Diego, CA, December 1995. IEEE Computer Society Press.

[391] Bill Segall and David Arnold. Elvin Has Left the Building: A Publish/-Subscribe Notification Service with Quenching. In *Proc. of the AUUG (Australian users group for Unix and Open Systems) 1997 Conference*, September 1997.

[392] Gautam Shah, Jarek Nieplocha, Jamshed Mirza, Chulho Kim, Robert Harrison, Rama K. Govindaraju, Kevin Gildea, Paul Dinicola, and Carl Bender. Performance and experience with LAPI - a new high-performance communication library for the IBM RS/6000 SP. In *In Proceedings of the International Parallel Processing Symposium*, pages 260–266, 1998.

[393] Hemal Shah, James Pinkerton, et al. Direct Data Placement over Reliable Transports. Technical report, IETF RFC 5041, October 2007.

[394] Niraj Shah and Kurt Keutzer. Network Processors: Origin of Species. In *Proceedings of the 17th International Symposium on Computer and Information Science (ISCIS XVII)*, 2002.

[395] Leah Shalev, Vadim Makhervaks, Zorik Machulsky, Giora Biran, Julian Satran, Muli Ben-Yehuda, and Ilan Shimony. Loosely Coupled TCP Acceleration Architecture. In *Proceedings of Hot Interconnects*, 2006.

[396] Shashank Shanbhag and Tilman Wolf. Massively Parallel Anomaly Detection in Online Network Measurement. In *Proc. of Seventeenth IEEE International Conference on Computer Communications and Networks (ICCCN)*, St. Thomas, USVI, August 2008.

[397] P. Shivam, P. Wyckoff, and D. Panda. Can User Level Protocols Take Advantage of Multi-CPU NICs? In *the Proceedings of International Parallel and Distributed Processing Symposium '02*, April 2002.

[398] Piyush Shivam and Jeffrey Chase. Promises and Reality: On the Elusive Benefits of Protocol Offload. In *Proceedings of the ACM SigComm Workshop on Network-IO Convergence (NICELI)*, August 2003.

[399] Piyush Shivam, Pete Wyckoff, and Dhabaleswar K. Panda. EMP: Zero-copy OS-bypass NIC-driven Gigabit Ethernet Message Passing. In *Int'l Conference on Supercomputing (SC '01)*, November 2001.

[400] Lance Shuler, Chu Jong, Rolf Riesen, David van Dresser, Arthur B. Maccabe, Lee Ann Fisk, and T. Mack Stallcup. The Puma operating system for massively parallel computers. In *Proceeding of the 1995 Intel Supercomputer User's Group Conference*. Intel Supercomputer User's Group, 1995.

[401] SIAC. FAST for OPRA - v1, 2007.

[402] SIAC. Data Recipient Specification, 2008.

[403] SIAC. FAST for OPRA - v2, 2008.

[404] SIAC. National Market Systems - Common IP Multicast Distribution Network - Recipient Interface Specification, 2008.

[405] Dale Skeen. The enterprise-capable publish-subscribe server. www.vitria.com.

[406] David Slogsnat, Alexander Giese, Mondrian Nüssle, and Ulrich Brüning. An Open-Source HyperTransport Core. *ACM Trans. Reconfigurable Technol. Syst.*, 1(3):1–21, 2008.

[407] SNIA-OSD Working Group. www.snia.org/osd.

[408] Marc Snir, Steve W. Otto, Steve Huss-Lederman, David W. Walker, and Jack Dongarra. *MPI: The Complete Reference.* MIT Press, Cambridge, MA, 1996.

[409] Solace Systems. At http://www.solacesystems.com/.

[410] Robert J. Souza and Steven P. Miller. UNIX and Remote Procedure Calls: A Peaceful Coexistence? In *In Proceedings of the 6th International Conference on Distributed Computing Systems*, pages 268–277, Cambridge, Massachusetts, May 1986.

[411] Tammo Spalink, Scott Karlin, Larry Peterson, and Yitzchak Gottlieb. Building a Robust Software-Based Router Using Network Processors. In *Proc. of 18th SOSP'01*, Chateau Lake Louise, Banff, Canada, October 2001.

[412] Raj Srinivasan. RPC: Remote Procedure Call Protocol Specification Version 2, 1995.

[413] Lawrence C. Stewart and David Gingold. A new generation of cluster interconnect. White paper, SiCortex, Inc., Maynard, Massachusetts, April 2008.

[414] Randall Stewart and Qiaobing Xie. Stream Control Transmission Protocol, 2000. IETF 2960.

[415] Ivan Stojmenovic. Honeycomb Networks: Topological Properties and Communication Algorithms. *IEEE Transactions on Parallel and Distributed Systems*, 8(10):1036–1042, October 1997.

[416] Harold S. Stone. Parallel Processing with the Perfect Shuffle. *IEEE Transactions on Computers*, C-20(2):153–161, February 1971.

[417] Harold S. Stone. Dynamic Memories with Enhanced Data Access. *IEEE Transactions on Computers*, C-21(4):359–366, April 1972.

[418] StorageReview.com Drive Performance Resource Center. 2008. www.storagereview.com.

[419] Robert Strom, Guruduth Banavar, Tushar Chandra, Marc Kaplan, Kevan Miller, Bodhi Mukherjee, Daniel Sturman, and Michael Ward. Gryphon: An Information Flow Based Approach to Message Brokering. In *International Symposium on Software Reliability Engineering '98 Fast Abstract*, 1998.

[420] Shinji Sumimoto, Hiroshi Tezuka, Atsushi Hori, Hiroshi Harada, Toshiyuki Takahashi, and Yutaka Ishikawa. High Performance Communication using a Gigabit Ethernet. Technical Report LA-UR-05-2635, Los Alamos National Laboratory, April 2005.

[421] Yuzhong Sun, Paul Y. S. Cheung, and Xiaola Lin. Recursive Cube of Rings: A New Topology for Interconnection Networks. *IEEE Transactions on Parallel and Distributed Systems*, 11(3):275–286, March 2000.

[422] TABB Group. Trading At Light Speed: Analyzing Low Latency Data Market Data Infrastructure, 2007.

[423] Kun Tan, Jingmin Song, Qian Zhang, and Murari Sridharan. A Compound TCP Approach for High-Speed and Long-Distance Networks. In *Proceedings of IEEE INFOCOM*, 2006.

[424] Tarari Content Processors. www.lsi.com.

[425] The BlueGene/L Team. An overview of the BlueGene/L supercomputer. In *Proceedings of SC2003: High Performance Networking and Computing*, Baltimore, MD, November 2002.

[426] TelX - The Interconnecton Company. At http://www.telx.com/.

[427] Priyanka Tembey, Anish Bhatt, Dulloor Rao, Ada Gavrilovska, and Karsten Schwan. Flexible Classification on Heterogeneous Multicore Appliance Platforms. In *Proceedings of the International Conference on Computer and Communication Networks*, St. Thomas, VI, August 2008.

[428] Tervela. At http://www.tervela.com/.

[429] Samuel Thibault and Tim Deegan. Improving Performance by Embedding HPC Applications in Lightweight Xen Domains. In *HPCVirt '08: Proceedings of the 2nd workshop on System-level virtualization for high performance computing*, pages 9–15, 2008.

[430] Tibco. TIB/Rendezvous. http://www.rv.tibco.com/rvwhitepaper.html.

[431] Brian L. Tierney, Tom Dunigan, Jason R. Lee, Dan Gunter, and Martin Stoufer. Improving distributed application performance using TCP instrumentation. Technical report, Lawrence Berkeley National Laboratory, May 2003.

[432] Intel EP80579 Integrated Processor Product Line. www.intel.com/-design/intarch/ep80579/.

[433] Don Tolmie and John Renwick. HIPPI: Simplicity Yields Success. *IEEE Network*, 7(1):28–32, January 1993.

[434] TOP 500 Supercomputer Sites. http://www.top500.org.

[435] Lewis W. Tucker and George G. Robertson. Architecture and applications of the Connection Machine. *Computer*, 21(8):26–38, August 1988.

[436] Keith D. Underwood, K. Scott Hemmert, Arun Rodrigues, Richard Murphy, and Ron Brightwell. A Hardware Acceleration Unit for MPI Queue Processing. In *19th International Parallel and Distributed Processing Symposium (IPDPS'05)*, April 2005.

[437] Keith D. Underwood, Michael Levenhagen, K. Scott Hemmert, and Ron Brightwell. High Message Rate, NIC-Based Atomics: Design and Performance Considerations. In *Proceedings of the 2008 IEEE International Conference on Cluster Computing*, September 2008.

[438] Keith D. Underwood, Arun Rodrigues, and K. Scott Hemmert. Accelerating List Management for MPI. In *Proceedings of the 2005 IEEE International Conference on Cluster Computing*, September 2005.

[439] U.S. Environmental Protection Agency ENERGY STAR Program. Report to Congress on Server and Data Center Energy Efficiency Public Law 109-431, 2007. www.energystar.gov/ia/partners/prod_development/-downloads/EPA_Datacenter_Report_Congress_Final1.pdf.

[440] Jeffrey S. Vetter, Sadaf R. Alam, Jr. Thomas H. Dunigan, Mark R. Fahey, Philip C. Roth, and Patrick H. Worley. Early evaluation of the Cray XT3. In *20th International Parallel and Distributed Processing Symposium*, pages 25–29, April 2006.

[441] Oreste Villa, Daniele Scarpazza, and Fabrizio Petrini. Accelerating Real-Time String Searching with Multicore Processors. *IEEE Computer*, 41(4):42–50, April 2008.

[442] Oreste Villa, Daniele Paolo Scarpazza, Fabrizio Petrini, and Juan Fernandez Peinador. Challenges in Mapping Graph Exploration Algorithms on Advanced Multi-core Processors. In *In Proc. IPDPS '07*, 2007.

[443] Abhinav Vishnu, Matthew Koop, Adam Moody, Amith Mamidala, Sundeep Narravula, and Dhabaleswar K. Panda. Hot-Spot Avoidance With Multi-Pathing Over InfiniBand: An MPI Perspective. In *7th IEEE Int'l Symposium on Cluster Computing and the Grid (CCGrid07)*, Rio de Janeiro, Brazil, May 2007.

[444] Intel Virtual Machine Device Queues. software.intel.com/file/1919.

[445] The VMWare ESX Server. http://www.vmware.com/products/esx/.

[446] Thorsten von Eicken, Anindya Basu, Vineet Buch, and Werner Vogels. U-Net: A User-level Network Interface for Parallel and Distributed Computing. In *Proceedings of the 15th ACM Symposium on Operating Systems Principles (SOSP)*, pages 40–53, December 1995.

[447] Thorsten von Eicken, David E. Culler, Seth Copen Goldstein, and Klaus Erik Schauser. Active messages: a mechanism for integrated communication and computation. In *Proceedings of the 19th annual International Symposium on Computer Architecture*, pages 256–266, May 1992.

[448] Thorsten von Eicken and Werner Vogels. Evolution of the Virtual Interface Architecture. *IEEE Computer*, 31(11):61–68, November 1998.

[449] Adam Wagner, Darius Buntinas, Ron Brightwell, and Dhabaleswar K. Panda. Application-bypass reduction for large-scale clusters. In *International Journal of High Performance Computing and Networking, Special Issue Cluster 2003*. In press.

[450] Robert P. Weaver and Michael L. Gittings. Massively parallel simulations with DOE's ASCI supercomputers: An overview of the Los Alamos Crestone project. In Tomasz Plewa, Timur Linde, and V. Gregory Weirs, editors, *Adaptive Mesh Refinement—Theory and Applications. Proceedings of the Chicago Workshop on Adaptive Mesh Refinement Methods*, volume 41 of *Lecture Notes in Computational Science and Engineering*, pages 29–56, Chicago, Illinois, September 3–5, 2003. Springer.

[451] David X. Wei, Cheng Jin, Steven H. Low, and Sanjay Hegde. FAST TCP: Motivation, Architecture, Algorithms, Performance. *IEEE/ACM Transactions on Networking*, 14(6):1246–1259, 2006.

[452] Eric Weigle and Wu-chun Feng. A Comparison of TCP Automatic-Tuning Techniques for Distributed Computing. In *Proceedings of the IEEE Symposium on High-Performance Distributed Computing (HPDC'02)*, Edinburgh, Scotland, 2002.

[453] Sage A. Weil, Scott A. Brandt, Ethan L. Miller, Darrell D. E. Long, and Carlos Maltzahn. Ceph: A scalable, high-performance distributed file system. In *Proceedings of the 2006 Symposium on Operating Systems Design and Implementation*, pages 307–320. University of California, Santa Cruz, 2006.

[454] Brent Welch, Marc Unangst, Zainul Abbasi, Garth A. Gibson, Brian Mueller, Jason Small, Jim Zelenka, and Bin Zhou. Scalable performance of the panasas parallel file system. In Mary Baker and Erik Riedel, editors, *Proceedings of the USENIX FAST'08 Conference on File and Storage Technologies*, pages 17–33. USENIX, February 2008.

[455] Bob Wheeler and Lenley Gwennap. *The Guide to Network Processors*. The Linley Group, 2008.

[456] P. Willmann, H. Kim, S. Rixner, and V.S. Pai. An Efficient Programmable 10 Gigabit Ethernet Network Interface Card. In *Proc. of HPCA*, 2005.

[457] Paul Willmann, Jeffrey Shafer, David Carr, Aravind Menon, Scott Rixner, Alan L. Cox, and Willy Zwaenepoel. Concurrent Direct Network Access for Virtual Machine Monitors. In *HPCA '07: Proceedings of the 2007 IEEE 13th International Symposium on High Performance Computer Architecture*, pages 306–317, 2007.

[458] Charlie Wiseman, Jonathan S. Turner, Michela Becchi, Patrick Crowley, John D. DeHart, Mart Haitjema, Shakir James, Fred Kuhns, Jing Lu, Jyoti Parwatikar, Ritun Patney, Michael Wilson, Ken Wong, and David Zar. A Remotely Accessible Network Processor-based Router for Network Experimentation. In *Proceedings of the Symposium on Architectures for Networking and Communications Systems*, 2008.

[459] Matthew Wolf, Hasan Abbasi, Benjamin Collins, David Spain, and Karsten Schwan. Service Augmentation for High End Interactive Data Services. In *Proceedings of Cluster 2005*, 2005.

[460] Michael Woodacre, Derek Robb, Dean Roe, and Karl Feind. The SGI Altix 3000 global shared-memory architecture. White Paper 3474, Silicon Graphics, Inc., Mountain View, California, October 21, 2003.

[461] Jiesheng Wu, Pete Wyckoff, and Dhabaleswar Panda. PVFS over Infini-Band: design and performance evaluation. In *Proceedings of ICPP '03*, Kaohsiung, Taiwan, October 2003.

[462] Qishi Wu and Nageswara S. V. Rao. A Class of Reliable UDP-based Transport Protocols Based on Stochastic Approximation. In *Proceedings of INFOCOM*, 2005.

[463] Qishi Wu and Nagi Rao. A Protocol for High Speed Transport Over Dedicated Channels. In *Proceedings of the 3rd International Workshop on Protocols for Fast Long-Distance Networks*, Lyon, France, 2005.

[464] William A. Wulf and Sally A. McKee. Hitting the memory wall: implications of the obvious. Technical Report CS-94-48, University of Virginia, 1995.

[465] Ben Wun, Jeremy Buhler, and Patrick Crowley. Exploiting Coarse-Grained Parallelism to Accelerate Protein Motif Finding with a Network Processor. In *Proceedings of the 2005 International Conference on Parallel Architectures and Compilation Techniques (PACT)*, 2005.

[466] Pete Wyckoff and Jiesheng Wu. Memory registration caching correctness. In *Proceedings of CCGrid'05*, Cardiff, UK, May 2005.

[467] Xsigo Virtual I/O Overview. white Paper at www.xsigo.com.

[468] Lisong Xu, Khaled Harfoush, and Injong Rhee. Binary Increase Congestion Control for Fast, Long Distance Networks. In *Proceedings of IEEE INFOCOM*, 2004.

[469] Lisong Xu, Khaled Harfoush, and Injong Rhee. Binary Increase Congestion Control for Fast, Long Distance Networks. In *Proceedings of INFOCOM*, 2004.

[470] Sudhakar Yalamanchili. ADEPT: An Architecture Description Language for Polymorphic Embedded Systems. In *Presentation at the Center for High Performance Embedded Systems, School of Computer Engineering. Nanyang Technological University*, Singapore, 2003.

[471] Eric Yeh, Herman Chao, Venu Mannem, Joe Gervais, and Bradley Booth. Introduction to TCP/IP Offload Engines (TOE). White Paper, May 2002.

[472] Katherine Yelick, Arvind Krishnamurthy, Steve G. Steinberg, David E. Culler, and Remzi H. Arpaci. Empirical Evaluation of the CRAY-T3D: A Compiler Perspective. In *Proceedings of International Symposium on Computer Architecture*, 1995.

[473] Ken Yocum and Jeff Chase. Payload Caching: High-Speed Data Forwarding for Network Intermediaries. In *Proc. of USENIX Technical Conference (USENIX'01)*, Boston, Massachusetts, June 2001.

[474] Takeshi Yoshino, Yutaka Sugawara, Katsusi Inagami, Junji Tamatsukuri, Mary Inaba, and Kei Hiraki. Performance Optimization of TCP/IP over 10 Gigabit Ethernet by Precise Instrumentation. In *Proceedings of Supercomputing'08*, Austin, TX, 2008.

[475] Jeffrey Young, Sudhakar Yalamanchili, Federico Silla, and Jose Duato. A HyperTransport-Enabled Global Memory Model For Improved Memory Efficiency. In *First International Workshop on HyperTransport Research and Applications*, 2008.

[476] Weikuan Yu, Shuang Liang, and Dhabaleswar K. Panda. High Performance Support of Parallel Virtual File System (PVFS2) over Quadrics. In *International Conference on Supercomputing (ICS '05)*, pages 323–331, 2005.

[477] Weikuan Yu, R. Noronha, Shuang Liang, and D. K. Panda. Benefits of High Speed Interconnects to Cluster File Systems: A Case Study with Lustre. In *Parallel and Distributed Processing Symposium, 2006. IPDPS 2006. 20th International*, 2006.

[478] Weikuan Yu, Jeffrey S. Vetter, Shane Canon, and Song Jiang. Exploiting Lustre File Joining for Effective Collective I/O. In *Proceedings of the*

Seventh IEEE/ACM International Symposium on Cluster Computing and the Grid, pages 267–274. IEEE Computer Society, May 2007.

[479] Rumi Zahir. Tolapai A System on a Chip with Integrated Accelerators. In *Proceedings of HotChips*, 2007.

[480] Li Zhao, Yan Luo, Laxmi Bhuyan, and Ravi Iyer. A Network Processor-Based Content Aware Switch. *IEEE Micro, Special Issue on High-Performance Interconnects*, 2006.

Index